A DICTIONARY OF GEOLOGY

D. G. A. Whitten was born in London in 1921. He went to Westminster City School and won a scholarship in 1941 to Imperial College. Graduating in geology, he joined ICI in 1943. In 1946 he became Lecturer in Geology at Kingston Technical College, the first person to hold such an appointment. He remained at Kingston and is now Head of the School of Geology in the Polytechnic. He has been a member of the Council of the Geologists' Association and is on the advisory committee of the Associated Examining Board and on the Council of the Geological Society of London. His hobbies are photography, science fiction and geological field work. He is married with one daughter.

J. R. V. Brooks was born in Richmond, Surrey, educated at Kingston Grammar School and graduated from the University of London in Geology. He spent over ten years in the oil industry before joining what is now the Petroleum Engineering Division of the Department of Energy, where he is Head of the Exploration and Appraisal Branch.

He is a Fellow of the Geological Society of London, a member of the American Association of Petroleum Geologists and a member of the Institution of Geologists.

He is interested in the education of geologists at all levels and has recently spent a five-year term as Chief Examiner in Ordinary and Advanced Levels of GCE in geology for the Associated Examining Board; he continues to examine the subject at GCE level. From time to time he lectures in petroleum geology and on the deep geology underlying the UK Continental Shelf, and writes book reviews.

J. R. V. Brooks is married with two children and enjoys photography and all things medieval.

KT-144-307

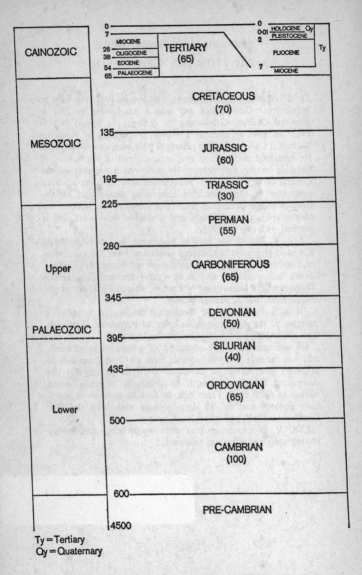

| CAINOZOIC | 0 ——— 7 ——— MIOCENE 26 ——— OLIGOCENE 38 ——— EOCENE 54 ——— PALAEOCENE 65 | TERTIARY (65) | 0 ——— 0·01 ——— HOLOCENE Qy PLEISTOCENE 2 ——— PLIOCENE 7 ——— MIOCENE | Ty |

CRETACEOUS
(70)

135———

MESOZOIC

JURASSIC
(60)

195———

TRIASSIC
(30)

225———

PERMIAN
(55)

280———

Upper

CARBONIFEROUS
(65)

345———

DEVONIAN
(50)

PALAEOZOIC

395———

SILURIAN
(40)

435———

ORDOVICIAN
(65)

500———

Lower

CAMBRIAN
(100)

600———

PRE-CAMBRIAN

4500

Ty = Tertiary
Qy = Quaternary

The geological time-scale in millions (10^6) of years.

A Dictionary of

GEOLOGY

D. G. A. WHITTEN

with

J. R. V. BROOKS

PENGUIN BOOKS

Penguin Books Ltd, Harmondsworth, Middlesex, England
Penguin Books, 625 Madison Avenue, New York, New York 10022, U.S.A.
Penguin Books Australia Ltd, Ringwood, Victoria, Australia
Penguin Books Canada Ltd, 2801 John Street, Markham, Ontario, Canada L3R 1B4
Penguin Books (N.Z.) Ltd, 182–190 Wairau Road, Auckland 10, New Zealand

—

First published 1972
Reprinted 1973, 1974, 1975, 1976, 1977, 1978, 1979 (twice), 1981

—

Copyright © D. G. A. Whitten and J. R. V. Brooks, 1972
All rights reserved

—

Made and printed in Great Britain by
Hazell Watson & Viney Ltd,
Aylesbury, Bucks
Set in Linotype Times

To Iris and Christine

CONTENTS

INTRODUCTION

'When *I* use a word,' Humpty Dumpty said, in rather a scornful tone, 'it means just what I choose it to mean – neither more nor less.'

'The question is,' said Alice, 'whether you *can* make words mean so many different things.'

'The question is,' said Humpty Dumpty, 'which is to be master – that's all.'

Through the Looking Glass

Within the last few years, there has been a great upsurge of interest in geology, both by laymen and in schools. This has coincided with a period of great development in the Earth sciences, especially in the application of new techniques and the study in depth of many topics, both new and old.

All this has meant that new terms, and modifications of old terms, have appeared in vast numbers, to confuse the beginner and expert alike. Geological nomenclature and terminology is a curious mixture of classical (and pseudo-classical) derivatives, common words used in a special way, terms borrowed from other disciplines (especially mining and quarrying) and words derived from names of persons and places, and even some words 'constructed' in a purely arbitrary way. There is little which is systematic in the naming of rocks, minerals, fossils, stratigraphic units and the like, and the entrenched position of many existing terms is likely to prevent any systematisation in the future.

In compiling the dictionary, the authors have concentrated on terms which seem to them to be in the widest use; special and local terms have been excluded. In a few cases, notably the stratigraphic entries, the temptation to include some local British detail has proved irresistible! In a number of cases the deciding factor has been simply 'Does the word occur in several commonly available books?' A number of obsolete and 'popular' terms have been included, since the former may be encountered in older works frequently still in use, and the latter in some of the books and articles which purport to explain geology to the layman.

Nobody is likely to agree completely with the authors' selection of terms – somebody's favourite word is sure to have been omitted;

likewise, there will be disagreements over some of the definitions, particularly where the authors have had to be arbitrary in their choice between conflicting authorities. Nevertheless they would welcome comments.

Geology remains one of the few sciences where amateurs are still able to make significant contributions, and it is hoped that this dictionary will be especially useful to them. For the aspiring geologist it may offer some helpful ideas and guidance, whilst for any other scientist who needs a quick reference to the vocabulary of geology it should be a useful starting point.

In the course of writing we have consulted many texts. The reference works used include *Glossary of Geological Terms* of the American Geological Institute, and the *Nomenclature of Petrology* by Arthur Holmes. Any original papers referred to are noted at the end of the relevant entry and a selected bibliography is included at the end of the book.

Throughout, we have been helped by the expert knowledge of our colleagues, in particular Mr G. Jenkins (structure and general topics), Mr A. Gardner (geophysics and well-logging), Mr P. Gurr (Palaeontology), Dr G. R. Parslow (optics), and Dr R. C. L. Wilson (sedimentary structures and limestones); their guidance and encouragement has been invaluable to us. The authors alone are responsible for the accuracy of the entries. We are also indebted to Mrs Elizabeth Sergent and Mrs Alex Gray, who typed the manuscript, and to Miss Jennifer Chadwick, who coped with the problem of drawing ellipses. The authors would also like to thank their friends and colleagues in the geological fraternity with whom they have, from time to time, discussed a variety of topics which greatly helped in the formulation of their ideas. The assistance of David Duguid in preparing the manuscript for publication is greatly appreciated.

Our wives have read the manuscript at all stages of its preparation and acted as guinea-pigs for some of the definitions. Anyone who has ever attempted to write a work of this kind will understand the magnitude of their contribution.

D.G.A.W.
J.R.V.B.

Surbiton, Surrey, August 1971

NOTES ON USE

Cross-references are indicated by ⬦ (= see) and ⬥ (= see also). In order to keep the number of such references down to a reasonable amount, certain very frequently occurring terms (e.g. quartz, feldspar, igneous, metamorphic) are not always given a cross-reference.

Stratigraphical divisions (other than periods) are restricted to stage names. No local stratigraphic names are included.

The minerals listed in the Appendix Table of Minerals have entries in the text, and both Appendix entry and text entry should be referred to.

Definitions of terms in related sciences, e.g. geomorphology, are given only if they have some specifically geological meaning.

NOTES ON USE

Aa (Hawaiian). ◊ *Lava* with blocky structure. (Cf. ◊ *Pahoehoe*.)

Aalenian. A stratigraphic stage name for the top of the Lower to the base of the Middle European ◊ *Jurassic*.

Ablation. The removal of rock debris by wind action.

Aboral. ◊ *Echinodermata* (Echinoids) and Fig. 44.

Abrasion. Reduction in size of rock particles by mechanical wearing-away, especially by the rubbing of two particles together. (◊ *Roundness.*)

Abrasion platform. Not a surface formed by ◊ *abrasion*, but generally any horizontal surface cut into a slope; commonly of marine origin, e.g. a wave-cut platform (◊ *Marine erosion*).

Absarokite. A porphyritic ◊ *basalt* containing a small amount of ◊ *orthoclase* in the ◊ *groundmass*.

Absolute age. Age expressed in numbers of years. (◊ *Radioactive dating*.)

Abstraction. The complete absorption of one river by another. In abstraction, a series of smaller valleys are progressively incorporated into a larger one. Most commonly seen in streams radiating from a cone, e.g. a volcano.

Abyssal. Belonging to the environment below 4,000 m., in which the only sediments are red clay and the deep-sea oozes (◊ *Abyssal deposits*). At this depth the temperature does not exceed 4°C. and only specialised animals are found. (Cf. ◊ *Bathyal*.)

Abyssal deposits. Sediments formed at depths greater than 2,000 m., in the deeper part of the ◊ *bathyal* zone and the ◊ *abyssal* zone itself. Two main types of sediment occur – the biogenic oozes and the non-biogenic sediments (red clay). The following are the main types recognised:

CALCAREOUS OOZES: Formed between 2,000 and 3,900 m., and consisting chiefly of the minute calcareous skeletons of ◊ *pelagic* animals such as Foraminifera (◊ *Protozoa*) and pteropods (highly adapted gastropods: ◊ *Mollusca*). The commonest Foraminifera are Globigerina, and Globigerina ooze covers perhaps 130×10^6 sq. km. (50×10^6 sq. miles) of the ocean bottom. Pteropod oozes are much less widespread. Both types commonly contain as much as 60% clay material derived from colloidal matter.

SILICEOUS OOZES: At depths greater than 3,900 m. calcium carbonate is more soluble than silica, and hence only the latter survives in sediments deposited below this depth. The main constituents of siliceous oozes are Radiolaria (◊ *Protozoa*), which are found mainly in tropical regions, and diatoms (single-celled plants secreting a siliceous skeleton), which are especially abundant in polar regions. Siliceous oozes cover 38×10^6 sq. km. (15×10^6 sq. miles) of ocean bottom.

RED CLAY: At depths greater than 5,000 m. much of the silica of radiolarian skeletons is also dissolved and the only deposit at greater depths is the red clay. This consists very largely of ultrafine particles, the finest wind-blown volcanic ash, and meteoritic material. The red colour is mainly due to ferric iron and manganese compounds. In some areas there are abundant sharks' teeth and whales' otoliths (ear bones), including those of extinct species. This tends to confirm the view that the rate of deposition of the red clay is no more than one to two mm. per 1,000 years. The red clay covers 100×10^6 sq. km. (40×10^6 sq. miles) mainly in the Pacific and Indian Ocean areas.

Manganese oxide nodules occasionally occur in vast quantities in association with all oozes, and may in the future become of economic importance.

Abyssal rocks. Rocks formed at very great depth. A synonym of ◊ *plutonic* rocks.

Acadian. Obsolete stratigraphic stage name for the North American Middle ◊ *Cambrian.* 'Albertan' is the modern equivalent.

Accessory minerals. Minerals present in a rock in such small amounts that their presence or absence is not significant when considering the mineral composition for classification purposes.

Accessory plate. A specially cut slice of a mineral mounted for use in a polarising microscope. The three commonest ones in use are: the quartz wedge, a wedge-shaped slice of quartz ground so as to produce the range of ◊ *polarisation colours* from zero to the end of the third or fourth order; the mica plate (or quarter-wave plate), which is of a thickness designed to give a standard retardation of a quarter of the wavelength of sodium light; and the gypsum plate (sometimes called a sensitive tint), designed to give a standard retardation of 590mμ, the 'sensitive purple' at the end of the first order.

All the above are marked with their slow or fast directions. Their principal use is for determining the fast or slow direction in a mineral, and hence, in appropriate cases, whether the mineral is positive or negative (◊ *Axes, optic*). They may also be used to determine ◊ *extinction* positions. For exact determination of extinction positions and birefringences, calibrated plates (compensators) are used. (◊ *Indicatrix*; *Polarisation colours* (Newton's Scale); *Vibration directions*.)

Accommodation structures. Small structures, such as ◊ *folds* and ◊ *faults*, which permit beds to occupy completely the available space during major ◊ *tectonic* activity. During major folding or faulting it is impossible for 'holes' to develop in the structure and any such tendency is prevented by the development of accommodation structures. Similarly, in the cores of tight folds structures develop which allow the excess material in the core to be accommodated.

Accretion theory. A theory which suggests that the planets grew by the condensation of small cosmic dust particles into large bodies. One theory suggests that the 'dust' arose from the supernova explosion of a twin star of the present sun. It can be shown that there will be a maximum tendency for 'clots' of dust to form at certain positions (Bode's law); it can also be shown that once a 'clot' has reached a diameter of about 100 miles, the rate of growth increases sharply because of the increased gravitational attraction of the mass. 'Clots' with a diameter of less than 100 miles are just as likely to break up because of weak gravitational attraction as they are to attract additional particles. It is very likely that a few 'clots' would grow very rapidly at the expense of the rest, and might even grow too large for their rate of rotation around the Sun and rate of spin about their axes. This may then produce partial break-up, giving rise to satellites.

ACF diagram. A ◊ *triangular diagram* in which the three apices are Al_2O_3, CaO, and FeO plus MgO. Used in the study of chemical variation in igneous and metamorphic rocks.

Acheulian. A stratigraphic stage name for the European Lower Pleistocene (◊ *Tertiary System*).

Achondrites. One of the two classes of stony ◊ *meteorites*; that in which ◊ *chondrules* are absent. (Cf. ◊ *Chondrites*.)

Achroite. A white potassium-rich variety of ◊ *tourmaline*.

Acicular (mineralogy). Needle-like.

Acid rock. An igneous rock with 10% or more free quartz. The term arises from the concept of silica as an acidic oxide, i.e. in theory, together with water, it can form a range of 'silicic acids', and thus the minerals forming the rocks were regarded as salts of these acids. If therefore a rock contained excess silica, it was regarded as having an excess of the acidic 'principle'. The name persists although the theory behind it is certainly untenable. Acid rocks were formerly defined in terms of silica percentage – any rock containing more than 66% silica being termed acid. (◊ *Granite*; *Granodiorite*; *Microgranite*; *Pitchstone*; *Rhyolite*.)

Acrozone (Range zone). A stratum defined by the time-range of a species. (◊ *Stratigraphic nomenclature*.)

Actinolite. ◊ *Tremolite–actinolite*.

Acute bisectrix. ◊ *Axes, optic*.

Adambulacrum (Echinodermata). ◊ Fig. 46.

Adamellite. A variety of ◊ *granite* containing a calcium-bearing plagioclase, usually oligoclase, and a potassium ◊ *feldspar*, in roughly equal amounts. For microadamellite, ◊ *Microgranite*.

Adaptive radiation. The evolution of a group of animals to fill a large number of ecological environments; e.g. for mammals:

Habitat	Mammal type evolved
Air	Bats
Grasslands	
(water meadow)	Cattle
(dry)	Sheep
(moist)	Horses
(burrowing)	Rabbits
Mountain regions (scrub)	Goats
Trees (leaf-canopy)	Squirrel, monkey
Trees (forest floor)	Deer
Desert	Camel
Aquatic	
(fresh water)	Otter
(swamp)	Swamp wallaby
(estuarine)	Manatee, dugong
(marine)	Seals, dolphins, whales

Adductor muscles. Muscles which draw together the two shells of a bivalved organism. (◊ *Brachiopoda* and Fig. 11; (Mollusca), Fig. 101.)

Adinole. An ◊ *albitised* slate.

Admission. One of the ways in which a trace element occurs. (◊ *Geochemistry*.)

Adobe. A ◊ *loess*-like clay.

Adularia. A very low-temperature monoclinic potassium ◊ *feldspar*.

Aegirine. A mineral of the ◊ pyroxene group, $NaFe'''Si_2O_6$, found in soda-rich igneous rocks. (◊ Appendix.)

Aeolian deposits. Sediments deposited after transport by wind. (◊ *Dunes*.)

Aeon. The use of this term for an indefinite but lengthy period of time is being replaced by a precise usage for 10^9 years. On this basis the ◊ *age of the Earth* is about 4·7 aeons, and the ◊ *Phanerozoic* represents just over half an aeon.

Aerolites. A class of ◊ *meteorite* composed dominantly of silicate material.

Affine deformation (Homogeneous strain). Deformation of rocks in which the particles move uniformly with respect to each other. Straight lines and planes are undistorted, but circles become ellipses.

Affluent. A tributary stream, especially a smaller stream emptying into a larger. (Cf. ◊ *Confluent*.)

Agalmatolite. ◊ *Pyrophyllite*.

Agassiz, Jean Louis (1807–73). Swiss zoologist and palaeontologist, famed in his lifetime for his classification of fossil and living fish, but now remembered more for advancing the theory that vast areas of the Earth had once been covered by ice. Although most of his work on glacier movement and structure was carried out in the Alps, he also travelled widely in search of further evidence for his theory. Amongst indications of past glaciation which he demonstrated during his visit to Britain were the Parallel Roads of Glen Roy, which had previously caused much speculation as to their origin.

Agate. Banded chalcedonic silica. (◊ *Silica group of minerals*.)

Age. ◊ *Stratigraphic nomenclature*; *Radioactive dating*.

Age of the Earth. The earliest estimate not dependent on Biblical evidence was that of 100,000,000 years suggested by Lord Kelvin in 1883, on the basis of: (1) the rate at which the Earth's rotational speed declined; (2) the time for which the Sun's energy output could have been maintained; (3) the time taken for the Earth to cool to its present temperature state from a molten condition.

In 1900, Joly produced supporting evidence for Kelvin's figure by considering the rate at which salt was being carried into the sea by rivers, and the time necessary for an initially 'freshwater' ocean to acquire its present salt content. The discovery of radioactivity in 1896 and the subsequent realisation that radioactivity is widespread in the rocks of the Crust invalidated Kelvin's arguments and provided a 'clock' by which rocks could be dated (◊ *Radioactive dating*). The oldest dated rocks (3,900 million years old) are found in West Greenland; they consist of ◊ *anorthosites* which are intruded into an older series of rocks. These rocks, intruded by the anorthosites, must be older than them by an unknown period of time, perhaps 100 million years or more. Thus it is effectively certain that rocks exist with an age of greater than 4,000 million years. The origin of the Earth, if it can be dated at all, must ante-date the oldest rocks by some period at which we can only guess. Another method based on the relative abundance of the various isotopes of lead in galena (PbS) yields figures of 5,000 to 5,400 million years. Astronomical data suggest an age for the Solar System of $5,000 \pm 1,000$ million years, which is in reasonable agreement with geochemical evidence. Rock samples from the Moon have been dated at 4,700 million years, also in good agreement with these estimates.

Agglomerate. ◊ *Pyroclastic rocks* consisting mainly of fragments larger than 2 cm. in diameter.

Agmatite. A ◊ *migmatite* in which the introduced material appears to form a network of veins throughout the host rock.

Agnatha (lit. 'without jaws'). ◊ *Chordata* (Pisces).

Agricola, Georgius (1494–1555). Latin name adopted by the German Georg Bauer. Sometimes referred to as the father of mineralogy, he produced the first detailed descriptions of minerals according to their physical properties, and the distribution of ores and their modes of occurrence, as well as the mining methods of his time.

Ahermatypic. Ecological term applied to a non-colonial assemblage or individual organism.

Akerite. An ◊ *oversaturated* ◊ *syenite*.

Alar septum (Coelenterata). ◊ Fig. 21.

Alaskites. ◊ *Granites* consisting only of quartz and alkali ◊ *feldspar*.

Albertan. A stratigraphic stage name for the North American Middle ◊ *Cambrian*.

Albertite. A ◊ *hydrocarbon* mineral.

Albian. A stratigraphic stage name for the base of British Upper ◊ *Cretaceous*.

Albion. A stratigraphic stage name for the North American Lower ◊ *Silurian*. A synonym of Medinian.

Albite. A variety of plagioclase ◊ *feldspar*.

Albitisation. The development of ◊ *albite* in a rock at the expense of a pre-existing mineral, usually another ◊ *feldspar*, by the introduction of sodium ions into the rock (◊ *Metasomatism*). A typical example is the albitised slate known as adinole.

Albitite. A ◊ *syenite* consisting almost entirely of sodic alkali ◊ *feldspar*.

Algoman. A stratigraphic division of the Canadian ◊ *Precambrian*, Pre-◊*Algonkian*.

Algonkian. The youngest stratigraphic division of the Canadian ◊ *Precambrian*.

Alkali, Alkaline. A term applied to igneous rocks in which the ◊ *feldspar* is dominantly sodic and/or potassic. The term is also used for rocks which contain essential ◊ *feldspathoid*, e.g. alkali basalt. It should be noted that the opposite of alkali is ◊ *calc-alkali*, not ◊ *acid*. Alkaline rocks commonly contain alkaline ferromagnesian minerals, e.g. ◊ *micas*, sodic ◊ *amphiboles*, and sodic ◊ *pyroxenes*, etc. The term is also applied to silicate minerals containing essentially Na and/or K, e.g. alkali ◊ *feldspars*.

Alkali basalt. The term is here used to cover all igneous rocks of generally basaltic character which contain ◊ *feldspathoids*. (For saturated alkali basalts, ◊ *Trachybasalts*.)

	Feldspar and feldspathoid		Feldspathoid only	
	Leucite	Nepheline	Leucite	Nepheline
Olivine-bearing	Leucite basanite	Nepheline basanite	Olivine leucitite	Olivine nephelinite
Olivine-free	Leucite tephrite	Nepheline tephrite	Leucitite	Nephelinite

Appropriate names may be constructed for analcite-bearing and kalsilite-bearing varieties, e.g. analcitite, kalsilitite. In older literature 'leucite basalt' and 'nepheline basalt' are used for rocks now termed olivine leucitite and olivine nephelinite respectively. It has been suggested that the feldspar-free varieties are not strictly basalts, but it is convenient to consider them with the undersaturated basalts, since both types occur in association. Melilite basalt is a feldspar-free type related to ◊ *alnöite*.

Many of these rocks have an exceptionally low silica content, and Na_2O and K_2O are notably high. The feldspathoidal basalts occur in a few restricted areas, notably Italy, the East African Rift region, Wyoming, U.S.A., and West Australia.

Alkali basalts are the volcanic equivalents of ◊ *alkali gabbros*.

Various attempts have been made to extract potassium salts from the leucite-bearing lavas where the leucite occurs as abundant ◊ *phenocrysts*.

Alkali feldspar. ◊ *Feldspars*.

Alkali gabbro and alkali dolerite (syenogabbro and syenodolerite are close synonyms). Igneous rocks which, though gabbroic and doleritic in general character, contain alkali feldspar and/or feldspathoids in addition to normal minerals. The rock names found in this group are used indiscriminately for doleritic and gabbroic types, although in many instances the finer grain size could be distinguished by using the prefix 'micro'. The typical minerals found are titanium-bearing ◊ *augite* and ◊ *labradorite* together with ◊ *analcite* and/or ◊ *nepheline*, with or without an alkali feldspar, and sometimes with ◊ *olivine*. Common accessory minerals are ◊ *ilmenite* and ◊ *apatite*. ◊ *Aegirine* or aegirine-augite may sometimes be present to the exclusion of titan-augite. Chemically, these rocks resemble gabbros except that they are notably higher in their sodium and potassium content. They are closely related to certain types of melanocratic ◊ *alkali syenites* and the distinction between the two groups is largely based on the presence of labradorite (and olivine) in the alkali gabbros.

Varieties showing ophitic or porphyritic textures are common. Kentallenite is sometimes included with this group but is here regarded as an olivine ◊ *monzonite*. ◊ *Essexite* is a labradorite, titan-augite, olivine, nepheline and/or analcite variety, with the occasional presence of alkali feldspar. ◊ *Theralite* is a variety similar to essexite, but without olivine and containing nepheline to the exclusion of analcite. ◊ *Teschenite* is a variety similar to essexite, but containing analcite to the exclusion of nepheline. Minverite is albite gabbro or dolerite containing barkevicite but no feldspathoids.

The volcanic equivalent of alkali gabbros and alkali dolerites are ◊ *basanites* and tephrites (◊ *Alkali basalt*). These alkali gabbros and dolerites are most commonly found as differentiates in basic sills, and in plugs and small dykes, associated with basic and ◊ *ultrabasic rocks*.

Alkali syenite. Alkali ◊ *syenite* is here taken to be synonymous with feldspathoidal syenite. These igneous rocks are divided into two major groups: those containing feldspar and feldspathoid, and those with no feldspar – the latter being termed 'syenoids' by Shand. (For several types containing nepheline in accessory amounts, ◊ *Syenite*.)

The commonest mineral present is ◊ *nepheline*, and there are few alkali syenites which do not contain it. Other feldspathoids which occur include sodalite, cancrinite, and analcite. The ferro-magnesian minerals commonly present are ◊ *aegirine*, sodic ◊ *amphiboles*, and, rarely, melanite ◊ *garnet* and ◊ *olivine*. ◊ *Corundum* rarely occurs, while ◊ *apatite* is an abundant accessory.

It will be noticed that ◊ *leucite* is absent from the above list, implying that there are no plutonic equivalents of leucitophyres (◊ *Trachyte*); in fact, leucite syenites are practically unknown, but a number of rocks contain polygonal-to-rounded masses, consisting of orthoclase and nepheline. These are referred to as 'pseudo-leucites', although it is doubtful whether they were leucite at any stage in their history.

Alkali syenites are characterised by SiO_2 contents lower than for other syenites, and higher contents of K_2O, Na_2O, and Al_2O_3.

The following types contain both feldspar and feldspathoid:

BOROLANITE. Pseudo-leucite, orthoclase, melanite, aegirine, and biotite.

DITROITE. Nepheline, ◊ *perthite*, sodalite, aegirine, and soda-amphibole. (The term sodalite syenite refers to a rock which contains sodalite to the exclusion of nepheline: similarly for analcite and cancrinite syenites.)

FERGUSITE. Pseudo-leucite and aegirine. (Unaltered leucite has been recorded from one occurrence of this rock.)

FOYAITE. The most widespread alkaline syenite, containing nepheline, perthite, and aegirine.

LITCHFIELDITE. Similar to foyaite, but contains dominant albite and microcline. Sodalite and/or cancrinite may occur in small quantities.

MALIGNITE. Nepheline, orthoclase, and aegirine: nepheline-rich shonkinite (◊ *Syenite*).

MARIUPOLITE AND MONMOUTHITE. Both of these rocks consist essentially of albite, nepheline, and aegirine or soda-amphibole, but whereas albite is dominant in mariupolite, nepheline is dominant in monmouthite.

MISSOURITE. An olivine-bearing melanocratic fergusite.

The following types, containing feldspathoids only, consist essentially of nepheline, aegirine and/or titan-augite. The ◊ *colour index* is used to delimit four main types:

URTITE. Colour index less than 30.

IJOLITE. Colour index between 30 and 70.

MELTEIGITE. Colour index between 70 and 90.

JACUPIRANGITE. Colour index more than 90 (syn. nepheline pyroxenite).

Pure nepheline rocks (note that nephelinite is pre-empted for an ◊ *alkali basalt* lava) have colour indexes of less than 10.

For the occurrence of alkali syenite, ◊ *Syenite*. Nepheline syenite ◊ *pegmatites* occur and are noted for the rare minerals they contain. Micro-foyaites and micro-ijolites have been recorded, but other medium-grained types are rare.

The volcanic equivalent of the alkali syenites are the phonolites (◊ *Trachyte*).

Nepheline-apatite rocks, which occur locally, have been worked on a small scale as a source of phosphate and as raw material for the glass industry.

Allanite. ◊ *Epidotes.*

Allivalite. ◊ *Gabbro.*

Allochthonous. A term applied to the material forming rocks which have been transported to the site of deposition. (Cf. ◊ *Autochthonous*; ◊ *Detrital*.)

Allogenic (Allothigenous). The term applied to the derived portion of a sediment, whether minerals or other constituents (e.g. the pebbles in a conglomerate, the grains of quartz making up a sand), which originated away from the area of sedimentation and is transported to it for final deposition. (Cf. ◊ *Authigenic*.)

Allotrio- (prefix). Alien, foreign.

Allotriomorphic. A term applied to mineral grains showing no development of crystal form whatsoever. (◊ *Texture*.)

Alluvial deposits, alluvium. ◊ *Detrital* material which is transported by a river and deposited – usually temporarily – at points along the ◊ *flood plain* of a river. Commonly composed of sands and gravels. (◊ *River terrace*; *Rejuvenation*; *Colluvial deposits*; *Eluvial deposits*.) Many important ore minerals, e.g. gold, platinum, diamonds, cassiterite (SnO_2), are locally found concentrated in alluvial deposits.

Alluvial fan. A mass of sediment deposited at a point along a river where there is a decrease in gradient, e.g. from a mountain to a plain. The mass is thickest at its point of origin, and thins rapidly in a downstream direction. In time adjacent fans may coalesce and extend for many kilometres away from the mountain front. Such an alluvial plain may be termed a ◊ *bajada*.

Almandine. ◊ *Garnet.*

Alnöite. ◊ *Alkali basalt*; *Lamprophyre.*

Alsbachite. An uncommon porphyritic variety of ◊ *aplite*.

Alstonite. An orthorhombic ◊ *carbonate* mineral, $CaBa(CO_3)_2$.

Alternation of generations. An alternation between a planktonic, reproductive stage (medusoid) and a sedentary, vegetative stage typical of ◊ *coelenterata*.

Aluminium silicates. ◊ *Silicates* containing aluminium as the only cation. The four main members of the group, having the formula Al_2SiO_5, are all found in metamorphic rocks, and the study of their phase relationships has thrown important light upon the temperatures and pressures which can be assigned to various grades of ◊ *metamorphism*:

ANDALUSITE (orthorhombic) occurs in nearly square-section prismatic crystals; the variety containing dark inclusions in a cruciform shape is known as chiastolite. Andalusite typically occurs in low-grade thermally metamorphosed ◊ *argillaceous rocks*. (◊ Appendix).

SILLIMANITE (orthorhombic), also known as fibrolite because of its common occurrence in fibrous masses, is found mainly in high-grade regionally metamorphosed ◊ *pelitic* schists and gneisses. It occurs occasionally in very high-temperature, thermally metamorphosed pelitic rocks.

KYANITE (= cyanite, though this spelling is not used owing to the possibility of confusion with syenite, the rock) (triclinic). Disthene is a synonym and relates to the mineral's widely differing ◊ *hardnesses* in different directions ($5\frac{1}{2}$–6 parallel to the c-axis; $6\frac{1}{2}$ parallel to the b-axis; and $6\frac{1}{2}$–7 parallel to the a-axis). The name of the mineral relates to its common occurrence in blue crystals. It is found characteristically in regionally metamorphosed pelitic rocks of slightly lower grade than rocks containing sillimanite. Hydrothermal quartz-kyanite veins also occur. (◊ Appendix.)

MULLITE (orthorhombic). The formula is commonly given as $Al_6Si_2O_{13}$ and this represents the natural material fairly well. However, synthetic mullite made under controlled conditions has the formula $Al_6Si_3O_{15} = 3(Al_2SiO_5)$ and the SiO_2 deficiency in natural material is presumably accounted for by a defect lattice. Mullite occurs as fine needles, commonly with corundum, in ◊ *xenoliths* which have been heated to very high temperatures (1,300°C.) enclosed in basic igneous rocks. It is a common constituent of the so-called 'sillimanite' refractory bricks.

Other minerals closely related to the aluminium silicates include topaz and staurolite:

TOPAZ (orthorhombic) may be regarded as being derived from the Al_2SiO_5 composition by replacement of an oxygen ion by two OH and/or F ions, giving a formula $Al_2SiO_4(OH, F)_2$. The mineral is formed almost exclusively as a result of the fluorine metasomatism of granitic rocks, and more rarely of other aluminium-rich rocks. It also occurs in quartz and ◊ *pegmatite* veins and occasionally in cavities in ◊ *rhyolites*. Detrital topaz also occurs. It is occasionally used as a gemstone despite a perfect basal cleavage. (◊ *Pneumatolysis*; Appendix.)

STAUROLITE (monoclinic – pseudo-orthorhombic – with β angle = 90°). It is impossible to demonstrate the monoclinic nature of staurolite by ordinary techniques. Cruciform twins are well known. Staurolite has a 'sandwich' structure consisting of kyanite layers between ferrobrucite $Fe''(OH)_2$ layers (cf. brucite, $Mg(OH)_2$), giving a formula $FeAl_4Si_2O_{10}(OH)_2$. It is formed in medium-grade regionally metamorphosed ◊ *argillaceous rocks*. A number of cases where the staurolite metamorphic ◊ *zone* appears to be absent are due to the absence of rocks of the appropriate chemical composition. (◊ Appendix.)

Amber. A resin found in Oligocene (◊ *Tertiary*) estuarine deposits around the Baltic coast. (◊ *Hydrocarbon minerals*.)

Ambitus (Echinodermata). ◊ Fig. 44.

Ambulacrum (Echinodermata). ◊ Fig. 44.

Amethyst. A violet/purple variety of quartz (◊ *Silica group of minerals*).

Ammanian. A stratigraphic stage name for the British Lower ◊ *Westphalian* (Carboniferous System).

Ammonites, Ammonoidea, Ammonitina. ◊ *Mollusca* (Cephalopoda).

Amorphous. Term applied to material having no regular arrangement of atoms. (◊ *Crystallinity*.)

Amphibia. ◊ *Chordata*.

Amphiboles. A group of inosilicates (◊ *Silicates*) characterised by a double chain (or ribbon) of linked SiO_4 tetrahedra. They are usually monoclinic, but orthorhombic forms are known. The general formula is $X_{2-3} Y_5 Z_8 O_{22} (OH)_2$, where X may be Ca, Na, or K, Y may be Mg, Fe''', Fe'', Al, or Ti, Z may be Si or Al (maximum of two Al). The hydroxyl may be partially replaced by F, Cl, or O.

ORTHORHOMBIC: anthophyllite $(MgFe'')_7(Si_8O_{22})(OH,F)_2$; gedrite $(Fe'',Mg)_7(Si_8O_{22})(OH,F)_2$ (both these minerals commonly contain some Al, replacing Mg and/or Si).

MONOCLINIC: cummingtonite $(Mg,Fe'')_7(Si_8O_{22})(OH,F)_2$; grunerite $(Fe'')_4(Fe'',Mg)_3(Si_8O_{22})(OH,F_2)$; tremolite $Ca_2Mg_5(Si_8O_{22})$ $(OH,F)_2$; actinolite $Ca_2(Mg,Fe'')_5(Si_8O_{22})(OH,F)_2$; hornblende $NaCa_2$ $(MgFe'')_4(Al,Fe''')(Si,Al)_8O_{22}(OH,F)_2$ (the term basaltic or oxyhornblende may be used when Ti and Fe''' are dominant over Mg, Al and Fe''); edenite, pargasite, and barkevicite (varieties in which Na is dominant over Ca); glaucophane $Na_2(Mg,Fe'')_3(Al,Fe''')_2(Si_8O_{22})(OH,F)_2$; riebeckite $Na_2Fe_3''Fe_2'''$ $Si_8O_{22}(OH,F)_2$. Where Fe'' completely replaces Mg in some of the above, the prefix 'ferro-' is used, e.g. ferro-actinolite. 'Magnesio-' is used where Mg replaces Fe.

The amphiboles are a group similar to the ◊ *pyroxenes* in many respects, the main point of difference being the 124° prismatic cleavage angle in the amphiboles compared with the 90° angle of the pyroxenes.

Many amphiboles occur in characteristic fibrous or acicular forms. Amphiboles are widespread in igneous and metamorphic rocks, and certain types occur as detrital grains in sediments. Certain of the fibrous forms belong to the group of minerals collectively known as ◊ *asbestos*.

Amphibolite. A ◊ *metamorphic* rock composed mainly of ◊ *amphibole*, generally with an orientated ◊ *fabric* (◊ *Hornblende schist*). The term should not be used for igneous rocks consisting entirely of amphibole.

Amphineura. ◊ *Mollusca*.

Amygdale, Amygdule (literally 'almond-shaped'). A name for the cavities in lavas formed by the evolution of gas by the lava, causing frothing. They range in size from 1mm. to 30cm. in diameter. The term is often applied only to cavities subsequently filled by a mineral, e.g. ◊ *zeolites*, calcite, or quartz. A lava with empty cavities is termed scoriaceous. Occasionally elongated cylindrical cavities (pipe amygdales) occur at right angles to the surfaces of the lava flow. They are formed when lava flows over a patch of wet ground, e.g. a spring, the steam so evolved penetrating the lava, leaving tubes behind. The top part of lava flows is usually the most amygdaloidal portion. Amygdaloidal structure should not be confused with ◊ *spherulitic texture*. (◊ *Way-up criteria* and Fig. 155.)

Ana- (prefix). Towards, up to.

Anaerobic. Literally 'without air', but implying absence of oxygen.

Analcite. ◊ *Feldspathoids* and Appendix.

Analcitite. ◊ *Alkali basalt*.

Analyser. The name given to the polaroid (or Nicol prism) in a petrological microscope which is situated in the microscope tube above the nose-piece and arranged to allow only ◊ *polarised light* vibrating parallel to one of the cross-wires, at right angles to the vibration plane of the ◊ *polariser*, to pass. The analyser is removable so that the microscopic slide can be viewed in either plane polarised light (analyser out) or between crossed polarisers (analyser in). Two conventions are in use for orientating polariser and analyser: in Britain, polarisers are orientated to give an E–W vibration and the analyser N–S, while in several other countries the reverse arrangement is adopted.

Anastomosing. Of streams, branching and rejoining irregularly to produce a net-like pattern. (Cf. ◊ *Dichotomous.*)

Anatase. A titanium mineral, TiO_2, found on joint planes and in veins in schists and gneisses. (◊ Appendix.)

Anatexis. Regeneration of magma by the fluxing of pre-existing rocks. (◊ *Palingenesis; Syntexis.*)

Anatexite. An ◊ *ultra-metamorphic rock* showing evidence of ◊ *anatexis.* (Cf. ◊ *Migmatite.*)

Anchi- (prefix). Almost.

Andalusite. ◊ *Aluminium silicates* and Appendix.

Andesine. A variety of plagioclase ◊ *feldspar.*

Andesinite. A ◊ *diorite* containing 5% or less of ◊ *ferro-magnesian minerals*, the remainder being andesine.

Andesite. A fine-grained ◊ *intermediate* volcanic igneous rock characterised by the presence of ◊ *oligoclase* or ◊ *andesine.* Their chemistry and mineralogy are closely similar to those of the ◊ *diorites.* ◊ *Porphyritic* varieties are fairly common, both ◊ *ferro-magnesian* minerals and feldspars occurring as ◊ *phenocrysts* – the latter commonly showing zoning. ◊ *Hypersthene* and enstatite are more common in andesites than in diorites. There is little doubt that some pyroxene andesites have been called olivine-free basalts and vice-versa. Pure glassy andesites are rare, but glass is of frequent occurrence in the groundmass of andesites, commonly in a devitrified state.

Rocks which display characteristic basaltic features (especially by containing olivine), except that they contain oligoclase or andesine, may be called mugearite and hawaiite respectively, but they are more properly classified as ◊ *trachyandesites.* With increasing silica content and the development of free quartz, andesites pass into dacites (◊ *Rhyolite*). Andesites occur as extensive lava flows always associated with continental masses. The 'Andesite

Line' is the boundary between the 'continental' andesitic association and the 'oceanic' basaltic association.

Andradite. ◊ *Garnet.*

Angoumian. A stratigraphic stage name for the European Upper ◊ *Turonian* (Cretaceous System).

Anhedral. A term applied to mineral grains showing no development of crystal form whatsoever. (◊ *Texture.*)

Anhydrite. An ◊ *evaporite* mineral, $CaSO_4$, found in sedimentary rocks associated with ◊ *gypsum.* (◊ Appendix.)

Anisian. A stratigraphic stage name for the East European and Russian Middle ◊ *Triassic.*

Anisotropic. Literally 'not ◊ *isotropic*'. Applied to a substance having different physical properties when measured in different directions, e.g. the hardness of topaz varies markedly between basal plane and prism; the thermal conductivity of quartz is least along the c-axis. However, the term is generally applied to optical properties such as variation of refractive index or colour absorption. All crystalline substances, other than those belonging to the cubic system, are anisotropic, but optically, basal sections, i.e. sections perpendicular to the c-crystallographic axis, of tetragonal and hexagonal minerals are isotropic.

Ankaramite. An augite-rich olivine mela-basalt. (◊ *Basalt.*)

Ankaratrite. Olivine nephelinite. (◊ *Alkali basalt.*)

Ankerite. A hexagonal ◊ *carbonate* mineral $(Ca,Mg,Fe)CO_3$.

Annelida. This phylum, the worms, is of little geological importance. The only fossil remains consist of chitinous jaw structures comparable to those found in the modern polychaete worms (the Bristle Worms), which are usually referred to as scolecodonts, and are of no stratigraphic importance. The only other fossil traces of this phylum consist of worm tubes and trails (◊ *Fossils*). It is thought that the phylum ◊ *Arthropoda* emerged from this group.

Anorthite. A calcic plagioclase ◊ *feldspar.*

Anorthoclase. A high-temperature sodium-potassium ◊ *feldspar.*

Anorthosite. A coarse-grained plutonic igneous rock consisting of more than 90% plagioclase ◊ *feldspar*, the remainder being made up of other gabbroic minerals. The term anorthosite is generally confined to rocks made up of basic plagioclase, i.e. labradorite, bytownite, or anorthite. Rocks consisting of the more sodium-rich plagioclases are generally referred to as ◊ *andesinites*, ◊ *oligoclasites*, and ◊ *albitites*, and these types may contain small amounts of different ◊ *ferromagnesian minerals.*

Anorthosites, in the strict sense, occur mainly as: (1) differentiation layers in layered basic complexes (◊ *Layered igneous structures*); (2) isolated plug-like intrusions, often in the same region as similar masses of ◊ *peridotite*; and (3) patches in gabbroic masses. With increasing ferromagnesian content they grade into leuco-gabbros. No hypabyssal or volcanic equivalents of anorthosites are known.

Antecedent drainage. ◊ *Inconsequent drainage.*

Anthophyllite. An orthorhombic mineral of the ◊ *amphibole* group.

Anthozoa. ◊ *Coelenterata.*

Anthracite. ◊ *Coal.*

Antian. A stratigraphic stage name for the top of the British Lower Pleistocene (◊ *Tertiary System*).

Anticline, Anticlinorium, Antiform. A ◊ *fold* or fold system in the form of an arch.

Antidune. ◊ *Cross-bedding* (Small-scale).

Antigorite. The laminar variety of ◊ *serpentine.*

Antiperthite. ◊ *Perthite.*

Anti-stress minerals. Minerals formed under conditions of metamorphism where ◊ *stress* was entirely absent. It was supposed that 'stress' inhibited the growth of such minerals. Doubt has been cast on the validity of the concept, as numerous examples of typical anti-stress minerals occurring in 'stress' environments have been recorded. (Cf. ◊ *Stress mineral.*)

Antithetic. A term applied to ◊ *fault* planes which dip in the opposite direction to the bedding.

Anus. ◊ *Echinodermata* and Figs. 42–45.

Apatite. A mineral, $Ca_5(F,Cl)(PO_4)_3$, found in igneous rocks, especially ◊ *pegmatites*, and in metamorphosed limestones. Varieties form bones and teeth. It is the main source of phosphates. (◊ Appendix.)

Aperture. (Mollusca) ◊ Figs. 100 and 105; (Polyzoa) ◊ Fig. 119.

Apex (Mollusca). ◊ Fig. 100.

Aphanitic. Textural term used to describe igneous rocks having such a fine grain that individual constituents are not visible to the naked eye, but are visible under suitable magnification. (Cf. ◊ *Hyaline*; ◊ *Phanerocrystalline.*)

Aphebian. A Canadian ◊ *Precambrian* ◊ *orogeny.*

Aphotic. An environmental term meaning 'without light'. In the sea, below approximately 200 m. (◊ Fig. 49.)

Apical disc, apical system (*Echinodermata*). ◊ Figs. 44 and 45.

Aplite. Without modification, the term implies an igneous rock of ◊

granitic affinities, but ◊ *syenite*, ◊ *diorite*, and ◊ *gabbro* aplites occur. Granite aplites are light-coloured fine-grained rocks, having the same mineralogy as granites, although ◊ *biotite* is rarely, if ever, present. They often contain ◊ *pneumatolytic* minerals such as tourmaline, topaz, fluorite, etc., and concentrations of rare elements such as beryllium and lithium. This suggests a relationship to the ◊ *pegmatites*, and in the field aplites and pegmatites are commonly found in association, mainly as dykes or veins (◊ Fig. 8). It has been suggested that whereas a pegmatite forms from a residual magma rich in ◊ *volatiles*, an aplite forms from a residual magma low in volatiles, thus promoting fineness of grain. Aplites have a characteristic sugary texture; the ◊ *porphyritic* variety alsbachite is uncommon. Gabbro aplite may be termed beerbachite.

Apo- (prefix). Derived from.

Apophyses. Thin branches or offshoots – a term applied to bodies of igneous rock.

Appinite. A group of coarse-grained igneous rocks similar in character to ◊ *syenites*, ◊ *monzonites*, or ◊ *diorites* but rich in ferromagnesian minerals, e.g. ◊ *hornblende*, ◊ *biotite*, and sometimes ◊ *pyroxene*. Generally regarded as the plutonic equivalent of the hypabyssal ◊ *lamprophyres*, they are found in small ◊ *stocks*, ◊ *plugs*, ◊ *dykes*, etc., or as marginal varieties of other rock types. The term is derived from the Appin district of Argyllshire, Scotland, and might well be replaced by such terms as mela-syenite, mela-diorite, etc.

Aptian. A stratigraphic stage name for the top of the British Lower ◊ *Cretaceous*.

Aptychus (Mollusca). ◊ Fig. 105.

Aquamarine. The blue ◊ *gemstone* variety of the mineral beryl (◊ *Cyclosilicates*).

Aquifer. A water-bearing bed of strata, either by virtue of its ◊ *porosity* or because it is ◊ *pervious*. (◊ *Artesian structure*.)

Aquitanian. A stratigraphic stage name for the top of the Oligocene to the base of the Miocene (◊ *Tertiary System*) in Europe.

Aragonite. An orthorhombic ◊ *carbonate* mineral, $CaCO_3$, found in sedimentary rocks. (◊ Appendix.)

Archaeocyatha. A group of extinct marine ◊ *benthonic* forms confined to the Lower Palaeozoic, and thought to be intermediate in form between the ◊ *Porifera* and ◊ *Coelenterata*. The archaeocyathids are classified on the character of the structure of their walls, and have a calcareous skeleton, usually an inverted cone shape (cornute). In primitive forms the skeleton consists of an outer wall supporting internal radial septa called parietes; in the advanced forms a second inner wall appears and is attached to the circlet of septa. The whole structure in both forms is perforated throughout. (◊ Fig. 1.)

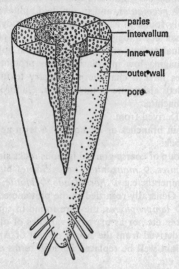

paries
intervallum
inner wall
outer wall
pore

1. A generalised archaeocyathid.

The archaeocyathids range from Lower ◊ *Cambrian* to Middle ◊ *Ordovician* and are not used as zonal indices, although a number of genera are taken as characteristic of certain horizons, and may be present in sufficient numbers to form ◊ *reefs*, as in the Cambrian. Archaeocyathids appear to be confined to clear, shallow-water areas of ◊ *kratonic* sedimentation.

Archean. Regarded as synonymous with ◊ *Precambrian*, but sometimes used to refer to older Precambrian rocks.

Arduino, Giovanni (1713–95). Italian professor and Director of Mines who was the first to make a division of rock-succession into Primary non-fossiliferous, Secondary fossiliferous, and Tertiary arenaceous fossiliferous, with a subsidiary division for alluvial material.

Arenaceous rocks. A group of detrital sedimentary rocks, typically sandstones, in which the particles range in size from 1/16 mm. to 2 mm. The term psammitic, although precisely equivalent to arenaceous, is now used almost entirely to describe metamorphosed arenaceous rocks. The term arenite is a convenient one for all arenaceous rocks regardless of their individual characteristics. It should be noted that the term sandstone is often used as a synonym for arenaceous rocks.

Arenaceous rocks may be either accumulated by wind action or deposited by water action, and in the latter case they may form in marine, brackish, or freshwater environments. They may be sub-divided according to the following criteria:

(1) PARTICLE SIZE:

2 mm.

Very coarse

1 mm.

Coarse

$\frac{1}{2}$ mm.

Medium

$\frac{1}{4}$ mm.

Fine

$\frac{1}{8}$ mm.

Very fine

$\frac{1}{16}$ mm.

(2) SORTING: Arenites are termed well sorted or well ◊ graded if the particle size range is small and poorly sorted or poorly graded if the particle size range is large.

(3) PARTICLE SHAPE: Arenaceous rocks in which the ◊ particle shape is angular to sub-angular are termed ◊ grits, while the remainder, with rounded to sub-rounded grains, are termed sandstones.

(4) MINERALOGY OF THE GRAINS: In most arenaceous rocks the grains are predominantly quartz, and unless some qualifying term is used it may be assumed that a quartzose sandstone is meant. The following are some of the more important minerals other than quartz which may occur in significant quantities in sandstone: (a) Feldspars, in feldspathic sandstone or arkose (see below). (b) Mica (usually white mica), in micaceous

sandstones or flagstones. (c) ◊ *Glauconite*, in glauconitic sandstones (greensands – see below), commonly an ◊ *authigenic* mineral. (d) Iron oxides (◊ *magnetite*, ◊ *ilmenite*, etc.), in black sands. These minerals, together with ◊ *garnet*, ◊ *zircon*, ◊ *rutile*, ◊ *monazite*, etc., may make up the bulk of the sediment. Modern black sands (India, Brazil) are of economic importance. (e) Rock particles, in lithic sandstones and greywackes (see below).

(5) MINERALOGY OF THE CEMENT: For loose grains to become sandstone, a cement is necessary (◊ *Diagenesis*). The following are the commonest: (a) Calcareous, including dolomitic. (b) Siliceous, giving rise to orthoquartzites; cementation by opal and chalcedonic silica is known. (c) Ferruginous, mainly ◊ *limonite* or ◊ *haematite*. ◊ *Siderite* is rare. (d) Argillaceous. (e) ◊ *Pyrite* (or other sulphides); especially significant of ◊ *anaerobic* environments. (f) ◊ *Fluorite*, ◊ *barytes*, copper minerals (mainly carbonates); probably of ◊ *hydrothermal* origin. (g) ◊ *Anhydrite* and ◊ *gypsum*; probably secondarily derived from local ◊ *evaporite* deposits.
For detrital limestones of sand grade (calcarenites), ◊ *Limestone* (Clastic).
Most sandstones contain a small proportion of non-quartz minerals other than those mentioned above. These are commonly referred to as heavy minerals, since they are denser than bromoform (density 2·85). They consist of resistates (◊ *Sedimentary rocks*), derived from the source rocks, and their identification is used in determining the ◊ *provenance* of the material. Occasionally sediments may be found in which the 'heavy minerals' are unusually abundant, commonly in close proximity to the source. These sometimes contain unstable minerals which are destroyed during transport or ◊ *diagenesis*, e.g. the olivine sands on beaches adjacent to basic igneous masses.
Maturity of a sandstone is a measure of how nearly it approaches an ideal stable state. This implies both compositional maturity (i.e. the elimination of all unstable minerals, leaving quartz), and textural maturity, where the grains are of uniform size and perfectly rounded. It should be noted that any one of these requirements (composition, sorting, roundness) may be fulfilled without necessarily involving the others.
ARKOSES contain notable quantities of feldspar grains, in addition to quartz; other minerals, e.g. micas, are often present in lesser amounts. The use of this term exclusively for sandstones containing more than 25% feldspar (those with less than

25% being termed feldspathic sandstones) is a refinement. Arkoses are commonly regarded as indicative of erosion under arid conditions, and rapid burial.

GREENSANDS contain a sufficient quantity of the mineral glauconite (◊ *Micas*) to give a green colour to the fresh unweathered rock. Weathered or otherwise oxidised 'greensands' are brown. 'Greensand' is also used as a stratigraphic term, especially for the Cretaceous Lower and Upper Greensands of Britain and adjoining parts of North-West Europe.

GREYWACKES consist of fine to coarse, angular to sub-angular particles which are mainly rock fragments (lithic fragments). They are usually poorly sorted and may even be locally pebbly. The cementing material is generally argillaceous, which under conditions of low-grade metamorphism (high-grade diagenesis) may become chloritic. The term lithic sandstones is preferred for rocks of more normal sandstone character, with rounded rock particles instead of quartz grains. Greywackes are a common rock type of ◊ *geosynclines* and are remarkable for the range of ◊ *sedimentary structures* which they display. (◊◊ *Turbidite*.)

Arenigian. A stratigraphic stage name for the European Lower ◊ *Ordovician*.

Arenite. Any ◊ *arenaceous rock*.

Arête. ◊ *Cirque*.

Argilla- (prefix). Clay.

Argillaceous rocks. A group of detrital sedimentary rocks, commonly clays, shales, mudstones, siltstones, and marls. Two grades of particle size are recognised, the silt grade, in which the particles range in size from 1/16–1/256 mm., and clay grade, with particles of less than 1/256 mm. The term pelitic, although precisely equivalent to argillaceous, is now used almost entirely in relation to metamorphosed argillaceous rocks. (For the minerals occurring in argillaceous rocks, ◊ *Clay minerals*.)

In addition to the clay minerals, argillaceous rocks may contain colloidal material, very finely divided quartz, carbonate dust, finely divided carbon and iron pyrites – the latter two minerals being characteristic of rocks formed under ◊ *anaerobic* conditions. Argillaceous rocks are almost always laid down in water – fresh, brackish, or marine. Their mineralogy is to some extent controlled by their environment of deposition.

CLAY is the term generally reserved for material which is plastic when wet and has no well-developed parting along the bedding planes, although it may display banding.

SHALE has a well-marked ◊ *bedding-plane* fissility, primarily due to the orientation of the clay mineral particles parallel to the bedding planes. Shales do not form a plastic mass when wet, although they may disintegrate when immersed in water.

MUDSTONE is a term used for rocks which are similar to shales in their non-plasticity, cohesion, and lower water content, but lack the bedding plane fissility.

SILTSTONES are similar to mudstones but consist of a predominance of silt-grade material.

MARL is a calcareous mudstone.

The above constitute the most important general types. Special types are : ◊ *ball clay* (pipe clay); boulder clay (◊ *Drift* (3)); ◊ *china clay*; fault gouge (◊ *Rock flour*); ◊ *fireclay*; and ◊ *fullers' earth*. Numerous other names are used in the ceramic industry for clays having certain properties or specialised uses.

Because of their extremely fine grain, the study of clays presents difficulties. Modern techniques include electron microscopy, differential thermal analysis (DTA), X-ray analysis, absorption, and staining.

The term 'lutite' is used for any rock consisting entirely of particles in the silt/clay grades of ◊ *particle size*, e.g. calci-lutite – a limestone in which the particles are less than 1/16 mm. in diameter.

Argovian. A stratigraphic stage name for the European Lower ◊ *Corallian* (Jurassic System).

Arikareean. A stratigraphic stage name for the North American equivalent of the European ◊ *Aquitanian*, approximately synonymous with the ◊ *Burdigalian* (Tertiary System).

Arkose. ◊ *Arenaceous rocks*.

Arm (*Echinodermata*). ◊ Figs. 43 and 47.

Armoured clay balls. Rolled masses of clay, the surfaces of which are covered with sand, gravel, and even pebbles, so that the whole mass behaves as a unit which may be transported for some distance without disintegration. Armoured clay balls range in size from a few centimetres to a few tens of centimetres in diameter and their occurrence in a sediment is usually taken as evidence of the proximity of a shore line. They vary in shape from almost spherical bodies to flattened bun-shaped masses, and this is clearly a function of the plasticity of the clay.

Unarmoured clay balls are occasionally found and are usually regarded as produced by the break-up of a clay bed *in situ* with an absolute minimum amount of transportation – sometimes only a matter of centimetres or tens of centimetres. (Cf. ◊ *Nodule*; ◊ *Clay gall*.)

Arsenopyrite. An arsenic mineral, FeAsS, found in hydrothermal veins. (◊ Appendix.)

Artesian structure. A series of sedimentary rocks disposed in such a way that an ◊ *aquifer* holds water under a pressure head between two layers of impermeable strata (◊ Fig. 2). When a

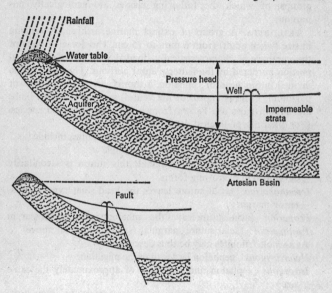

2. An artesian basin. Impermeable strata are unshaded.

well is sunk into the aquifer, water rises to the surface by virtue of the pressure head and may in fact rise as a fountain if unconfined. The term takes its name from the basin of Artois in France, and many similar structures are known, e.g. the London Basin. Since the rate of extraction of water from the London Basin has exceeded the rate at which water penetrates into the rocks, the water no longer rises to ground level. Artesian structures not involving basins are known, but are uncommon.

Arthropoda. A phylum of segmented animals which have an external shell (exoskeleton or carapace) consisting of chitin, strengthened in some cases by $CaCO_3$, and which have jointed

35

limbs. Growth is accompanied by the shedding of the carapace (ecdysis), during which a rapid increase in size precedes the hardening of the new shell.

The phylum Arthropoda contains the largest number of known species, which have colonised every conceivable ecological environment, ranging from abyssal through terrestrial to pools of crude oil (petroleum). It is divided into a very large number of groups, of which the following classes are geologically important:

TRILOBITA. A group of extinct marine arthropods ranging in size (when adult) from 6 mm. to 75 cm. The dorsal surface of the carapace is characteristically divided into an inflated axial portion bordered by the flatter pleural portions. The body is also divided into a head or cephalon, a thorax, and a tail portion or pygidium. Little is known of the ventral surface, although limbs and mouth parts are known from a few well-preserved species. (For other morphological details, ◊ Fig. 3.)

Terms used to describe the facial suture of trilobites include:

Gonatoparian   facial suture cuts genal angle

Hypoparian   facial suture marginal; this suture is secondarily developed in burrowing forms

Opisthoparian   facial suture leaves the head shield via the posterior margin

Proparian   facial suture leaves the cephalon via the lateral margin

Protoparian   facial suture marginal (see Hypoparian, above)

As a whole trilobites may be thus described:

Heteropygous   cephalon larger than the pygidium

Isopygous   cephalon and pygidium of approximately the same size

Classification of the Trilobita is surrounded with difficulties, mainly because no single characteristic has yet been found which exhibits a gradual evolutionary change within the group without being adaptational in one way or another. The classification is further complicated by the fact that ◊ *neotony* occurs within the group. The original classification was based on the character of the facial suture, but this is no longer tenable and has been superseded by a division of the Class into Six Orders, which group together genera of similar morphology without reference to any single character.

The Trilobita range from Lower ◊ *Cambrian* to ◊ *Permian*, although there must have been an extensive Precambrian history to this group. They have been used as zonal indices in the Cam-

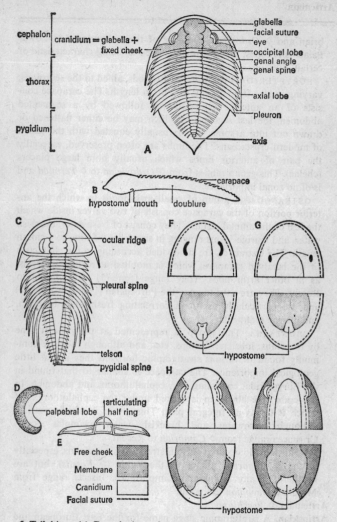

3. Trilobites. (A) Dorsal view of a generalised trilobite; (B) Longitudinal section of a generalised trilobite; (C) Dorsal view of a trilobite; (D) Detail of an eye, showing the palpebral lobe; (E) Transverse thoracic section, showing the articulating half ring; (F–I) Dorsal (upper) and ventral (lower) views of stylised cephalons, showing various relationships of the eye, free cheek, cranidium, facial suture, and hypostome.

brian and ◊ *Ordovician*. Because of their abundance in Lower Palaeozoic times, many forms are used as being characteristic of certain horizons.

EURYPTERIDA. Freshwater arthropods, allied to the scorpions, varying in size from 10 cm. to 2 m. in length. The carapace consists of an anterior cephalothorax followed by a segmented abdomen, the last segment of which may be either flattened or drawn out into a spine, and is usually equated with the telson of modern crustaceans. The limbs are often preserved, especially the pair of anterior limbs which usually bore large pincers (chelae). This group ranges from ◊ *Ordovician* to ◊ *Permian* and is of no zonal importance.

OSTRACODA. A group of small arthropods in which the anterior portion of the carapace consists of two valves inside which the rest of the animal resides. They consist of freshwater, brackish water, and marine forms ranging in size from 0·5 mm. to 1 cm. in length. The Ostracoda are subdivided according to the characters of the bivalved carapace, which is moulted at regular intervals, as in other arthropods. They range from (?) ◊ *Cambrian* to Recent. They are used to zone the Purbeckian (Upper ◊ *Jurassic*) and are particularly useful in correlating freshwater, brackish, and marine sediments.

CRUSTACEA. This group is represented at the present time by the crabs, lobsters, shrimps, etc., and although they are commonly found in various stratigraphic horizons they are of little geological importance. The carapace is similar to that found in the Eurypterida, consisting of a cephalothorax and abdomen. In the crabs, the abdomen is tucked under the cephalothorax, of which it forms an integral part. This group is mainly marine, although freshwater and brackish water examples occur. Crustacea range from ◊ *Cambrian* to Recent.

Fossil insects do occur at various stratigraphic horizons, especially in the ◊ *Oligocene* amber of the Baltic (◊ *Fossils*), but are nowhere of any stratigraphic importance. Insects range from Middle ◊ *Devonian* to Recent.

Articulata. ◊ *Brachiopoda*.

Artinskian. A stratigraphic stage name for the East European and Russian Lower ◊ *Permian*.

Asbestos. The name given to fibrous varieties of several distinct mineral species. All are silicates and common varieties are ◊ *tremolite*, crocidolite (a fibrous ◊ *riebeckite*), and chrysotile (fibrous ◊ *serpentine*). They are important as they can be felted and woven in the same way as any other fibre, the resulting product being 'fireproof'. (◊ *Amphiboles*; *Layer-lattice minerals*.)

Ash. The unconsolidated fine-grained material formed as a result of volcanic explosions. (◊ *Pyroclastic rocks*.)

Ashgillian. A stratigraphic stage name for the European Upper ◊ *Ordovician*.

Asphalt. A naturally occurring hydrocarbon of very viscous character. Some asphalts are just pourable under normal conditions, whereas others are virtually solid. They may occur as the high-boiling-point residuum of an oil pool after the lighter fraction has evaporated. The most noted occurrences are the Trinidad Pitch Lake and the Athabasca Tar Sands. (◊ *Tar pits*; *Hydrocarbon minerals*.)

Assimilation. A process by which material is incorporated into an igneous rock by a process of melting and solution. If this proceeds to any large extent, the final composition of the rock may well be significantly altered. Assimilation is often regarded as a marginal feature of igneous rocks, but may well be of greater importance than is generally realised, especially as a process for producing unusual rock types. (◊ *Batholith*; *Granitisation*.)

Astartian. A synonym of ◊ *Sequanian*.

Asteroidea. ◊ *Echinodermata*.

Asthenosphere. A name used for the lower part of the sima (◊ *Crust of the Earth*), which is presumed to have low strength and little rigidity when compared with the ◊ *lithosphere*.

Astian. A stratigraphic stage name for the Upper Pliocene (◊ *Tertiary System*) in Europe.

Astreoid. ◊ *Coelenterata* and Fig. 23.

Atacamite. A copper ◊ *ore mineral*, $CuCl_2.3Cu(OH)_2$, found in the oxidised zones of copper lodes. (◊ Appendix.)

Atmophile. A term in ◊ *geochemistry* for the elements which are concentrated in the atmosphere.

Atokan. A stratigraphic stage name for the Lower ◊ *Pennsylvanian* in the U.S.A. (Carboniferous System).

Atoll. ◊ *Reef*.

Atomic structure. The position of the atoms constituting a substance relative to one another. The structural pattern so developed is called the atomic lattice, and consists of a number of repeated units of pattern. (◊ *Unit cell*; *Silicates*.)

Augen (German, 'eyes'). Clots of very coarse crystals which develop locally in the quartzo-feldspathic bands of some ◊ *gneisses*. Rocks containing them are known as augen-gneisses.

Augite. A mineral of the ◊ *pyroxene* group, $(Ca,Mg,Fe,Al)_2$ $(Al,Si)_2O_6$, found in ultrabasic and basic volcanic and plutonic rocks. (◊ Appendix.)

Augitite. A ◊ *basalt* consisting of ◊ *augite* phenocrysts, sometimes with ◊ *biotite* or ◊ *hornblende*, set in a glassy ◊ *groundmass* which is usually soda-rich.

Aureole. ◊ *Metamorphic aureole.*

Authigenic, Authigenous. A term implying a development in place during or after deposition. It is mainly applied to sedimentary material, e.g. secondary overgrowths of quartz on sand grains, and overgrowths of calcite on shell fragments. (Cf. ◊ *Allogenic.*)

Auto- (prefix). Self.

Autochthonous. A term used to imply 'not transported', i.e. 'in place'. For example, an autochthonous conglomerate is one formed by the break-up of an underlying bed without transport being involved. An autochthonous sediment is one in which the main constituents have formed *in situ*, e.g. ◊ *evaporites.* 'Autochthon' is a term applied in Alpine geology to rocks which, although they have been faulted and folded, have not been tectonically transported to any great extent, i.e. recumbently ◊ *folded* on a large scale. However an autochthonous ◊ *nappe* is a recumbent fold which can be traced back to its root zone. In this sense the opposite of autochthonous is not the usual ◊ *allochthonous* but exotic, implying a completely disconnected nappe. (Cf. ◊ *Allochthonous;* ◊ *Granite series.*)

Automorphic. A term used to describe the ◊ *texture* of a rock in which the mineral grains display fully developed crystal form.

Autotheca. ◊ *Chordata* and Figs. 17 and 18.

Autunian. A stratigraphic stage name for the European Lower ◊ *Permian.*

Auversian. A synonym of ◊ *Ledian* (Tertiary System).

Aves. ◊ *Chordata.*

Avonian. A synonym of ◊ *Dinantian.*

Axes, crystallographic. Directions in space selected so that crystals belonging to the various systems may be conveniently referred to them by means of ◊ *intercepts.* In normal circumstances, crystallographic axes are selected from amongst the axes of ◊ *symmetry* of a crystal; where this is not possible, arbitrary decisions have to be made.

Axes may be divided into two groups: the orthogonal, in which the three directions chosen are mutually at right angles, and the non-orthogonal, in which one or more directions is not at right angles to the other directions.

The cubic, tetragonal, and orthorhombic systems are referred to orthogonal axes and the monoclinic and triclinic systems to non-orthogonal axes. The hexagonal and the trigonal systems are referred to a special set of axes: three horizontal axes, arranged mutually at 120°, at right angles to a vertical axis.

Since axes are directions in space, their lengths are infinite and thus the phrase 'The orthorhombic system is referred to three axes

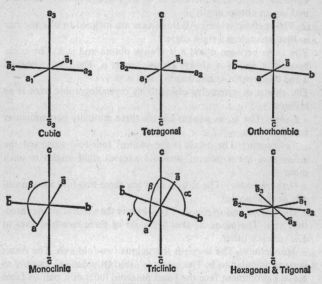

4. Crystallographic axes.

mutually at right angles, but of different lengths' is strictly inaccurate (although commonly used). The correct wording should be 'The orthorhombic system is referred to three axes mutually at right angles and having unequal ◊ *parameters*'.

When the crystal is held in the 'reading' position, the back-to-front axis is labelled 'a', the left-to-right axis 'b', and the vertical axis 'c'. The front, right hand, and upper ends respectively of these axes are positive and the opposite ends negative. (◊ Fig. 4.) In the cubic system, the parameters are all equal, the three axes become interchangeable and are labelled a_1, a_2, a_3 respectively. In the tetragonal system, the two horizontal axes have equal parameters and are labelled a_1, a_2, while in the hexagonal system all three horizontal axes have equal parameters and are correspondingly labelled a_1, a_2, a_3. The special convention used in the notation of horizontal hexagonal axes is shown in the diagram. The orthorhombic system is defined above.

For the non-orthogonal axes two cases are recognised:

(1) The monoclinic system. Two axes, b and c, are at right angles while the third, a, is inclined to c, the positive parts of a and c making an obtuse angle, β.

(2) The triclinic system. All three axes are inclined to one another so that no angle is a right angle.

The angle between c and a is always obtuse and is β. The angle between c and b is always obtuse and is α. The angle between a and b is greater or less than 90° and is γ.

The choice of symmetry elements as crystallographic axes is as follows:

Cubic: The a_1, a_2, a_3 axes are the three mutually perpendicular four-fold axes.

Tetragonal: The c-axis is the vertical four-fold axis, and the a_1, a_2 axes are a pair of two-fold axes at right angles to each other.

Orthorhombic: The a, b, c, axes are three two-fold axes mutually at right angles.

Hexagonal and trigonal: The c-axis is the vertical six or three-fold axis. The a_1, a_2, a_3 axes are a set of three two-fold axes at 120° to each other.

Monoclinic: The b-axis is the unique two-fold axis. The c-axis is perpendicular to b. The a-axis is arbitrarily defined, usually to make a prominent face the basal pinacoid (001) or a pair of faces the clino-domes (011 and 0$\bar{1}$1).

Triclinic: All axes are arbitrarily selected so as to be parallel to edges between suitable pairs of prominent faces.

Axes, optic. ◊ *Indicatrix, Ellipsoid*. The direction perpendicular to a circular cross-section of an indicatrix. For a uniaxial mineral, the unique circular cross-section has either X or Z perpendicular to it and hence a single optic axis. Sections perpendicular to this optic axis are ◊ *isotropic* but yield uniaxial ◊ *interference figures*. In biaxial crystals, two circular cross-sections exist, which are symmetrically arranged relative to the axes of the indicatrix. If one considers the Y–Z plane of the indicatrix, two points can be found on the curves between X and Z where the refractive index equals that in the Y direction. These positions will define circular cross-sections and hence the optic axes perpendicular to them. (◊ Fig. 5.)

The acute angle between the two possible optic axes is the optic axial angle, 2V, and the bisector of this angle is the acute bisectrix. When 2V = O a biaxial mineral has degenerated into

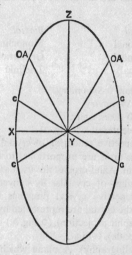

5. Optic axes. X, Y, Z, axes of ellipsoid (X, obtuse bisectrix, Z, acute bisectrix); c–c, circular cross-sections; OA, optic axes.

a uniaxial one. If 2V = 90° there is no acute bisectrix and the mineral is neutral. Owing to refraction, the emergence of the optic axes from the mineral into air results in the angle between them appearing to be greater than 2V; this emergent angle is called 2E:

$$\text{Sin } E = (\sin V) \times (\text{refractive index in the Y direction})$$

A section perpendicular to the acute bisectrix yields a biaxial interference figure; a section perpendicular to an optic axis yields an optic axis figure of a single ('compass needle') ◊ *isogyre*; a section perpendicular to the obtuse bisectrix yields the so-called 'flash figure' – a weak biaxial figure, visible over a very narrow angle of observation as compared with a normal acute bisectrix figure.

If Z is the acute bisectrix in a biaxial mineral, or the optic axis direction in a uniaxial mineral, then the mineral is said to be positive; if X is the acute bisectrix in a biaxial mineral, or the optic axis direction in a uniaxial mineral, then the mineral is said to be negative.

The crystal systems are divided between uniaxial and biaxial as follows:

Uniaxial	*Biaxial*
Hexagonal	Monoclinic
Tetragonal	Orthorhombic
(Trigonal)	Triclinic

Axes, rotary inversion. ◊ *Symmetry.*

Axes, symmetry. ◊ *Symmetry.*

Axial angle, optic. ◊ *Axes, optic.*

Axial complex (Coelenterata). ◊ Fig. 21.

Axial cross. A perspective rendering of crystal axes drawn so that the lengths of the arms are proportional to the ◊ *parameters* of a particular crystal. Axial crosses are used as an aid in preparing perspective drawings of crystals. By convention, the vanishing point of all perspective crystal drawings is infinity; thus all parallel edges in the crystal are represented by parallel lines in the drawing (clinographic projection). (◊ Fig. 4.)

Axial lobe (Arthropoda). ◊ Fig. 3.

Axial plane (crystallography). A plane which contains two crystallographic axes.

Axial-plane cleavage. ◊ *Cleavage, rock.*

Axial ratios. ◊ *Parameters.*

Axinite. A mineral of the ◊ *cyclosilicate* group, $Ca_2(Mn,Fe'')Al_2 BO_3(Si_4O_{12})(OH)$, formed by the contact ◊ *metamorphism* of limestone or by ◊ *pneumatolysis*. (◊ Appendix.)

Axis (Arthropoda). ◊ Fig. 3.

Axis, optic. ◊ *Axes, optic.*

Axis, tectonic. Tectonic features can be related to a set of three-dimensional rectangular axes which are chosen so as to bear a convenient relationship to the forces involved, or supposed to be involved. The a-axis is the direction of movement (= the direction of tectonic transport) – shear or compressional; b is the normal to a in the most prominent s-plane present; and c is perpendicular to the ab-plane. The special importance of these axes is in relating petrofabric analysis to the macroscopic structures. It should be noted that the b-tectonic axis (or ◊ *fabric* axis) is parallel to the b-axis of the strain ◊ *ellipsoid*. (◊ Fig. 6.)

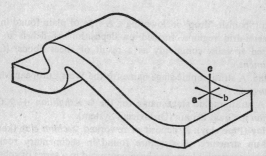

6. Tectonic axes.

Axis, zone. ◊ *Zone, crystallographic.*

Azoic (lit. 'no life'). A name given to ◊ *Precambrian* strata presumed to have been formed before life appeared on the Earth. The name is of doubtful validity and is obsolescent.

Azurite. A copper ◊ *ore mineral*, $2 CuCO_3Cu(OH)_2$, found associated with other oxidised copper minerals. (◊ *Carbonates* and Appendix.)

Bajada (Spanish, 'drop' or 'lowering'). A type of plain found in arid or semi-arid regions, formed by deposition of debris in fan-shaped spreads, commonly as a result of sheet floods. (Cf. ◊ *Pediment*.)

Bajocian. A stratigraphic stage name for the West European Middle ◊ *Jurassic*.

Bala. A stratigraphic stage name for the ◊ *Ashgillian* + ◊ *Caradocian* in Great Britain (Ordovician System).

Ball clay (Pipe clay). A deposit of reworked ◊ *china clay* (kaolin).

Balled-up structure. A structure found in sedimentary rocks in which a particular layer has been converted to a series of spherical or near-spherical masses (pseudonodules), by a process of local deformation of the layer while in an unconsolidated state. ◊ *Slumping* and ◊ *collapse* are two processes which commonly give rise to balling-up. (◊ Fig. 24.)

Ballstone. A sedimentary rock consisting essentially of an ◊ *argillaceous* matrix surrounding numerous nearly spherical nodular masses, usually calcareous in nature. They can often be traced laterally into continuous layers of limestone and are commonly associated with reef limestones. A secondary origin has been claimed in some cases. (◊ *Concretions*.)

Banakite. One of a variety of saturated or oversaturated ◊ *trachybasalts* (sometimes inaccurately classified as trachyandesites).

Banatite. Quartz ◊ *monzonite*.

Banded structure. A term applied to a rock showing 'striping', i.e. linear structure. A rock is said to have banded structure when the layers have varying physical properties and/or chemical compositions. Individual 'stripes' (bands) should have a measurable thickness; where the bands are extremely thin the term ◊ *lamination* or laminar structure is more appropriate.

Banket. The name given to a ◊ *conglomerate* consisting mainly of quartz pebbles with ◊ *pyrite*, found in the Precambrian of South Africa and Ghana. It is important in South Africa as the source of the gold in the Witwatersrand goldfield, where the individual beds of banket are referred to as 'reefs'.

Bar. A more or less linear deposit of sand and/or gravel generally found in the sea, which is parallel or sub-parallel to the coastline. Bars are usually formed as a result of ◊ *longshore drift* and may extend across the mouths of bays and inlets. The term 'bar' is also used for any rock mass which produces a shallow-water zone between two deeper water regions, e.g. the rock bars at the mouths of some ◊ *fiords*. (◊ *Spit*; *Tombolo*.)

Barchan. ◊ *Dunes.*

Barite. ◊ *Barytes.*

Barkevicite. A monoclinic mineral of the ◊ *amphibole* group.

Barrande, Joachim (1799–1883). French-born geologist who spent his life producing a detailed account of the ◊ *Silurian System* in Bohemia, and amassing a famous collection of fossils which is still regarded as one of the most important sources of information about Silurian faunas.

Barremian. A stratigraphic stage name for the European mid Lower ◊ *Cretaceous.*

Barrow's (Barrovian) metamorphic zones. ◊ *Zone, metamorphic.*

Barstovian. A stratigraphic stage name, the North American equivalent of the European Upper ◊ *Vindobonian* (Tertiary System).

Bartonian. A stratigraphic stage name for the top of the European Upper Eocene (◊ *Tertiary System*).

Barytes (Barite). A barium mineral, $BaSO_4$, found in ◊ *hydrothermal* veins and ◊ *replacements* and in nodular masses. (◊ Appendix.)

Baryto-calcite. A monoclinic ◊ *carbonate* mineral, $BaCa(CO_3)_2$.

Basal conglomerate. A ◊ *conglomerate* formed at the beginning, i.e. the earliest portion, of a stratigraphical unit.

Basalt. A fine-grained, sometimes glassy basic igneous rock. The essential minerals are a calcic ◊ *plagioclase* and ◊ *pyroxene* (usually augite), with or without ◊ *olivine*. ◊ *Magnetite* is an important accessory, while ◊ *quartz*, ◊ *hornblende*, and ◊ *hypersthene* are sometimes present in significant amounts. (For basaltic rocks containing feldspathoids and/or feldspar, ◊ *Alkali basalt*.)

Glassy basalt is called tachylite; basalt glass containing olivine and augite phenocrysts is called limburgite, but analysis suggests that limburgites are commonly ◊ *undersaturated*. Amygdaloidal and porphyritic basalts are common, the ◊ *amygdales* often being filled with ◊ *zeolites*.

Basalts are characterised by low SiO_2 content (45–50%), tholeiites (see below) having the higher content. FeO, MgO, and CaO are generally high, while Na_2O and K_2O are low, especially in the olivine basalts.

Tholeiite is an important type of basalt consisting of basic plagioclase and pigeonite (a pyroxene), with interstitial glass or quartz-alkali feldspar intergrowths. 'Big feldspar basalt' contains large plagioclase ◊ *phenocrysts*, sometimes two or more inches in length. Absarokite is a porphyritic basalt containing a small amount of ◊ *orthoclase* in the groundmass. Augitite consists of augite phenocrysts, sometimes with biotite or hornblende, set in a glassy groundmass, which is usually soda-rich. Ankaramite is an augite-rich olivine mela-basalt, and oceanite an olivine-rich mela-basalt.

With decrease in feldspar content, basalts grade into ◊ *ultrabasic* types, while an increase in the soda content of the plagioclase, to give andesine or oligoclase, coupled with the development of hornblende, produces ◊ *andesite*. Increase in alkali feldspar produces ◊ *trachybasalts* and ◊ *alkali basalts*.

Basalts are generally found in the form of lava flows which may be extensive and are often erupted from fissures and sometimes from central-type vents. Small-scale ◊ *dykes* and ◊ *sills* of basic rock are often fine-grained enough to warrant the use of the term basalt. In many parts of the world basalt flows form great piles, tens of thousands of feet thick, as for example in Iceland, Mull, the Deccan of India, and the Columbia River basalts. A particular type of ◊ *columnar* jointing, which produces hexagonal 'prisms' of basalt forming at right-angles to the surfaces of flows, dykes, or sills, is fairly common, and forms such well-known features as the Giant's Causeway in Antrim, and Fingal's Cave on the island of Staffa in the Hebrides. More than 90% of volcanic rocks are basalt, and more than 90% of basic igneous rocks are basalt.

In regions where basalt flows are developed, small quantities of other lavas are also found.

Basalts are the fine-grained (volcanic) equivalent of the coarse-grained (plutonic) ◊ *gabbros*, and the medium-grained (hypabyssal) ◊ *dolerites*. (For 'oligoclase basalt' and 'andesine basalt', ◊ *Andesite*.)

Basanite. An olivine-bearing ◊ *alkali basalt* containing feldspar and feldspathoid.

Base exchange. In certain ◊ *clay minerals*, the exchange of loosely bonded cations for others as a result of changes in the local concentration of cations.

Base level. The surface to which ◊ *long-profiles* of rivers are related. (◊ Fig. 7.) The base level of a stream is usually the sea, but if there are irregularities in the long-profiles local base levels will

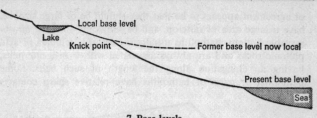

7. Base levels.

be temporarily created, thus defining a series of sub-profiles, each one cutting down to its particular base level. Examples of these local base levels include surfaces of lakes, tributary streams, waterfalls, or ◊ *knick points*. (◊ *Rejuvenation.*)

Basement complex. Usually synonymous with the ◊ *Precambrian*, but the term may also be applied to any widespread association of igneous and metamorphic rocks which are covered unconformably by unmetamorphosed sediments.

Bashkirian. A stratigraphic stage name for the Upper ◊ *Namurian* to the Lower ◊ *Westphalian* in Russia (Carboniferous System).

Basic front. ◊ *Granitisation.*

Basic rocks. A quartz-free igneous rock containing feldspars which are generally more calcic than sodic. The term was used originally as the antithesis of ◊ *acid rock*, and does not imply the presence of free bases in the chemical sense. Rocks in this group, formerly defined in terms of silica percentage, contain between 45 and 55% of silica. Pyroxene and olivine are the common ◊ *ferromagnesian* minerals although ◊ *hornblende* and ◊ *biotite* may occur in small quantities.

'Basic' is not synonymous with ◊ *'alkaline'*. Basic rocks grade into ◊ *intermediate rocks* by an increase in the sodium content of the feldspar, and into ◊ *ultrabasic rocks* by decrease in the amount of feldspar. (Cf. ◊ *Acid rocks*. For typical basic rocks, ◊ *Basalt*; *Dolerite*; *Gabbro*.)

Basin. A depression of large size, which may be of structural or erosional origin.

Basin facies. The sediments and their associated fauna which are found in the deeper parts of a ◊ *basin*. (Cf. ◊ *Shelf facies*.)

Batholith, Bathylith. The term is applied to any large intrusive mass of igneous rock (almost always ◊ *granite*, in its widest sense). Definitions vary greatly from author to author, mainly on the basis of individual ideas as to supposed origin. The main point of agreement appears to be that they should be of large size, i.e. have a large area of outcrop, and be subjacent – having no observable bottom. Batholiths usually consist of a complex of acid plutonic rocks and are always associated with ◊ *orogenic* belts, having an elongation along the length of such belts. Some batholiths or parts of batholiths have relative sharp contacts

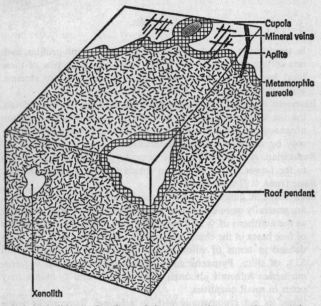

8. A batholith. ◊ *Cupolas* and ◊ *roof pendants* are characteristic of batholiths and other large igneous masses. ◊ *Aplite* veins and ◊ *xenoliths* may also occur.

with the country rocks, while others have diffuse, transitional contacts.

In many batholiths there is evidence that they have replaced the country rock (◊ *Granitisation*), but others have been emplaced by ◊ *stoping* or structural adjustment. Batholiths commonly have mineralised zones associated with them, together with a well-developed ◊ *metamorphic aureole*. (◊ Fig. 8.)

Bathonian. A stratigraphic stage name for part of the Middle ◊ *Jurassic* in Great Britain.

Bathy- (prefix). Depth.

Bathyal. A term used to describe the environment of the ◊ *continental slope* between 200 and 4,000 m. Conditions are much more variable than in the ◊ *abyssal* region, owing to fluctuations in sedimentation, oxygen content, current action, and changes brought about by slumping on the continental slope. The region between 2,000 and 4,000 m. is often one where oozes (◊ *Abyssal deposits*) are formed.

Bauer, Georg. ◊ *Agricola*.

Bauxite. ◊ *Laterite and bauxite*.

Baventian. A stratigraphic stage name for the top of the British Lower Pleistocene (◊ *Tertiary System*).

Bayou (American). An ◊ *ox-bow lake* developed in a delta.

Beach. ◊ *Littoral*.

Beak (Brachiopoda). ◊ Fig. 11.

Beaufort. A stratigraphic stage name for the top of the South African Middle ◊ *Karroo*.

Beaumont, Élie de (1798—1874). French geologist who worked with Dufrénoy over eighteen years to produce a geological map of France which brought him much fame in his lifetime. He also put forward a popular ◊ *orogenic* theory based on distortion of the crust during cooling by variation of pressure – these changes being violent, and sudden, in line with ◊ *Cuvier*'s catastrophic theory.

Beche, Henry de la (1796–1855). English geologist whose work in south-west England was the first in the world to be government-supported, and resulted in the establishment of the ◊ *Geological Survey* in 1835 with himself as first director. The practical knowledge and clarity of his *Manual of Geology* and other books made him widely respected internationally, and his work at the School of Mines did much to stimulate the science in this country. (Traditionally his name is pronounced as in 'Beech'.)

Becke lens. An auxiliary lens on a petrological microscope above the eyepiece, used to view ◊ *interference figures*. (Cf. ◊ *Bertrand lens*.)

Becke line and test. When the illumination of a thin section of rock being viewed under a microscope is reduced by means of the substage iris diaphragm, a bright line can be seen at the boundaries of grains which differ in refractive index from the adjoining material (either another grain or the mounting medium). If the microscope is defocused by racking *up* slightly, the bright line (Becke line) is observed to move *into* the material with the

51

higher refractive index. Conversely, on racking down, the line moves into the material with the lower refractive index. This test is widely used in identifying minerals in thin section, and grains immersed in liquids of varying refractive index. (Cf. ◊ *Shadow test*; ◊ Fig. 9.)

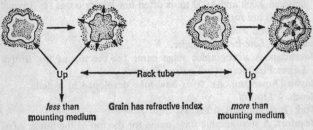

9. The Becke test.

Bed. ◊ *Stratigraphic nomenclature*; *Sedimentary structures*.

Bedding, Bedding plane. A bedding plane is a surface parallel to the surface of deposition, which may or may not have a physical expression. In shales, for example, the rock splits along planes which are bedding planes, whereas in some sandstones bedding planes are marked by changes of colour, grain size, etc., but no plane of preferred splitting occurs. The original attitude of a bedding plane should not be assumed to have been horizontal. (◊ *Cross-bedding*; *Sedimentary structures*; *Stratigraphic nomenclature*.)

Bedding-plane slip. Movement along a ◊ *bedding plane*, usually an ◊ *accommodation structure* during folding or faulting.

Bedoulian. A stratigraphic stage name for the European Lower ◊ *Aptian* (Cretaceous System).

Bedrock. A mining term for the unweathered rock below the soil and ◊ *drift* cover.

Beef. Fibrous ◊ *calcite* (rarely applied to other minerals), commonly showing ◊ *cone-in-cone structure*.

Beekite. A variety of chalcedony (◊ *Silica group of minerals*) often formed as a replacement of calcareous fossils.

Beerbachite. Gabbro ◊ *aplite*.

Beidellite. A ◊ *clay mineral* of the montmorillonite group.

Bekinkinite. ◊ *Teschenite*.

Belemnite, Belemnoidea. ◊ *Mollusca* (Cephalopoda).

Belonites. ◊ *Devitrification.*

Benitoite. A very rare mineral of the ◊ *cyclosilicate* (three-membered ring) group, BaTiSi₃O₉.

Benthos (adj. benthonic). Those animals which live on the sediments of the sea floor, including both mobile and non-mobile forms. The free-living (mobile) forms move to their food, whereas the sessile (non-mobile) forms (which may or may not be fixed) wait for their food to come to them. The characteristics of this second group are radial symmetry and complex food-gathering organs, e.g. Crinoidea (◊ *Echinodermata*). (◊ Fig. 49.)

Bentonite. A special assemblage of ◊ *clay minerals*, in many instances formed by weathering of ◊ *acid* ◊ *lavas* and ◊ *pyroclastic rocks.* (◊ *Argillaceous rocks.*)

Bergschrund. A crevasse at the back of a glacier in a ◊ *cirque.*

Beringer, Johannes (1667–1740). German professor who was the victim of probably the first geological practical joke. His love of strange and 'freakish' fossils led him to accept fireclay forgeries, even to the extent of producing a book about them, in the belief that they were 'divine miracles'.

Berriasian. A stratigraphical stage name for the basal European ◊ *Cretaceous.*

Bertrand lens. An auxiliary lens in a petrological microscope, between the eyepiece and the objective, used to view ◊ *interference figures.* (Cf. ◊ *Becke lens.*)

Beryl. ◊ *Cyclosilicates* and Appendix.

Berzelius, Jacob (1779–1848). Swedish chemist who first presented a mineral classification based upon chemical composition, in particular detailing the ◊ *silicates* as a group as a result of his discovery that silica behaved as an acidic oxide.

Biaxial. ◊ *Axes, optic.*

Billabong (Australian). A synonym for ◊ *ox-bow lake.*

Bioclastic. Term applied to sediments made up of broken fragments of organic skeletal material, the commonest type being bioclastic ◊ *limestones.*

Biocoenosis. A fossil assemblage of forms living together or closely associated before death and subsequent burial. This type of assemblage is more commonly found under conditions of quiet deposition.

Biofacies. A rock unit which contains an assemblage of fossils characteristic of a particular environment; thus there may be several contemporaneous biofacies adjacent to one another. (◊ *Facies.*)

Biogenic (lit. 'life-formed'). A term applied to material produced by the action of living organisms.

Bioherm. Synonymous with large organic ◊ *reef*. The term implies organisms *in situ*, in their position of growth. (◊ Fig. 125.)

Biolith. A rock formed by organic processes, usually of organic material, and including bedded structures.

Biolithite. A ◊ *limestone* composed entirely of the skeletons of reef-building organisms.

Biophile. The elements concentrated in living matter. (◊ *Geochemistry*.)

Bioseries. ◊ *Stratigraphic nomenclature.*

Biosphere. The sum total of all organic life living on or in the surface of the Earth.

Biostrome. Any sheet-like mass of purely organic material derived from sedentary organisms. (◊ *Limestone*.)

Biotite. ◊ *Micas* (Trioctahedral) and Appendix.

Biotope. A ◊ *palaeoecological* term meaning a distinct environmental unit.

Biozone. Rocks deposited during the total time range of a species, as evidenced by the presence of fossils of the species.

Birefringence (lit. 'double refraction'). All substances capable of transmitting light (excluding non-crystalline material and substances belonging to the cubic system) have the property of passing light through them at two different speeds, corresponding to two different refractive indices. The numerical value of birefringence is the difference between the greatest refractive index and the least refractive index of a mineral. A beam of light passing through a thin plate of a mineral will be split into two beams, one 'fast' and one 'slow', and the difference between the corresponding refractive indices will vary from a minimum to a maximum according to the direction in which the plate is cut relative to the crystallographic axes. A characteristic effect is the production of a double image, as seen when a spot is viewed through a calcite cleavage-rhomb. (◊ *Polarisation colours*; *Vibration directions*.)

Bischof, Karl (1792–1870). German chemistry professor who developed the study of chemical geology, in particular with his detailed analyses of rocks and his descriptions of the chemical action of water on rocks.

Bisectrix. ◊ *Axes, optic.*

Biserial. ◊ *Chordata* (Graptolithina).

Biserial pores. ◊ *Echinodermata* (Echinoidea) and Fig. 44.

Bitumen. Naturally occurring tar-like ◊ *hydrocarbon mineral* of indefinite composition. It ranges in consistency from a thick liquid to a brittle solid. (Cf. ◊ *Asphalt*.)

Bivalvia (lit. 'two valves'). A name for the class of ◊ *Mollusca* better known as Lamellibranchiata or Pelecypoda. A recent decision to revive the name Bivalvia on grounds of priority has hardly yet influenced text-books, although original papers relating to the group employ it. It is an unfortunate choice, since all 'bivalves' are not 'Bivalvia' – Brachiopoda and Ostracoda are both 'two-valved' – and confusion by inexperienced geologists seems inevitable.

Bivium (Echinodermata). ◊ Fig. 45.

Black-band ironstone. A sedimentary rock consisting mainly of iron carbonate (siderite) and coaly material, which is effectively self-smelting. The most famous deposit was the Staffordshire black-band ironstone, which reached a maximum thickness of 50 ft, but thinner seams are found in many parts of the Coal Measures. (◊ *Sedimentary iron ores*.)

Black earth. ◊ *Chernozem*.

Blackriverian. A stratigraphic stage name for the mid Middle ◊ *Ordovician* in North America.

Blancan. A stratigraphic stage name, the North American equivalent of the European ◊ *Villafranchian* (Tertiary System).

Blanket sand. A thin layer of sandstone horizontally extensive in two dimensions. (◊ *Sedimentary structures* (1).)

-blastic (suffix). Implies the growth *in situ* of crystals in a metamorphic rock, e.g. poikiloblastic – a ◊ *poikilitic*-type texture which develops during metamorphic recrystallisation.

Blasto- (prefix). Implies growth during metamorphic recrystallisation which partly destroys a pre-existing texture, e.g. blastoporphyritic – a metamorphosed ◊ *porphyritic* rock in which phenocrysts remain only as relics.

Blastoidea. ◊ *Echinodermata*.

Block lava. ◊ *Aa*.

Blue elvan ◊ *Elvan*.

Blue ground (Kimberlite). A thoroughly brecciated and altered ◊ *ultrabasic rock*. A ◊ *mica* (phlogopite) peridotite containing bronzite and chrome-diopside, formerly worked for diamonds.

Blue John. ◊ *Fluorite*.

Bode's law. ◊ *Accretion theory*.

Body chamber (Mollusca). ◊ Fig. 105.

Body whorl (Mollusca). ◊ Fig. 100.

Bog burst. When an area of peat or swampy ground is confined, either by some slight physical feature or by some organic agency (e.g. small bushes), the bog may build up within the confined space and subsequently become oversaturated with water and flow out as a kind of mud flow. The term is often used as a synonym for a mud flow derived from an organic swamp, however caused.

Boghead, Boghead coal. A type of ◊ *coal* consisting of algal material with some fungal matter. Torbanite, the well-known oil shale of the Midland Valley of Scotland, is a typical boghead coal.

Bog iron ore. Iron hydroxide deposited in swamps and lakes, probably by bacterial action. Such deposits are of some commercial importance, as they may have a low content of phosphorus and sulphur and can yield very pure iron. (◊ *Sedimentary iron ores.*)

Bojite. A ◊ *gabbro* containing primary ◊ *amphibole* – either barkevicite or hornblende – as opposed to many hornblende gabbros in which the amphibole is secondary.

Bole. A fossil ◊ *laterite* interbedded with lava flows, usually basalts.

Bombs, volcanic. Large discrete masses of molten ◊ *lava* ejected into the air from a volcano. (◊ *Pyroclastic rocks.*)

Bone bed. A sedimentary layer characterised by a high proportion of fossil bones, scales, teeth, ◊ *coprolites*, etc. (◊ *Phosphatic deposits*). Bone beds are commonly pyritic, and rarely very thick. (◊ *Bone breccia; Remanié fossils.*)

Bone breccia. A large unstratified mass of bones or bone fragments cemented together. The commonest type of bone breccia develops where bones accumulate in a cave, or in a sink hole in a limestone region, subsequently becoming cemented by the deposition of calcium carbonate in the form of stalagmitic material. (◊ *Tar pits.*)

Bononian. A stratigraphic stage name equivalent to ◊ *Portlandian* (Jurassic System).

Boreal (lit. 'northern'). Implying a northern element in a fauna. Since the term originated in the Northern Hemisphere it is effectively synonymous with 'cold', and is often used with reference to an Arctic element in a fauna, especially in ◊ *Tertiary* and ◊ *Quaternary* stratigraphy.

Bornite. A copper ◊ *ore mineral*, Cu_5FeS_4, found in ◊ *hydrothermal* copper veins. (◊ Appendix.)

Borolanite. ◊ *Alkali syenite.*

Bort. A granular mass of ◊ *anhedral* ◊ *diamonds*, used industrially for cutting and grinding. Black, low-density bort is called carbonado.

Boss. (1) A mass of plutonic igneous rock having a circular plan and steep contacts. (◊ *Stock*.)

(2) (Echinodermata). ◊ Fig. 44.

Bostonites. Certain microsyenites, often displaying ◊ *trachytic* features, consisting almost entirely of alkali ◊ *feldspars*. (◊ *Syenite*.)

Bothnian. A stratigraphic division of the Baltic ◊ *Precambrian* Post-◊*Svionian*, Pre-◊*Karelian*.

Bottomset bed. One of the three main types of ◊ *delta* bed. It forms on the sea bottom beyond the seaward face of the delta.

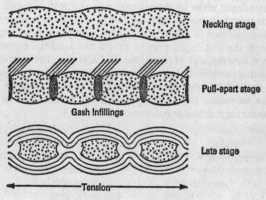

Necking stage

Pull-apart stage

Gash infillings

Late stage

◄─────Tension─────►

10. The formation of a boudinage.

Botryoidal. A term applied to minerals occurring as aggregates with rounded surfaces, resembling a bunch of grapes. The masses commonly consist of radiating acicular (needle-like) crystals. 'Reniform', named from the kidney-like character of the masses, implies a larger-scale variety. 'Mammilated' (from *mammae*, breasts) describes the same formation on an even larger scale.

Boudinage (French *boudin*, 'sausage'). A minor structure arising from tensional forces. It develops by the stretching of a ◊ *competent bed* along bedding planes, giving rise to pull-apart structures, tension cracks or necks, which may become filled with incompetent material from either side. The usual appearance in cross-section is that of a string of sausages. (◊ Fig. 10.)

Boulder bed. A sedimentary rock consisting of a high proportion of large blocks of rock (diameter greater than 256 mm.) together with fine-grained material. Many boulder beds consist of ice-transported material.

Boulder clay (Till). ◊ *Drift* (3).

Bounce marks. Tool marks produced by an object ('tool') such as a pebble, shell, or bone being bounced along the sea floor. They are especially associated with ◊ *turbidites*. (◊ *Sedimentary structures* and Fig. 134.)

Bourne. An intermittent or seasonal ◊ *spring*.

Bowen's reaction series. N. L. Bowen, an American petrologist who studied the crystallisation of silicate melts in the laboratory, postulated two sequences of minerals which he regarded as being in the normal order of crystallisation from a melt. He also suggested that if a mineral which formed at an early stage of crystallisation persisted to a later stage, there would then be a reaction between the still liquid portion of the melt and the crystal, to form a new mineral which would be the stable phase under the prevailing conditions. The two reaction series which he suggested are, respectively, the Discontinuous and Continuous series, and it can be shown that these are effectively independent of one another except at low temperatures.

Discontinuous	*Continuous*
Olivine	Calcic plagioclase
Orthopyroxene*	Calci-sodic plagioclase
Clino-pyroxene*	Sodi-calcic plagioclase
Amphibole	Sodic plagioclase
Biotite	Potash feldspar

Muscovite

Quartz

*More precisely Ca-free pyroxene, and calcic-pyroxene respectively.

It will be observed that the discontinuous reaction series consists of ◊ *ferromagnesian minerals* and that each 'step' is a distinct phase. The continuous reaction series consists mainly of the plagioclase feldspars, and here there are no separate 'steps', simply a continuous gradation from the calcium-rich end to the sodium-rich end of the plagioclase series. At high temperatures the gradation continues from sodium to potassium feldspar, but at low temperatures there is a gap between these two types. Both series represent falling energy and it will be observed that the progression corresponds approximately to the sequence basic–intermediate–acid. It is sometimes possible to observe the operation of the reaction series in rocks from the development of such phenomena as zoned ◊ *feldspars*, ◊ *zoned crystals* and/or ◊ *reaction rims* of a later mineral enveloping an earlier one.

Box-stones. Hollow ◊ *concretions*.

Boxwork. The pattern of ◊ *limonite* left after sulphide minerals containing iron have been oxidised and other cations removed.

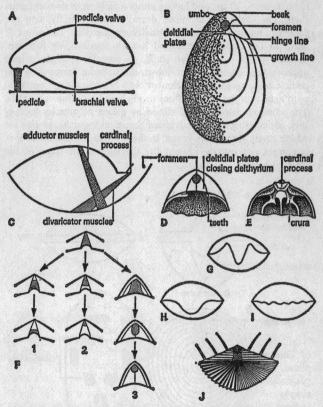

11. Articulate brachiopods. (A) Side view; (B) Pedicle view; (C) Longitudinal section; (D) Detail of pedicle view; (E) Detail of brachial valve; (F) Evolution of various types of delthyrium closure; (G–I) Types of margin; (G) Plications; (H) Sulcus; (I) Crenulations; (J) Exterior morphology (the left-hand side shows coarse ribbing, the right, fine ribbing; the horizontal shading represents the interarea).

The various sulphide minerals give rise to characteristic boxworks. (◊ *Gossan*.)

Brachial valve. ◊ *Brachiopoda* and Figs. 11, 12 and 13.

Brachidium (Brachiopoda). ◊ Fig. 13.

Brachiole (Echinodermata). ◊ Fig. 41.

Brachiopoda. A group of bivalved marine animals varying in size from 5 mm. to 20 cm. and having either a calcareous or a chitinous shell. The shells are arranged dorsally and ventrally, and are generally unequal in size and usually symmetrical. The larger ◊ *valve* in the typical brachiopod has an aperture, or foramen, through which a muscular stalk, or pedicle, emerges, which anchors the animal to the sub-stratum. Some forms lose the pedicle as a means of attachment, and become either cemented onto debris on the stratum floor or anchored by means of spines. The valves are opened and closed by means of muscles (adductor), articulating in the advanced forms along a hinge line on which a pair of teeth are borne. In these forms internal calcareous structures are usually present for the support of some of the internal organs.

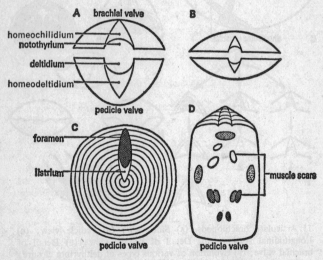

12. Inarticulate brachiopods. (A) Hinge views; (B) Hinge view of an equivalve inarticulate; (C) *Orbiculoidea*; (D) Interior of pedicle valve showing muscle scars.

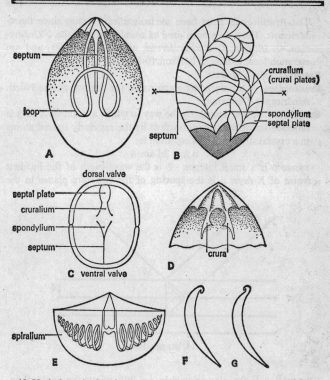

13. Variations in brachiopod morphology. (A) Internal view of a
brachial valve showing loop of a Terebratulid; (B) Longitudinal
section of a Pentamerid; (C) Section along x–x of (B); (D) Interior
view of the ventral valve of a Rhynchonellid; (E) Interior of spiri-
ferid brachiopod showing the spiralium; (F) Convexi-concave shell
of a Strophomenid (brachial valve convex, pedicle valve gently
concave); (G) Concavo-convex shell of a Strophomenid (brachial
valve concave, pedicle valve convex). A, D, E are types of brachidium.

The Brachiopods are divided into two classes, the Inarticulata
and the Articulata. The former are primitive forms, having no
hinge structure, the shell being commonly chitinous. They range
from Lower ◊ *Cambrian* to Recent. Articulate forms have a hinge
structure and commonly possess a calcareous shell: they also range
from Lower Cambrian to Recent.

61

The Brachiopods have been an insignificant group since the ◊ *Mesozoic*. They have been used as zonal indices in the ◊ *Ordovician*, ◊ *Silurian*, ◊ *Carboniferous*, and ◊ *Cretaceous*, and are common local indices of horizon. (◊ Figs. 11–13.)

Brachy- (prefix). Short.

Bradfordian. A stratigraphic stage name for the top of the North American ◊ *Devonian*.

Bragg's law. A law describing the way in which a beam of X-rays is reflected (more strictly, diffracted) by the regularly spaced atoms in a crystal structure. The equation is:

$$n \lambda = 2d \sin \theta$$

where n is a small integer; λ is the wavelength of the incident beam of X-rays; d is the spacing of the structure planes in the

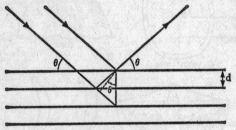

14. Bragg's law.

crystal; θ is the glancing angle, i.e. 90°, minus the angle of incidence as usually defined. (◊ Fig. 14.)

Braided stream. A stream consisting of interwoven channels constantly shifting through islands of alluvium and sandbanks. Although it develops like a ◊ *meander* as the stream tries to become ◊ *graded*, conditions for its development are different. Braided streams occur where gradients are steep, the stream has a high rate of discharge, and the banks are composed of soft sediments (often self-deposited) which are easily eroded. The accumulation of debris may also be helped by quick growth of vegetation due to climatic conditions. The stream beds are wider and shallower than where meanders occur, and are more liable to flood, which in itself helps the constant shifting, widening process.

Bravais lattice. ◊ *Unit cell.*

Breadcrust bombs. ◊ *Pyroclastic rocks.*

Breccia. One of the two main types of ◊ *rudaceous rocks* (◊ *Conglomerates*), consisting of angular fragments implying minimum transport of material. Breccias are generally poorly sorted and commonly contain rock fragments derived from a restricted source. They are of various types: (a) Cemented ◊ *scree*/talus deposits: breccias of this sort are often sub-aerial; (b) ◊ *Fault breccias*; (c) Volcanic breccia (◊ *Agglomerate*); (d) ◊ *Intraformational*; (e) ◊ *Bone breccia*; (f) Collapse/fissure. These are kinds of breccias typical of limestone regions. Collapse breccias derive from material broken up during the collapse of a solution cavity roof, whilst fissure breccias are accumulations of fragments in a solution fissure (cf. ◊ *Neptunean dyke*).

Breccio-conglomerate. A ◊ *rudaceous rock* containing both angular and rounded pebbles, intermediate between ◊ *breccia* and ◊ *conglomerate* in character.

Brickearth. ◊ *Loess* reworked by river action.

Bridgerian. A stratigraphic stage name for the North American equivalent of the ◊ *Auversian* + ◊ *Lutetian* (Eocene-Tertiary System).

Bringewoodian. A stratigraphic stage name for the British Middle ◊ *Ludlovian* (Silurian System).

Briovenian. A stratigraphic division of the late ◊ *Precambrian* in Brittany.

Brockram. Name given to a Permo-Trias ◊ *breccia* (Northern England).

Bronn, Heinrich (1800–62). German zoologist who attempted the first complete chronological survey of known fossil types.

Bronzitite. An *ultrabasic rock* consisting essentially of orthorhombic ◊ *pyroxene*-bronzite.

Brookite. A titanium mineral, TiO_2, found in ◊ *hydrothermal* veins in ◊ *schists* and ◊ *gneisses.* (◊ Appendix.)

Brown coal. Lignite. (◊ *Coal.*)

Brown soil. A term covering several varieties of ◊ *soil* characterised by a higher humus content than ◊ *podsol.*

Brucite. A magnesium mineral, $Mg(OH)_2$, found in ◊ *contact-metamorphosed* dolomitic ◊ *limestones* and in ◊ *serpentine.* (◊ Appendix.)

Bruxellian. A stratigraphic stage name for the European Lower ◊ *Lutetian* (Tertiary System).

Bryozoa. ◊ *Polyzoa.*

Buch, Leopold von (1774–1853). German geologist and pupil of ◊

Werner whose adherence to his teacher's Neptunean theory was eventually broken by his own exhaustive fieldwork. He was responsible for a complete geological map of Germany and many papers concerned with various European areas.

Buchan metamorphic zones. A series of metamorphic ◊ *zones* characterised by the occurrence of ◊ *andalusite,* ◊ *cordierite,* ◊ *staurolite,* and ◊ *sillimanite,* in the Buchan district of north-east Scotland. (◊ Fig. 160.)

Buchite. ◊ *Hornfels.*

Buckland, William (1784–1856). The first Professor of Geology at Oxford and later Dean of Westminster, who was much esteemed by his contemporaries. He was one of the last British advocates of a universal deluge, though in later years under the influence of ◊ *Agassiz* he relinquished it in favour of a similarly destructive ice age. He worked particularly on Jurassic and Tertiary strata and his name is remembered in certain fossils. In 1824 he displayed the first remains of a great dinosaur under the name Megalosaurus.

Buffon, Georges Louis Leclerc de (1707–88). Influential French naturalist who believed that the abundance of fossils was evidence that the earth was once completely covered by ocean which was swallowed up into the earth when the crust broke up. He divided the history of the earth into six epochs, and made the earliest recorded attempts to work out the age of the earth and the planets independent of Genesis.

Bunsen, Robert von (1811–99). German chemist with many notable inventions to his credit. He pioneered the chemical analysis of rocks, particularly in his study of the volcanic rocks of Iceland, and went on to work on the origins and classification of the eruptive rocks.

Bunter. A stratigraphic stage name for the European Lower ◊ *Triassic.*

Burdigalian. A stratigraphic stage name for the European Lower Miocene (◊ *Tertiary System.*)

Burnet, Thomas (1635–1715). English theologian who published a *Sacred Theory of the Earth* to try to satisfy both Church and Science. He suggested that before the Flood the earth was an oceanless, mountainless ball, but with man's wickedness in this paradise the sun opened up great crustal cracks which God used to release the Flood. The present surface was all that was left when the flood subsided.

Bustamite. ◊ *Pyroxenoids.*

Butleyian. Obsolete stratigraphic stage name for the British Lower Pleistocene (part ◊ *Ludhamian*) (◊ *Tertiary System*).

Butte. A small flat-topped hill caused by the continual erosion of a ◊ *mesa*.

Bysmalith. An approximate synonym of ◊ *plug*.

Bytownite. A variety of plagioclase ◊ *feldspar*. The name is derived from the old name for Ottawa, Canada.

Cainozoic, Kainozoic. The division of geological time which succeeds the ◊ *Mesozoic* and ends at the ◊ *Quaternary*. The duration is approximately 63 m.y. from 65 m.y. to 2 m.y. It is commonly used as a synonym for ◊ *Tertiary*.

Cairngorm. Brown variety of quartz (◊ *Silica group of minerals*).

Calabrian. A stratigraphic stage name for the European Lower Pleistocene (◊ *Tertiary System*).

Calamine. Obsolete name for the mineral ◊ *smithsonite*, $ZnCO_3$.

Calc- (prefix). Calcium-bearing, containing $CaCO_3$.

Calc-alkali. Applied to igneous rocks in which the dominant feldspar is calcium-rich. The converse of calc-alkali is ◊ *alkali*. Calc-alkaline rocks tend to contain calcium-bearing ◊ *ferromagnesian minerals* – hornblende, augite, etc. Also applied to calcium-bearing minerals, e.g. calc-alkali feldspar.

Calcarenite. Clastic ◊ *limestones* of grain-size 1/16 mm.–2 mm.

Calceoloid. ◊ *Coelenterata*.

Calc-flinta. A ◊ *calc-silicate hornfels* which is ultra-fine grained and which has a flat, smooth ('flinty') appearance on broken surfaces. It is commonly light in colour, whitish to greenish.

Calci-lutite. Clastic ◊ *limestones* of grain-size less than 1/16 mm.

Calci-rudite. Clastic ◊ *limestones* of grain-size 2 mm. and above.

Calcite. ◊ *Carbonates* and Appendix.

Calcrete. Superficial gravels cemented by ◊ *calc tufa*.

Calc-silicate hornfels. A thermally metamorphosed impure limestone or dolomite. The rock has been recrystallised and the bulk of it is in the form of various calcium and/or magnesium-containing silicates. Examples are ◊ *wollastonite*, forsterite (◊ *Olivine*), grossularite (◊ *Garnet*), ◊ *diopside*, ◊ *idocrase*. (◊ *Marble*; *Skarn*.)

Calc tufa. A general name for deposits of $CaCO_3$ formed by deposition from solutions of calcium bicarbonate, $Ca(HCO_3)_2$:

$$Ca(HCO_3)_2 \rightleftharpoons CaCO_3 + H_2O + CO_2$$

The solubility of calcium carbonate in water is a function of the quantity of dissolved carbon dioxide, and this in turn is a function of temperature and pressure – low temperature and high pressure increasing the amount of carbon dioxide in solution.

If water saturated with carbon dioxide at a particular temperature and pressure is also saturated with calcium carbonate, any increase in the temperature or decrease in pressure will cause calcium carbonate to be precipitated so as to restore equilibrium. Similarly, loss of water by evaporation will also cause deposition of the calcium carbonate.

Calc tufa is found mainly in limestone regions, filling cracks, joints, fissures, and cavities in the rocks, and around springs and resurgences of water which have traversed limestone strata. The calc tufa formed in these cases is often spongy or cellular in character and may enclose fragments of rock, plants, or animal remains. Calc tufa sometimes cements superficial gravels to produce material known as calcrete.

In underground caverns a more massive type of deposit is formed, usually banded in character, which is known as stalagmite or dripstone. This material commonly coats the floor and walls of limestone caves. Where water containing carbonate in solution drips from the ceiling, long, more or less cylindrical, pendant concretions may build up, known as stalactites. A corresponding upward projection from the floor is known as a stalagmite, which may ultimately link up with the corresponding stalactite. Certain hot springs in volcanic regions also deposit a kind of calc tufa known as travertine.

Caldera. A very large ◊ *crater*, which may arise by (a) the coalescence of several small craters, (b) repeated explosion, (c) collapse, or (d) ◊ the *stoping* of surface rocks by a large underground magma chamber. Crater Lake, Oregon, U.S.A., is a typical example of a collapse caldera. A 'super-caldera' in north-west Sumatra has an area of 1,810 sq. km. (700 sq. miles), and may represent the effects of a very large mass of intrusive igneous rocks perforating the surface by explosion and subsequently collapse.

Caledonian, Caledonides. A period of ◊ *orogeny* which in north-west Europe seems to have extended from Middle ◊ *Ordovician* to Middle ◊ *Devonian* times. During this period the Lower Palaeozoic Geosyncline was folded, metamorphosed, and injected by granites, and the dominant N.E.–S.W. 'grain' (Caledonoid direction) was developed in Scotland, Northern Ireland, the Lake District, and Wales.

It can be shown that the various effects of the orogeny developed at different times in different localities, e.g. in Northern Scotland the orogeny seems to have been at a maximum in the Middle Ordovician, whereas further south, in the Southern Uplands, it came in late Silurian times, while in Wales and the Welsh borderlands the maximum was probably in the Middle Devonian Period.

Callovian. A stratigraphic stage name for the base of the West European Upper ◊ *Jurassic*.

Callus (Mollusca). ◊ Fig. 100.

Calyx. ◊ *Coelenterata* (Anthozoa) and Fig. 21; *Echinodermata* (Blastoidea, Crinoidea) and Figs. 42 and 43.

Cambering. ◊ *Fold*.

Cambrian (from the Roman name for Wales – Cambria). The oldest system of rocks in which fossils can be used for dating and correlation. The period commenced at least 530 ± 40 m.y. ago, and had a duration of at least 70 m.y. Generally speaking, the base of the Cambrian shows a marked ◊ *unconformity* with the underlying sediments, and contains the first unequivocal shelled fossil remains, although in some places there is evidence that Cambrian sediments accumulated in a basin of sedimentation formed in ◊ *Precambrian* times, with little or no disconformity between the two. In Great Britain there appears to be everywhere a marked unconformity between the Cambrian and the Precambrian, even where the Precambrian has proved to be fossiliferous, e.g. the Charnwood Forest Area of Great Britain. The upper limit of the Cambrian is taken to be the top of the Tremadoc Series, and in the type area of Tremadoc, in North Wales, these are overlain unconformably by the Arenig Grits of the Lower ◊ *Ordovician*. There has been much dispute over the exact stratigraphic position of the Tremadocian, many Continental stratigraphers preferring to place it at the base of the Ordovician. However, this procedure cannot be justified on the basis of historical priority, nor can it be justified in the type locality, where the fine-grained sediments of the Tremadocian follow on naturally from the Upper Cambrian sediments below.

The Cambrian sediments were deposited in two different sedimentary facies: (1) Unstable extra-◊ *kratonic* basin deposition, the characteristics of which are the massive accumulations of detrital muddy sediments. This facies was formerly referred to as the Atlantic Province. (2) Stable deposition of relatively shallow water sediments around the margins of the kratons, characteristically taking the form of calcareous sedimentation in clear seas, often with an abundance of colloidal silica. This facies was formerly referred to as the Pacific Province.

As would be expected, the faunas of the two areas differ in many respects. In Great Britain the 'Pacific' facies is represented only by the sediments in north-west Scotland, west of the Moine Thrust.

While most groups of invertebrates are represented in the Cambrian, only a few are sufficiently abundant to be of any geological importance. Trilobites (◊ *Arthropoda*) were abundant, especially the more primitive forms such as the Mesonacidae and the Ptychopariidae. ◊ *Brachiopoda* were common; inarticulate forms predominated throughout, with articulate forms being mainly represented by Orthids. Graptolites (◊ *Chordata*) first appear in the Tremadocian. The Cambrian is zoned by means of trilobites.

Camera (Mollusca). ◊ Fig. 105.

Camouflage. One of the ways in which trace elements occur. (◊ *Geochemistry*.)

Campanian. A stratigraphic stage name for part of the European Upper ◊ *Cretaceous*.

Camptonite. ◊ *Lamprophyre*.

Canada balsam. A naturally occurring resin almost universally used as a mounting medium for thin sections of rocks until recently, when it was replaced by more uniform and stable synthetic products. When properly cured it has a refractive index of 1·54. Its major disadvantage is that it turns yellow with age and becomes very brittle.

Canadian. A stratigraphic stage name for the North American Lower ◊ *Ordovician*.

Cancrinite. ◊ *Feldspathoids*.

Cannel. ◊ *Coal*.

Canyon. A deep valley with vertical sides excavated by a river, generally associated with ◊ *rejuvenation*. It is essential that the rocks should be near horizontal and fairly hard if a canyon is to be formed. Dipping rocks will make the valley asymmetrical, while soft rocks will rapidly collapse into the common V-shaped valley, although it is possible for ephemeral canyons to be produced by very rapid downcutting. The classic example of a canyon is the Grand Canyon, Arizona.

Capture. (1) One of the ways in which trace elements occur. (◊ *Geochemistry*.)

(2) ◊ *River capture*.

Caradocian. A stratigraphic stage name for the West European Upper ◊ *Ordovician*.

Carapace. ◊ *Arthropoda* and Fig. 3.

Carbonado. ◊ *Bort*.

Carbonates (◊ *Limestones*). Although the carbonate minerals do not form a homogeneous group, it is convenient to discuss them under a single heading in three major groups:

(1) HEXAGONAL CARBONATES. These are characterised by a dominant rhombohedral ◊ *cleavage*. The more important types are calcite $CaCO_3$; dolomite $CaMg(CO_3)_2$; magnesite $MgCO_3$; ankerite $(Ca,Mg,Fe)CO_3$; siderite $FeCO_3$; rhodocrosite $MnCO_3$; smithsonite $ZnCO_3$ (dolomite and ankerite have the ◊ *symmetry* $-\bar{3}$, whereas the others have a symmetry of $\bar{3}\ 2/m$). (◊ Appendix.) Extensive solid solution occurs between all the members listed above, and minerals intermediate in composition have been named. The ◊ *hardness* of calcite is 3, and its density is 2·71 (cf. aragonite, below); other hexagonal carbonates have hardnesses varying from $3\frac{1}{2}$ to $4\frac{1}{2}$.

(2) ORTHORHOMBIC CARBONATES. These are characterised by a moderate prismatic cleavage and belong to the orthorhombic holohedral (2/mm) class. They commonly form pseudo-hexagonal triplets (◊ *Twinned crystal*). The most common types are aragonite $CaCO_3$; witherite $BaCO_3$; strontianite $SrCO_3$; cerussite $PbCO_3$; alstonite $CaBa(CO_3)_2$. (Baryto-calcite – another calcium-barium carbonate – is monoclinic.)

Aragonite has a hardness of $3\frac{1}{2}$–4 and a density of 2·9–3 (cf. calcite) and is formed at higher temperatures than calcite.

(3) BASIC CARBONATES. ◊ *Malachite* $CuCO_3Cu(OH)_2$ (green); ◊ *azurite* $2CuCO_3.Cu(OH)_2$ (blue).

Various Na carbonates have been recorded from dried-up lakes in tropical regions.

Calcite, dolomite and siderite occur as primary sediments (◊ *Limestone*; *Dolomitisation*). Aragonite may be the original form in which much $CaCO_3$ is deposited, inverting in the course of time to the more stable calcite. Magnesite has also been claimed as a primary sediment or sedimentary ◊ *replacement*, but is most common as an alteration production of magnesian silicate rocks, e.g. in ◊ *serpentine* and certain ◊ *ultrabasic rocks*. Most of the minerals can occur as ◊ *gangue* minerals in ◊ *hydrothermal* veins, or as hydrothermal replacements in limestones. Cerussite, smithsonite, and the copper carbonates are most commonly found in the oxidised zone of hydrothermal deposits.

Carbonatite. A magmatic rock consisting of calcium carbonate, sometimes with magnesium carbonate and very rarely strontium carbonate, often accompanied by rare-earth minerals. They are almost inevitably associated with nepheline syenite (◊ *Alkali syenite*) ring complexes. Their origin is a matter of considerable dispute. (◊ *Cauldron subsidence; Ring structures.*)

Carboniferous System. A period named (by ◊ *Conybeare,* 1822) from the widespread occurrence of carbon in the form of coal in these beds. It extends from 345 to 280 m.y. and has a duration of 65 m.y. The Carboniferous includes two sub-systems – the Mississippian (Lower Carboniferous) and Pennsylvanian (Upper Carboniferous), the boundary being dated at 325 m.y. approximately. The lower limit of the Carboniferous is the horizon at which the Devonian faunas are replaced by a fauna of Productid ◊ *Brachiopoda* and corals (◊ *Coelenterata*); in terms of Goniatite zones, the boundary lies between the Clymenia zone of the Devonian and the Wocklumeria zone of the Carboniferous. The upper limit in most of Britain and Europe lies in continental ◊ *red bed* facies and is difficult to interpret; where a marine sequence occurs (as in Russia and North America), the incoming of the Foramenifera Pseudoschwagerina marks the beginning of the Permian.

In the Lower Carboniferous, marine sediments predominate in Great Britain, and generally speaking two faunal provinces can be recognised. The typical Lower Carboniferous facies is of a detrital organic limestone, often with a development of coral reefs, and with abundant crinoids (◊ *Echinodermata*) and brachiopods. The second environment is one of black shales which contain a reduced fauna of brachiopods, together with ◊ *Polyzoa* and locally abundant trilobites (◊ *Arthropoda*). These black shales also often contain goniatites (◊ *Mollusca*), especially in the upper part of the succession.

The Upper Carboniferous differs from the Lower Carboniferous in that everywhere in Britain it is represented by fresh water or lacustrine sediments, containing only occasional marine bands. The fauna consisted mainly of freshwater lamellibranchs (◊ *Mollusca*), fishes, and rare amphibians. The flora of the Upper Carboniferous was also highly characteristic, consisting mainly of the more primitive kinds of vascular plants, such as the club-mosses and horsetails. These often reached a height of 15–20 m. (45–60 ft), and contributed largely to the formation of ◊ *coal* seams. It should not, however, be assumed that these forms are common only in the Upper Carboniferous, as evidence from some coal seams found in the Lower Carboniferous indicates that these plants were already flourishing at that time.

The Carboniferous is perhaps the most important system economically, containing as it does the bulk of the world's coal reserves, together with important deposits of iron ore, oil shale, oil, fire clay, and gannister.

The period contains evidence of widespread vulcanicity and minor igneous intrusions. The end of it saw the commencement of the Variscan ◊ *Orogeny*. Towards the end widespread glaciation became established in the southern hemisphere, especially near the present-day equator. (◊ *Continental drift*; *Gondwanaland*.)

Carbonisation. The reduction of organic tissue to a carbon residue. An unusual kind of fossilisation in which the tissue is preserved as a carbon film. Plants are commonly preserved in this manner, soft-bodied animals more rarely.

Cardinal area (Mollusca). ◊ Fig. 101.

Cardinal process (Brachiopoda). ◊ Fig. 11.

Carnegieite. ◊ *Feldspathoids* (Simple).

Carnelian. A reddish or brownish variety of chalcedony. (◊ *Silica group of minerals*.)

Carnian. ◊ *Karnian*.

Carpoidea. ◊ *Echinodermata* (Cystoidea).

Carstone. A variety of sandstone with an exceptional amount of limonitic cement. Commonly seen as thin bands not necessarily coincident with bedding. (Cf. ◊ *Iron pan*.)

Cassadagan. A stratigraphic stage name for the middle of the North American mid Upper ◊ *Devonian*.

Casselian. A synonym of ◊ *Chattian* (Tertiary System).

Cassiterite. A tin ore, SnO_2, found in ◊ *hydrothermal* veins, alluvial deposits, and acid igneous rocks, especially ◊ *pegmatites*. (◊ Appendix; ◊◊ *Detrital*.)

Cast. The impression obtained from a ◊ *mould*, i.e. a replica of the original form. Casts and moulds are most commonly thought of in connexion with the preservation of fossils, but any three-dimensional feature may be so replicated, e.g. rain prints, ripple marks, footprints, etc. (◊ Fig. 15.)

Cata- (prefix). Deepest, at greatest depth.

Cataclasis. The process of mechanical fracture or break-up of rocks, usually associated with dynamic ◊ *metamorphism* or ◊ *faulting*. The term is used both for small-scale, i.e. microscopic, structures, and also for large-scale phenomena. Hence cataclastic (adjective), and cataclasite – any rock produced by cataclasis. (◊ *Mylonite*; *Crush breccia*; *Fault breccia*.)

Cataclastic rock. One that has undergone mechanical breakage, as

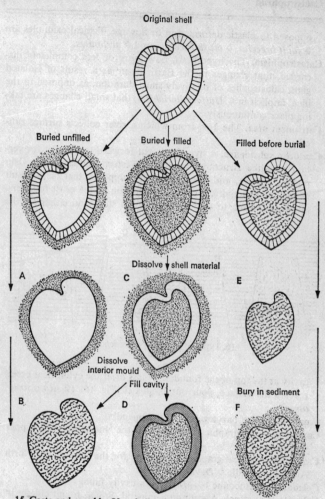

Original shell

Buried unfilled   Buried filled   Filled before burial

A   Dissolve shell material   C   E

Dissolve interior mould

Fill cavity   Bury in sediment

B   D   F

15. Casts and moulds. If a shell is buried unfilled and then dissolves, (A) is left – an external mould which if subsequently filled with material will yield (B), an external cast. If the shell is filled, internal and external moulds result (C), and if the cavity is filled the result is (D), an internal and external cast, i.e. a cast of the original shell. The third possibility is that the shell is filled before burial and (E) is left – an internal mould which may subsequently be buried (F) (note the absence of an external mould or cast in this latter case).

opposed to plastic deformation or flowage. Typical examples are ◊ *fault breccias*, ◊ *crush breccias*, and ◊ *mylonites*.

Catastrophism. The hypothesis, now more or less completely discarded, that changes in the Earth occur as a result of isolated giant catastrophes of relatively short duration, as opposed to the idea, implicit in ◊ *Uniformitarianism*, that small changes are taking place continuously.

Catchment area. The area from which a river collects surface run-off.

Cauldron subsidence. A hypothesis put forward to account for certain ring-like structures found in Western Scotland, the Oslo region, Nigeria, and elsewhere. It is supposed that, as a result of magmatic activity, a cylindrical portion of the crust founders, allowing magma to well up around the sides and either fill the

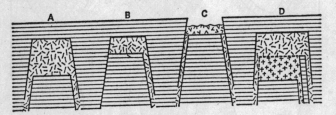

16. Types of cauldron subsidence.

cavity at the top of the foundered mass or, if the cylindrical break reaches the surface, pour out as lava (◊ Fig. 16). (◊ *Ring structures*.)

Caulk. A form of barytes, $BaSO_4$ (◊ Appendix).

Cayugan. A stratigraphic stage name for the North American Upper ◊ *Silurian*.

Cazenovian. A stratigraphic stage name for the base of the North American Middle ◊ *Devonian*.

Celadonite. Glauconite (◊ *Micas*) as a cavity filling.

Celestite (Celestine). A strontium mineral, $SrSO_4$, found chiefly in sedimentary rocks but also in ◊ *hydrothermal veins*. (◊ Appendix.)

Cement. Material binding particles together, usually in a sediment. (◊ *Arenaceous rocks*.)

Cementation. ◊ *Diagenesis*.

Cenomanian. A stratigraphic stage name for the basal European Upper ◊ *Cretaceous*.

Central eruption. A volcanic eruption from a more or less circular vent, effectively a point source, as opposed to the linear source of a fissure eruption (◊ *Volcano*). An ordinary volcano, in the usual sense of the word.

Cephalon. ◊ *Arthropoda* (Trilobita) and Fig. 3.

Cephalopoda. ◊ *Mollusca*.

Cerioid. ◊ *Coelenterata* and Fig. 23.

Cerussite. ◊ *Carbonates* and Appendix.

Ceylonite. A mineral of the ◊ *spinel* group.

Chadronian. A stratigraphical stage name, the North American equivalent of the European ◊ *Tongrian* (Tertiary System).

Chalcedony. ◊ *Silica group of minerals.*

Chalco- (prefix). Copper.

Chalcocite. A copper ore, Cu_2S, found mainly in the enriched zones of sulphide deposits. (◊ Appendix.)

Chalcophile. Elements having a strong affinity for sulphur, characterised by the sulphide ◊ *ore minerals*.

Chalcopyrite. The main copper ore, $CuFeS_2$, a widely occurring mineral found mainly in ◊ *hydrothermal* and ◊ *metasomatic* veins. (◊ Appendix.)

Chalk. Strictly, the very fine-grained pure-white ◊ *limestone* found in the Upper Cretaceous of Western Europe. However, the term is occasionally applied to similar fine-grained limestones which are not stratigraphically the same. The term is also used in a stratigraphic sense (with a capital C) as the equivalent of Upper Cretaceous, mainly in Western Europe where strata of this age are dominantly composed of this rock type.

Chamberlin, Thomas Chrowder (1843–1928). American geologist who was one of the first to suggest a dating of the Pleistocene from the advance and retreat of ice sheets. He also established the origin of ◊ *loess*, and discovered beneath the Greenland ice the fossil evidence of a former hot climate. He devoted much of his life to his planetismal theory of the origin and development of the Earth. He suggested that the Earth may have come into being when gatherings of matter and gases (planetismals) – blown into space by an explosion of the sun – gradually attracted further matter to them. The pressures caused by the cohesion of this matter forced all mobile and lighter material to the surface, shaping the Earth during the process of its growth.

Chamosite. A typical oxy-◊*chlorite* minerals. (◊ *Sedimentary iron ores*.)

Champlainian. A stratigraphic stage name for the North American Middle ◊ *Ordovician*.

Charmouthian. A stratigraphic stage name for the European mid Lower ◊ *Jurassic*.

Charnian. A stratigraphic division of the volcanic ◊ *Precambrian* in Leicestershire (England).

Charnockite. A coarse granular rock consisting mainly of quartz, feldspar, and hypersthene, plus other minerals, named after Job Charnock, an East India Company clerk and founder of Calcutta, whose tombstone is made of it. The minerals are characteristically fresh, undecomposed, and lacking in ◊ *inclusions*. Some charnockites may result from crystallisation of a magma at great depths under extremes of pressure. Others are thought to have arisen by ◊ *ultra-metamorphism* of basic *igneous* (rarely other) rocks. The term should not be used as an equivalent of hypersthene granite.

Chattian. A stratigraphic stage name for the European Upper Oligocene (◊ *Tertiary System*).

Chazyan. A stratigraphic stage name for the North American Lower ◊ *Champlainian* (Ordovician System).

Cheeks, fixed, free (Arthropoda). ◊ Fig. 3.

Chemungian. A stratigraphic stage name for the North American mid Upper ◊ *Devonian*.

Cheniers. Long sinuous ridges of sand deposited on top of the swamp deposits of a delta. They represent stationary phases of a regressing shore-line. They may form off-shore ◊ *bars* and islands when the sea advances and may later become incorporated into a pile of deltaic sediments. The type area for their development is the Mississippi Delta.

Chernozem (Black earth). Humus-rich ◊ *soils* developed as a result of low rainfall. The upper layers are relatively leached.

Chert. ◊ *Cryptocrystalline* silica which may be of organic or inorganic origin. It occurs as bands or layers of nodules in sedimentary rocks. It can be shown that it is sometimes a primary deposit, sometimes formed by the confluence of disseminated silica in a rock, and sometimes a secondary ◊ *replacement* material. Some cherts from geosynclinal deposits contain abundant radiolaria (◊ *Protozoa*) and are sometimes called radiolarites; these probably represent deep-water accumulations.

Flint is the variety of chert occurring primarily in the Upper Cretaceous, and as detrital pebbles in the Tertiary. It has a conchoidal fracture, as opposed to the flat fracture of chert.

Lyddite (Lydian stone) is a dense black variety of chert, formerly used as a ◊ *touchstone*. (For other chalcedony varieties, ◊ *Silica group of minerals*.)

Chesterian. A stratigraphic stage name for the North American Upper ◊ *Mississippian* (Carboniferous System).

Chiastolite. A variety of ◊ *andalusite* containing dark cruciform-shaped inclusions.

China clay. Deposits of kaolin produced by hydrothermal decomposition or weathering of ◊ *feldspars* in granites. (◊ *Argillaceous rocks*.)

Chitons. ◊ *Mollusca* (Amphineura).

Chlorite group (Greek *chloros*, 'green'). The name chlorite is applied to a somewhat heterogeneous group of ◊ *layer-lattice minerals*. Structurally they may be regarded as consisting of ◊ *talc* units – $Mg_6Si_8O_{20}(OH,F)_4$ – sandwiched between ◊ *brucite* layers – $Mg_6(OH)_{12}$. Al may replace Si in the talc layer, and ferrous iron and/or Al can replace Mg in both the talc and brucite units. The oxidised chlorites have ferric iron replacing ferrous iron. A typical chlorite formula is $(Mg,Fe'')_{10}Al_2(Si,Al)_8O_{20}(OH,F)_{16}$. (◊ Appendix.)

Most chlorites are green and have cleavage flakes which are flexible but not elastic (cf. ◊ *Micas*). They are often found as alteration products of ◊ *ferromagnesian minerals*, and in low-grade regionally metamorphosed rocks. Of the many special names used for varieties of chlorite, clinochlore and penninite may be mentioned as typical normal chlorites. Chamosite is a typical oxy-chlorite and occurs in sediments.

Chloritoid. A group of minerals with the formula $(Fe'' Mg)_2(AlFe''')Al_3O_2(SiO_4)_2(OH)_4$ having many features in common with the ◊ *chlorites*, including perfect basal cleavage, green colour, monoclinic pseudo-hexagonal crystallography, and certain optical properties. They are not, however, true ◊ *layer-lattice minerals*, as they consist of isolated SiO_4 tetrahedra and a brucite-type layer, $Mg_6(OH)_{12}$. Chloritoid is most commonly found in low-grade regionally metamorphosed rocks.

Chondrites. Stony ◊ *meteorites* containing ◊ *chondrules*. (Cf. ◊ *Achondrites*.)

Chondrodite. ◊ *Humite minerals*.

Chondrules. Small, globular masses of ◊ *pyroxene*, ◊ *olivine* and sometimes glass found in certain stony ◊ *meteorites*.

Chordata. The phylum Chordata consists of a number of sub-phyla of widely differing morphology, ranging from sessile sea squirts to man himself. The basic character of the phylum is that there is at some stage in the life history an internal skeletal rod, which may be of cartilage (gristle) or bone – the notochord. Of the many groups of the Chordata only two are of any geological importance: the graptolites (sub-phylum Hemichordata, class Graptolithina) and the vertebrates (sub-phylum Vertebrata).

GRAPTOLITHINA. Graptolites were colonial organisms consisting of one or more branches (stipes). The individuals of the colony (zooids) were situated in cups arranged in one or two rows along the stipe, the whole being covered by an external layer of chitinous material. The complete colony (rhabdosome)

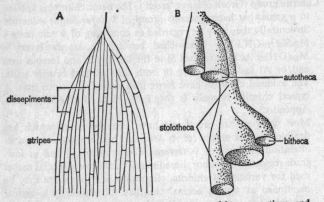

17. Dendroids. (A) Rhabdosome, showing multiramous stipes and connecting dissepiments; (B) Detail of stipe.

developed from a small cone-like structure – the sicula. The graptolites are divided into five orders, the two main ones being the Dendroidea and the Graptoloidea.

The dendroid graptolites (◊ Fig. 17) are the most primitive forms, showing a gradual evolutionary reduction from 64 to 16 stipes in the rhabdosome. They differ from the Graptoloidea in that three main types of thecae are present – a larger autotheca, a smaller bitheca, and a stolotheca which gives rise to the next set of thecae. The dendroids range from the Upper ◊ *Cambrian* (Tremadocian) to Lower ◊ *Carboniferous*.

The Graptoloidea (◊ Fig. 18) differ from the Dendroidea in that only autotheca are present. The number of stipes ranges from 8, in the most primitive, to 1, in the most advanced forms. Forms in which the rhabdosome consists of 1 stipe only may be primitive

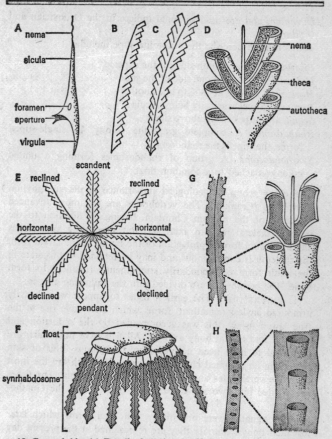

18. Graptoloids. (A) Detail of sicula; (B) Uniserial stipe; (C) Biserial stipe; (D) Detail of biserial stipe; (E) Successive positions of growth of a two-sided form, from early pendant to late scandent (the stipe positions of five individuals are represented); (F) Detail of synrhabdosome; (G) Detail of intratorted thecae; (H) Scalariform preservation of thecae.

biserial (two rows of thecae) forms, or advanced uniserial (one row of thecae) forms. Some graptoloids develop more than 8 stipes by thecal budding rather than growth from the thecae. The graptoloids ranged from Lower ◊ *Ordovician* to Lower ◊

Devonian, and are used as zonal indices in the Ordovician and ◊ *Silurian*.

Terms used to describe the graptolite stipe include:

Biserial Thecae on both sides of the stipe

Uniserial Thecae on one side of the stipe

Horizontal Stipes at 180° to one another

Pendant Stipes vertically below sicula

Scandent Stipes rising above sicula

Rhabdosome A complete graptolite colony. In single-stiped forms, the stipe is the rhabdosome

Synrhabdosome A group of rhabdosomes forming a united colony attached to a common float

The other group of geological importance is the sub-phylum Vertebrata (Craniata). The vertebrates are the most advanced members of the phylum Chordata, having, in addition to the internal skeletal rod, an axial skeleton supporting appendages and a cranium for the protection of the brain. In the larval form the skeleton is cartilaginous and may be retained in this state in the adult form or secondarily strengthened (ossified) to form bone. The vertebrates are divided into the following classes:

(1) PISCES (fish). The earliest fishes to appear were heavily armoured jawless (agnathan) forms which later gave rise to the true jawed fishes; this was accompanied by the reduction and eventual loss of the bony armour. All fish have fins which are presumed to have been derived from continuous folds of skin present in an assumed chordate ancestor. The fish are the most primitive vertebrates and gave rise to the Amphibia. The fishes are divided into two sub-classes, the Agnatha and the Gnathostomata:

(a) The Agnatha were primitively armoured forms which later lost their bony material: they are represented at the present day by the cartilaginous lampreys and hagfishes. They range from Middle ◊ *Ordovician* to Recent, and are used as zonal indices for the non-marine facies of the Lower ◊ *Devonian*.

(b) The Gnathostomata are a heterogeneous group of fish which were primitively armoured like their agnathan ancestors, but which later became more specialised to give rise to the cartilaginous sharks and the modern bony fishes, e.g. salmon and cod. They range from Lower ◊ *Devonian* to Recent and are used as zonal indices for the non-marine Middle and Upper Devonian.

(2) AMPHIBIA. The amphibians evolved from the fishes in

Upper Devonian times and although still living at the present day have never been particularly abundant. They are characterised by having a tetrapod (four-limbed) form in which the body is suspended between the limbs and not above them as in the more advanced tetrapods. The amphibians have never completely broken away from their aquatic ancestry, as they must return to water to breed, and they normally live in a moist environment as a safeguard against desiccation.

The amphibians are classified on their bone structure, using the skull and vertebral column. They range from the Upper ◊ *Devonian* to Recent.

(3) REPTILIA. The Reptilia were the first group of tetrapods to colonise the land successfully, owing to the evolutionary development of the shelled egg (which could withstand desiccation) and scales as protection for the skin. The Reptilia evolved from the Amphibia in Upper ◊ *Carboniferous* times and gave rise to the mammals early in the ◊ *Mesozoic* and the birds in the ◊ *Jurassic*. They are usually regarded as being cold-blooded forms (although some groups may have been primitively warm-blooded) and for this reason they are generally restricted to the warmer portions of the Earth's surface.

The reptiles are divided into a number of well-known groups, of which the dinosaurs (Archosaurs) are probably the best known. These forms evolved in the ◊ *Triassic* and became the dominant terrestrial animals in the Jurassic and ◊ *Cretaceous*, the largest dinosaurs reaching a length of about 35 m. (100 ft). These large forms were herbivorous; carnivorous forms rarely exceeded a length of 12 m. (35 ft) and a height of 7 m. (20 ft).

Reptiles range from the Upper Carboniferous to Recent, and are used as zonal indices in the non-marine sediments of the ◊ *Permian* and Triassic.

(4) AVES. The birds evolved from the Reptiles in Upper Jurassic times and have not lost many of their reptilian characters. Many authorities consider that had the birds become extinct before the present day, they would have been included within the class Reptilia. Feathers are only a modification of reptilian scales, and scaling is still present on the legs. Primitive birds had toothed beaks, but this is lost in advanced forms. Birds possess hollow bones, an adaptation towards lightening the skeleton for flight. Birds range from the Upper ◊ *Jurassic* to Recent, but fossil remains are extremely rare.

(5) MAMMALIA. Mammals evolved from an offshoot group of

the reptiles during Upper Triassic times. They are characterised by having hair, a dentition of different shaped (heterodont) teeth, and being warm-blooded (homeothermic).

The mammals can be divided into a primitive group, consisting mainly of (1) Marsupial forms in which the young after birth are kept within a pouch, and (2) Monotremes, the egg-laying mammals, and an advanced group, the Placentals. The marsupial mammals were abundant throughout the world until Miocene (◊ *Tertiary System*) times, after which they became largely restricted to Australia and South America. Mammals range from Upper ◊ *Triassic* to Recent, the placental mammals ranging from Upper ◊ *Cretaceous* to Recent; they are used as zonal indices for non-marine sediments in the Tertiary.

Chorology. The study of the geographical distribution of organisms.

Chromite. A chromium ore mineral, $FeCr_2O_4$, in the ◊ *spinel group* and found, often as small grains, in ◊ *ultrabasic* igneous rocks and ◊ *serpentine*. (◊ Appendix.)

Chrysocolla. A copper ore mineral, $CuSiO_3.2H_2O$, found in the oxidised zones of copper deposits. (◊ Appendix.)

Chrysolite. ◊ *Olivines.*

Chrysoprase. A green variety of chalcedony (◊ *Silica group of minerals*).

Chrysotile. A common variety of ◊ *asbestos.*

Cincinnatian. A stratigraphic stage name for the North American Upper ◊ *Ordovician.*

Cinnabar. HgS, the most important ore mineral of mercury, associated with volcanic activity. (◊ Appendix.)

Cinnamon stone (Hessonite). ◊ *Garnet.*

CIPW Classification. A system of classifying igneous rocks proposed by Cross, Iddings, Pirsson, and Washington in 1903. The system is based upon the chemical composition of the rock and requires the calculation of the ◊ *norm* as an essential first step. The relative proportions of the arbitrary minerals thus calculated are then used to establish a series of classes, orders, and rangs (with sub-classes, sub-orders, and sub-rangs). In this way a very large number of classificatory 'pigeon-holes' can be erected. Unfortunately many of the 'pigeon-holes' are unoccupied, while others contain several rocks which can be differentiated on other grounds. The process of determining a position in the classification is nowadays more or less obsolete, although there has been a revival of interest recently, especially in the calculation of the norm, since it lends itself to computer calculation.

Cirque, Corrie, Cwm. A landform of a glaciated highland region. A cirque originates as an ordinary valley head or other suitable hollow which becomes filled with snow. This snow becomes compacted into ◊ *firn* as the air is forced out and a thickness of about 33 m. is sufficient to initiate movement of the ice mass out of the cirque. The walls of the cirque are cut back by plucking of the back-wall which takes place across the ◊ *bergschrund*, and the cirque floor is gouged out by the moving ice. Some of the material so removed is deposited as moraine around the lip of the cirque. On slopes with a thinner patchy snow cover, nivation – the action of frost and thaw on the surrounding rocks – combined with the removal of the shattered material by solifluction (◊ *Gravity transport*) causes development of depressions known as nivation cirques or nivation hollows. These rarely develop into glaciers, and do not possess the steep walls, gouged floors and moraines associated with the stronger ice movement in true cirques.

In general cirques are circular in plan and in some cases may become the sites of small lakes called tarns. Where two cirques meet, a precipitous divide called an arête develops; the feature formed by the coalescence of more than two cirques is called a horn.

Cirrus (pl. cirri) (Echinodermata). ◊ Fig. 43.

Citrine. A yellow variety of quartz (◊ *Silica group of minerals*).

Clarain. ◊ *Coal*.

Clarendonian. A stratigraphic stage name, the North American equivalent of the European ◊ *Sarmatian* (Tertiary System).

Clarite. ◊ *Coal*.

Clarke, Rev. W. B. (1798–1878). English-born geologist who settled in New South Wales and is often thought of as the founder of Australian geology. He chiefly studied Silurian rocks, but he was also the first to discover gold *in situ*.

Clarke of concentration. The ratio of the amount of a particular element in a rock to the average amount of that element in the Earth's Crust. It is therefore a measure of the amount of concentration which the element has undergone in a particular environment. It is often applied to ore deposits.

	Average Concentration in Earth's Crust	Ore	Clarke of Concentration
Sn	·004%	1%	250
Fe	5%	30%	6
Pb	·0016%	4%	2500

It was named after F. W. Clarke, an American geochemist, and first introduced by the Russians W. Fersman and K. Vernadsky.

Clarkeforkian. A stratigraphic stage name for the North American Upper Thanetian (◊ *Tertiary System*).

Clastic rocks. Rocks built up of fragments of pre-existing rocks which have been produced by the processes of weathering and erosion, and in general transported to a point of deposition. Typically, ◊ *arenaceous rocks*, ◊ *conglomerates*, ◊ *breccias* etc. (◊ *Sedimentary rocks*.)

Clay. ◊ *Argillaceous rocks.*

Clay gall. When a patch of clay or mud dries out, the upper surface cracks and peels away from the under-layers. These thin 'leaves' of hardened clay may occasionally be transported a short distance and deposited in a sand or similar sediment in the form of flat or lens-shaped clay galls, generally orientated parallel to the bedding.

Clay ironstone. A sedimentary rock consisting of ◊ *siderite* ($FeCO_3$) and argillaceous impurity. It is often nodular, and was formerly of importance as an iron ore, e.g. in the Wadhurst Clay (Lower Cretaceous) of the Weald of Kent and Sussex. (◊ *Black-band ironstone.*)

Clay minerals. A term reserved for those constituents of a clay which give it its plastic properties. Their atomic structure is basically that of the ◊ *layer-lattice minerals* and they generally occur as minute, platy, more rarely fibrous, crystals. An important characteristic is their ability to lose or take up water according to the temperature and amount of water present in a system. Some clay minerals contain loosely bonded cations which can easily be exchanged for others, according to the local concentration ('base exchange'). Both two- and three-layer types are known. Clay particles range in size from the near-colloidal to those within the resolving power of an ordinary microscope. Clay minerals are produced by the degradation (◊ *weathering*, ◊ *hydrothermal processes*, etc.) of other ◊ *silicates* or silicate glasses. Five clay mineral groups are recognised:

(1) THE KAOLINITE GROUP, which includes kaolinite, dickite and nacrite, which are isochemical ($Al_4Si_4O_{10}(OH)_8$) but not isostructural. They contain no exchangeable cations and are mainly produced by the destruction of alkali feldspars under acidic conditions. Kaolin is the main constituent of ◊ *china clay*.

(2) THE ILLITE GROUP, which includes illite, the hydromicas and perhaps ◊ *glauconite*. The general formula is $K_{1-1.5}Al_4(Si,$

Al)$_8$O$_{20}$(OH)$_4$. Glauconite is often considered a ◊ *mica* and contains other cations besides K, including Na, Ca, Mg, Fe", and Fe"'. Illites are amongst the commonest clay minerals; they develop by the alteration of micas, alkali feldspars etc. under alkaline conditions.

(3) THE MONTMORILLONITE GROUP, which includes montmorillonite, nontronite and beidellite, and is sometimes called the fullers' earth group or smectite group. The general formula is

$$M^1{}_{\frac{2}{3}}(Y^3, Y^2)_{4-6}(Si, Al,)O_{20}OH_4.nH_2O$$

Where M^1 = Na or $\frac{1}{2}$ Ca, Y^2 = Mg or Fe" and Y^3 = Al or Fe"'. This group is especially notable for the way in which it takes up and loses water, and for its important base exchange properties. Montmorillonites are formed by the alteration of basic rocks, or other silicates low in K, under alkaline conditions, providing Ca and Mg are present.

(4) ◊ *VERMICULITE,* which is related both to the montmorillonites and to the ◊ *chlorites.* It appears as a constituent of clay in certain soils, and seems to form mainly as a result of the alteration of ◊ *biotite* flakes or, more rarely, chlorites, hornblende, etc.

(5) THE PALYGORSKITE GROUP, which are rare clay minerals possessing a chain structure rather than a layered one.

Clay-with-flints. The residual deposit left after the solution weathering of ◊ *chalk,* usually in the form of a red-brown sandy clay with numerous unrolled ◊ *flint* nodules. Probably formed during ◊ *periglacial* conditions in the Pleistocene (◊ *Tertiary System*).

Cleat. Jointing in ◊ *coal.*

Cleavage, mineral. Many minerals will, when broken, display a flat plane of breakage which is parallel to a possible crystal face. The mineral is then said to possess a cleavage parallel to 'hkl' (◊ *Form*). Cleavage planes are developed along planes of weakness in the atomic lattice and the perfection, or otherwise, of the cleavage depends upon the relative strength of the bonds in this plane; e.g. micas have a perfect cleavage parallel to 001; pyroxenes and amphiboles have cleavage parallel to 110 and 1$\bar{1}$0; while galena has three cleavages parallel to the cube faces (100, 010, 001).

Cleavage, rock. Rock cleavage may be one of the following types (◊ Fig. 19):

FRACTURE CLEAVAGE, which affects incompetent layers lying in a series of beds of varying degrees of competency (◊

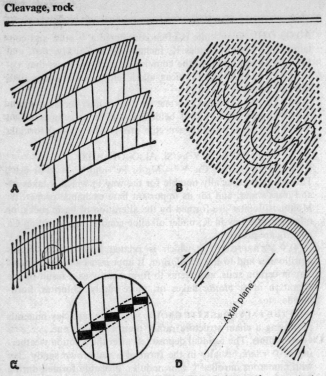

19. Types of rock cleavage; (A) 'Refraction' of cleavage in competent beds (the cleavage direction is refracted when it passes through the beds); (B) Slaty cleavage; (C) Flow cleavage; (D) Axial-plane cleavage.

Competent bed). In general, the term implies that the cleavage planes are not controlled by mineral particles in parallel orientation. Some authors use 'shear cleavage' as a synonym of fracture cleavage.

SLATY CLEAVAGE, which develops in fine-grained rocks as a result of intense deformation, producing a partial recrystallisation of platy minerals parallel to the axial planes of the ◊ *folds*, i.e. perpendicular to the compressive forces. In general, this recrystallisation does not destroy all traces of bedding.

FLOW CLEAVAGE, a further development of slaty cleavage, in which further recrystallisation has taken place and reduced the traces of bedding to mere vestiges. Schistosity is a further development of this process, whereby recrystallisation produces a fairly coarse crystalline, ◊ *foliated* rock with all traces of bedding destroyed. (◊ *Schist.*)

AXIAL-PLANE CLEAVAGE, the term applied to slaty cleavage occurring parallel to the axial plane of a fold.

STRAIN-SLIP CLEAVAGE, now used to describe a second cleavage. As originally used, it implied a second cleavage along which movement had taken place deforming the original cleavage planes. False cleavage is used in the same sense.

In assessing a type of cleavage, it is essential to take into account field relationships as well as microscopic structure. It will often be found difficult to determine which sort of cleavage is involved, in any particular case. Different authorities use the above terms (and occasionally others) with slight differences of meaning or emphasis.

Cleavage plane. The plane of mechanical fracture in a rock. Cleavage planes are normally sufficiently closely spaced to break the rock into parallel-sided slices. The term is also applied to *mineral* ◊ *cleavage.*

Clinochlore. A mineral of the ◊ *chlorite* group.

Clinozoisite. ◊ *Epidotes.*

Clintonite. ◊ *Micas* (Trioctahedral).

Clints. The ridges in the bare limestone rock surface of ◊ *karst scenery.*

Clone. A colony of animals or plants, all being descendants of the same individual, formed by vegetative reproduction (asexual reproduction), e.g. a coral colony.

Closed basin. A ◊ *basin* not connected to a larger body of water.

Closure. In studies of traps for ◊ *oil,* the concept of closure of an anticline (◊ *Fold*) has become significant in determining the maximum possible area over which fluid can be contained by the structure. It is most simply defined in terms of structure contours (◊ *Strike lines*) as the vertical distance between the highest part of the fold and the lowest continuous encircling structure contour. Synforms are, on this definition, said to have negative closure.

Coal. The general name given to stratified accumulations of carbonaceous material derived from vegetation. The starting point for coal formation is usually peat or some similar accumulation of partially decayed vegetable matter. By a process of compaction and slight heating during burial, the peat is converted to the familiar black coal. In general, a series of stages in the process of coalification corresponding to the amount of heating that the rock has undergone are recognised. These are:

> Peat
> Lignite (brown coal)
> Sub-bituminous
> Bituminous
> Sub-anthracite
> Anthracite

– collectively known as the humic coals. This series shows a progressive increase in carbon content and a decrease in volatiles. The percentage of carbon in dry mineral-free coal is called the 'rank'. The individual constituents of coals are known as 'macerals' (derived from the verb 'to macerate'). Three main groups of macerals (vitrinite, exinite, and inertinite) are recognised. These are in turn sub-divided according to the detailed character of the material. The various macerals are best studied in polished surfaces under an immersion medium. The name microlithotype is used for intimate associations of the various macerals. Three main microlithotypes are recognised: vitrite (vitrinite mainly), clarite (vitrinite and exinite) and durite (exinite and inertinite). Types containing all three macerals are also known and other combinations of two macerals occur. The terms 'vitrain', 'clarain', 'durain' and 'fusain' have been superseded by the above terms as a result of more detailed study.

Two other major types of coal are recognised, which may be referred to collectively as the sapropelic coals. They are:

(1) Cannel coal, which shows many similarities to the durite microlithotype, but is much finer grained. The main constituents appear to be finely divided vegetable matter, spores, algae, and fungal material. Resin globules are occasionally found.

(2) Boghead coals, which consist of algal material, together with some fungal matter. Torbanite, the well-known oil shale of the Midland Valley of Scotland, is a typical boghead coal.

Coals of various sorts occur throughout the stratigraphic column from the ◊ *Devonian* upwards, with a remarkable maximum in the ◊ *Carboniferous*. Bituminous and anthracite coals occur mainly in the Carboniferous and the lignites and brown coals in the ◊ *Mesozoic* and ◊ *Tertiary*.

Cobaltite. A cobalt ore mineral, CoAsS, found in ◊ *hydrothermal* veins with other cobalt and nickel minerals. (◊ Appendix.)

Cobble. A rock fragment with a diameter of between 64 and 256 mm., i.e. larger than a pebble and smaller than a boulder. Generally rounded to sub-rounded. (◊ *Particle size.*)

Coblenzian. A stratigraphic stage name for the European ◊ *Emsian* + ◊ *Siegenian* (Devonian System).

Coccolith. Microscopic calcareous plates secreted by certain organisms, probably algal Phytoplankton. An important constituent of chalk. (◊ *Limestone.*)

Coelenterata. A group of radially symmetrical aquatic animals consisting of two distinct layers of cells united by an acellular layer of jelly – the mesogloea. The name of the phylum is derived from the internal cavity or coelenteron, into which the food is passed from a central mouth, and in which digestion takes place. A typical coelenterate shows an alternation between a planktonic, reproductive stage (medusoid), and a sedentary, vegetative stage (polypoid). In the various groups of Coelenterata either generation may be dominant. The Coelenterata can be divided into three main classes, the Scyphozoa (jellyfish), the Hydrozoa, and the Anthozoa. Of these three groups the Anthozoa contain all the important fossil forms, especially the groups Zoantharia and the Tabulata.

SCYPHOZOA. These are the jellyfishes, which are marine and in which the medusoid stage is dominant. They are characterised by the enormous increase in thickness of the mesogloea which makes up the bulk of the animal. They range from ◊ *Precambrian* to Recent. They are very rarely found fossil and then only as impressions.

HYDROZOA. A group of small, usually insignificant, marine and freshwater forms, which may be solitary or colonial. The external skeleton, if present at all, is composed of chitinous material, although many authorities consider that the Stromatoporoids (a group within the Hydrozoa) are calcareous colonial Hydrozoa. Stromatoporoids are important reef-building forms, especially in the Silurian. They range from ◊ *Cambrian* to Recent (stromatoporoids are Cambrian to ◊ *Cretaceous.*)

ANTHOZOA. Marine benthonic Coelenterata, in which the polypoid stage is dominant, being characterised by a coelenteron divided by vertical radial mesenteries. This group includes sea anemones and the reef-building corals. Only the corals are of any geological importance, and they are divided into sub-classes:

(a) Tabulata (◊ Fig. 20). These are simple, colonial corals, in which the calcareous corallite (an individual coral unit) wall is the most important structural feature. Transverse plates of

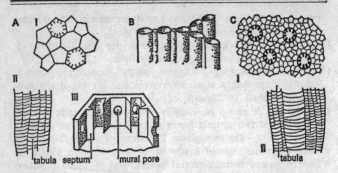

20. Tabulate corals. (A) *Favosites*: i, plan view, two corallites showing septa; ii, vertical section, showing tabulae; iii, enlarged corallite partly sectioned; (B) *Halysites* (chain coral); (C) *Heliolites*, corallites of two sizes: i, plan view; ii, vertical section.

CaCO₃ (tabulae) are usually present within the individual corallites. In some of the more advanced Tabulata, rudimentary vertical plates (septa) of skeletal material are arranged radially inside, and attached to, the corallite wall. The individual corallites of the colony may be separate from one another or they may unite in various ways to give rise to characteristic genera. They commonly range from ◊ *Ordovician* to ◊ *Carboniferous*, although the group is known to extend as far as the Middle ◊ *Jurassic*. The modern Octocorallia (Alcyonaria) may be a group of the Tabulata, or may have been derived from them. The Tabulata are not used as zonal indices, although some forms are characteristic of various horizons.

(b) Zoantharia. This sub-class may be further subdivided into (i) Rugosa or Tetracoralla and (ii) Scleractinia or Hexacoralla.

(i) The Rugosa or Tetracoralla (◊ Fig. 21) are a group of corals confined to the Palaeozoic in which the corallite wall is massive, and the septa are inserted during development in sets of four or multiples of four. They show a range of forms, from simple solitary to complex colonial types, and exhibit a number of features not found in the simpler Tabulata. These features include dissepiments – plates of skeletal material inserted between the septa and parallel to the corallite wall – and various columella structures in the centre of each corallite. These latter may be formed in a number of different ways, including the up-arching of the tabulae or the formation of a 'web' of discrete septa and

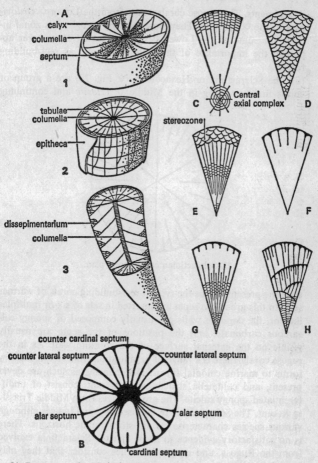

21. Rugose corals. (A) A simple rugose coral to show morphology; (B) Section; (C–H) Variations of septal plan and development of the dissepimentarium: (C) Outer dissepimentarium, central axis complex, and septa withdrawn from centre (Clisiophylloid); (D) All dissepimentarium, no septa (Cystiphylloid); (E) Septa with median dissepimentarium and outer stereozone; (F) No dissepiments, septa withdrawn from centre (Amplexoid); (G) Septa disappearing at margin, central and median dissepiment zone (Lonsdaleoid); (H) Strong development of tabulae with short discoidal septa; central dissepiment zone.

dissepiments. The rugose corals range from the Lower ◊ *Ordovician* to the ◊ *Permian*. They were used by Vaughan as zonal indices for some of the Lower ◊ *Carboniferous*. They never approach the importance of the following group as reef-building forms.

(ii) The Scleractinia or Hexacoralla (◊ Fig. 22) are a group of corals first appearing in the Middle ◊ *Triassic* and continuing

22. Scleractinian coral, septal plan.

until the present day as the typical reef-building corals of warmer seas. In this group the septa are inserted in sets of six or multiples thereof, the corallite walls are usually composed of spongy calcareous carbonate, and the positions of the septa are usually visible on the external surface of the corallite wall. As in the rugose corals, there is a range of organisation from simple solitary forms to marine colonial forms. Dissepiments as such are never present, and columella structures, if present, consist of undifferentiated spongy calcite. The group ranges from Middle Triassic to Recent. The Scleractinia are not used as zonal indices although various species characterise certain stratigraphic horizons. There is no satisfactory evidence to show that the Scleractinia evolved from the Rugosa, and many authorities consider that they may have arisen from an anemone-like ancestor.

Corals generally may be described as follows (◊ Fig. 23):

Astreoid Colonial corals in which individual epithecal walls have disappeared but the septa of individual corallites remain separate.

Calceoloid Thecal section D-shaped as in the genus Calceola.

Cerioid Colonial corals in which the corallites are in contact but still retain the individual epithecal walls.

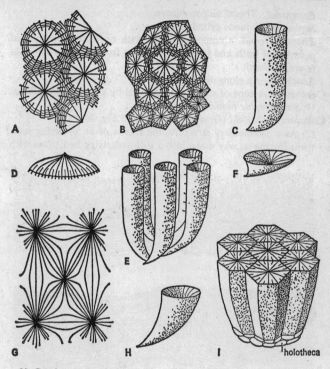

23. Corals – general descriptive terms. (A) Astreoid: epitheca replaced by stereozone; (B) Cerioid: epitheca retained; (C) Cylindrical; (D) Discoidal, with exsert septa; (E) Fasciculate (phacelloid); (F) Patellate; (G) Thamnasterioid: septa confluent; (H) Trochoid (cornute); (I) Compound massive.

Cylindrical Those corals which commence with a trochoid form and then develop a somewhat tubular form, i.e. they maintain a more or less constant diameter.

Discoidal A term applied to single corals in which the thecal wall is a flat disc.

Fasciculate Simply branched colonial corals in which neighbouring branches do not touch.

Patellate Depressed cone-shaped corallite, i.e. limpet-shaped. ⟡ *Mollusca* (Gastropoda).

93

Pyramidal Thecal section square.

Scolecoid Sinuous cylindrical cone.

Thamnasterioid Compound corals with no trace of individual epithecal walls and in which the septa from separate corallites are confluent.

Trochoid An elongated cone.

Coenozone. A zone of strata characterised by a faunal assemblage. (◊ *Stratigraphic nomenclature.*)

Collapse. (1) ◊ *Fold* (Gravity collapse) and Fig. 76.

(2) As a sedimentary structure, the results of an overlying sediment forcing its way down into a soft underlying bed, often with

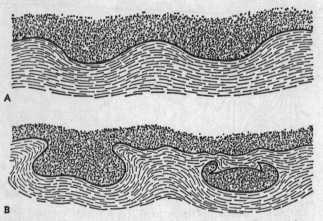

24. Collapse structures. (A) Downward bulging; (B) Exaggeration of downward bulge to form pseudonodule.

more or less complete breaking-up of the upper bed. This may be regarded as a development of ◊ *loadcasting* to an extreme extent. In some cases an overlying sandstone layer which collapses into a soft mud may develop a kind of ◊ *balled-up structure*, sometimes referred to as a pseudonodule. (◊ Fig. 24.)

(3) ◊ *Caldera*.

(4) A cavern with a collapsed roof, as found in limestone regions, is called a ◊ *polje*.

Colloid. A dispersion of ultra-fine particles (the disperse phase) suspended in a dispersion medium. A state which exists between a true solution and a suspension, and arises because the particles are so small (1–10 m μ) that the electrical and other forces which tend to support them in the dispersion medium are greater than the gravitational forces tending to cause them to settle. Glue and gelatin are typical more or less 'solid' colloids (gels). Egg-white and starch solution are 'liquid' colloids (sols). Colloids are coagulated (or flocculated) by heating or contact with electrolytes (e.g. sea water). Thus colloids transported by rivers are coagulated or flocculated on entering the sea and become part of the clay fraction of the sediment load. Opal (◊ *Silica group of minerals*) is a colloidal mineral.

Collophane. Apparently a ◊ *cryptocrystalline* or amorphous ◊ *phosphatic deposit*.

Colluvial deposits. Weathered material transported by gravity, e.g. ◊ *scree* (talus) slopes, etc. (◊ *Alluvial deposits*; *Eluvial deposits*.)

Colour index. The percentage of dark minerals in a rock, as determined from a thin section. The term is used almost exclusively of igneous rocks. Four subdivisions have been suggested, the limits of which vary according to different authors:

Johanssen	Shand	
0–5	0–30	◊ *Leucocratic*
5–50	30–60	◊ *Mesotype* (◊ *'Mesocratic'*)
50–95	60–90	◊ *Melanocratic*
95–100	90–100	◊ *Hypermelanic*

The term 'dark mineral' corresponds approximately to ◊ *ferromagnesian mineral*, i.e. in a normal igneous rock the colour index would be based upon the percentage of olivine, pyroxene, amphiboles, biotite, tourmaline and iron oxides present. However 'dark' the feldspar is in fact, it is not included with the 'dark' constituents. Shand objects to 'mesocratic' on semantic grounds.

Columella. ◊ *Coelenterata* (Anthozoa) and Fig. 21; (Mollusca) ◊ Fig. 100.

Columnal (Echinodermata). ◊ Figs. 42 and 43.

Columnar structure. A structure seen mainly in ◊ *lava* flows and ◊ *sills*, more rarely in ◊ *dykes*, and most commonly in basic rocks (◊ *basalts* and ◊ *dolerites*). The classic type – as seen in the Giant's Causeway, Antrim, Northern Ireland, and Fingal's Cave, Staffa, Inner Hebrides – consists of a close-packed series of hexagonal (sometimes pentagonal or heptagonal) 'prisms' lying perpendicular to the upper and lower surfaces of the flow or sill, or to the walls of the dyke (◊ Fig. 25). Sometimes the columnar part of

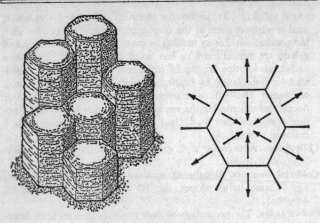

25. Columnar structure.

the rock is separated from the wall or floor by a massive, unjointed portion.

◊ *Joints* arise owing to contraction during the cooling of the lava, when a pattern of tensional forces acting towards a number of centres is set up in a layer. These forces tend to pull open a series of joints which ideally assume a hexagonal pattern, and, as cooling proceeds towards the central part of the mass, the joints develop in depth. A rather similar structure has been recorded as forming during the drying-out of clay masses; here the contractive forces are due to loss of water.

Comagmatic. A term applied to a series of igneous rocks which are assumed to have been derived from a common source (◊ *Magma*).

Comanchean. A stratigraphic stage name for the North American Lower ◊ *Cretaceous*.

Combe (coombe) rock (Head). An earthy mass containing angular fragments produced as a result of ◊ *solifluxion* in ◊ *periglacial* regions. The typical combe rock is to be found on the Chalk of south-east England.

Comminution. The breakdown of a solid to a fine powder; pulverisation, usually by mechanical means.

Compaction. ◊ *Diagenesis*. In the diagenetic formation of massive rock from loose sediment, the close-packing of the individual grains mainly by the elimination of pore-space and the expulsion of entrapped water, normally brought about by the weight of the overlying sediments.

Competent bed. A rock layer which, during folding, flexes without appreciable flow or internal shear. In metamorphosed rocks a competent bed will not develop slaty cleavage. (◊ *Incompetent bed.*) Competency of rocks is a relative term, depending to some extent on the surrounding rocks, i.e. a rock in one situation may act as a competent horizon while in other circumstances an identical rock may act incompetently. (◊ Fig. 26.)

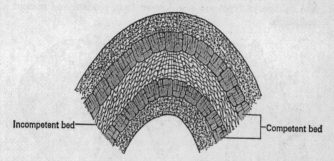

Incompetent bed ────────── ────── Competent bed

26. Competent and incompetent beds.

Composite. A term applied to igneous bodies, especially ◊ *dykes,* ◊ *sills,* ◊ *laccoliths,* etc., which show evidence of more than one phase of emplacement consisting of different materials, e.g. an original dolerite dyke may be subsequently intruded by a more acid type; a basic sill may have a more acid phase injected into it. (Cf. ◊ *Multiple.*)

Composition plane. ◊ *Twinned crystal.*

Conchoidal (lit. 'shell-like'). A term applied to the form of fracture of certain rocks and minerals (e.g. obsidian, quartz) which takes the form of a curved, concentrically ribbed surface, not unlike the shells of some lamellibranchs. (◊ *Mollusca.*)

Concordant. (1) A term applied to intrusive igneous rocks where the margins of the intrusion do not cut across the bedding or foliation of the country rock.

(2) A term applied to coastlines where the ◊ *strike* ('grain') of the country is parallel to the general trend of the coastline. Equivalent to Dalmatian or Pacific type.

Concretions. (1) During ◊ *diagenesis* certain constituents of sediments tend to concentrate in certain parts of the rock, often accumulating around a nucleus. The masses so formed may be rounded or irregular, and are known as concretions. Concretions composed of ◊ *clay ironstone,* ◊ *limestone,* ◊ *dolomite,* ◊ *chert* (flint), ◊ *gypsum,* ◊ *barytes,* and ◊ *limonite* have been recorded. (◊ *Ballstone; Dogger; Nodule; Incretion; Box-stones.*) (2) A term applied to primary aggregates developed during sedimentation, which form discrete masses enclosed in the main body of sediment. Manganese oxide concretions are forming today in the abyssal regions, while ◊ *chert,* calcium phosphate, and calcareous types are also known from modern and ancient sediments.

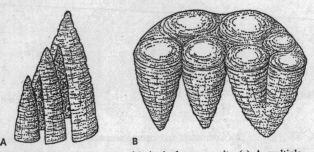

27. Cone-in-cone structures. (A) A single cone unit; (B) A multiple cone-in-cone.

Condensed sequence. A stratigraphic term used to describe a series of beds which show a relatively much thinner development than the equivalent beds elsewhere. All the beds are fully represented in the condensed sequence, although on a much reduced scale, i.e. the reduction in thickness is not produced by a series of non-sequences (◊ *Unconformity*) but by a reduced amount of sedimentation.

Cone-in-cone structure. A structure commonly found in fibrous ◊ *calcite* ('beef') and fibrous ◊ *gypsum* layers in sediments, and more rarely in ironstone and coal beds. It has the appearance of a series of cones packed one inside the other. Some occurrences may be concretionary, but in most cases the structures appear to be due to pressure. Detailed examination shows that in many cases the structure is actually a series not of 'nested' cones but of spirals. (◊ Fig. 27.)

Cone of depression. The conical form assumed by the de-watered zone surrounding a pumped well or artesian well.

Cone-sheet. An igneous ◊ *dyke* which has the form of part of a conical surface, dipping inward to a focus. Usually cone-sheets occur in swarms around a centre (e.g. Ardnamurchan in Argyll), and generally display an arcuate pattern of outcrop (◊ Fig. 28.) The term was used originally in the British Tertiary Volcanic Province. (◊ *Dyke* (Ring dyke); *Ring structure*.)

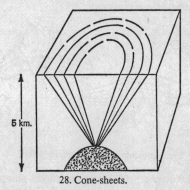

5 km.

28. Cone-sheets.

Conewangoan. A stratigraphic stage name for the top of the North American Upper ◊ *Devonian*.

Confining pressure. Equivalent to hydrostatic pressure (◊ *Pressure*).

Confluent. Two streams of approximately equal size which unite are said to be confluents. (◊ *Affluent*.)

Conformable. A sequence of beds are said to be conformable when they represent an unbroken period of deposition. (◊ *Unconformity*.)

Conglomerates. ◊ *Rudaceous rocks* consisting of rounded or subrounded fragments, implying rather more transport than ◊ *breccias*. The following are the main types of conglomerate: (a) Those consisting essentially of one type of pebble, usually quartz and/or chert pebbles – oligomict; (b) Those consisting essentially of a mixture of pebble types – petromict or polymict; (c) Glacial conglomerates, tillites, consisting of a mixture of rock pebbles and ◊ *rock flour* (◊ *Glaciers and glaciation*): (d) Those formed by the break-up, *in situ*, of a previously deposited layer of rock – ◊ *intraformational* or ◊ *autochthonous* conglomerates.

Modern pebble accumulations, such as beaches, flood plains, and outwash fans, may become conglomerates when cemented. (◊ *Banket*; *Limestone*, Clastic.)

Coniacian. A stratigraphic stage name for part of the European Upper ◊ *Cretaceous*.

Conjugate. A term used to describe two sets of structural features, e.g. ◊ *fractures*, ◊ *joints*, ◊ *faults*, ◊ *folds*, formed at the same time but aligned in differing directions.

Connate water. Water trapped in sediments at the time of deposition. (◊ *Juvenile*; *Meteoric water*.)

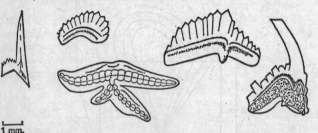

1 mm.

29. Representative conodont variations.

Conodonts. Microscopic (0·05–2 mm. in length) phosphatic tooth-like structures which range in time from ◊ *Cambrian* to ◊ *Jurassic*. They are used in the stratigraphic division of the ◊ *Devonian* and ◊ *Carboniferous*, although it is still not known to which group of animals they belong. The modern opinion is that they are vertebrate structures, because of their phosphatic nature. (◊ Fig. 29.)

Consequent. ◊ *Drainage pattern*.

Consolidation. The process of conversion of a loose or soft material to a compact, harder material – e.g. sand to sandstone (by cementation), mud to clay (by de-watering). (◊ *Diagenesis*.)

Contact metamorphism. Thermal ◊ *metamorphism* associated with igneous intrusions.

Contamination. A term used to imply the incorporation of ◊ *country rock* into ◊ *magma*. (◊ *Assimilation*.)

Continental. A term commonly used in geology in the sense of 'land'; e.g. 'continental conditions' implies conditions on a land mass as opposed to the oceans.

Continental drift. A hypothesis put forward to account for various major features of the Earth's surface – e.g. the distribution of the continents and ocean basins – in which the continents as we now see them are believed to have been formed by the break-up of one large land mass, and to have 'drifted' into their present positions. Amongst the evidence which has been put forward in support of the theory is the distribution of the Permo-Carboniferous glaciation in the Southern Hemisphere, the apparent wandering of the Poles, as revealed by palaeo-magnetic studies, the morphological 'fit' of South America and Africa, and the distribution of certain plants and animals, both as fossils and at the present time. (◊ *Gondwanaland*; *Laurasia*; *Pangaea*; *Palaeomagnetism*; *Plate tectonics*.)

Continental shelf, Continental slope. The continental shelf is that part of the sea floor adjoining a land mass over which the maximum depth of sea water is 200 m. (600 ft). The outer margin is marked by the continental slope, which extends to the ◊ *abyssal* region. (◊ Fig. 30.) Continental shelves are regarded as portions of

30. Convection currents in the Earth's mantle.

continental masses which are locally submerged. The edge of the shelf is also regarded as the edge of the continental sialic mass (◊ *Crust of the Earth*). Because of the relative shallowness of the water, the continental shelf and slope form a distinct environment with characteristic features. 'Shelf' sediments tend to be arenaceous and relatively thin, commonly containing local unconformities. ◊ *Reef* and ◊ *bioclastic* material is locally common. (◊ *Shelf facies*; Fig. 49.)

Convection currents. This term is used in its ordinary physical sense in geology, with two special applications:

(1) Convection currents occurring in the core and mantle of the Earth providing forces which produce certain structural effects at the Earth's surface – ◊ *geosynclines*, ◊ *orogenies*, and ◊ *continental drift*. (◊ Fig. 30.)

(2) Convection currents occurring in bodies of crystallising ◊ *magma*, resulting in their ◊ *differentiation*. (◊ *Layered igneous structures*.)

Convergence. (1) Metamorphic convergence occurs where two rocks of differing original character become closely similar after ◊

metamorphism, e.g. a basic igneous rock and an argillaceous limestone may yield similar amphibole-rich rocks on metamorphism. In some cases it is suggested that ◊ *metasomatic* changes may produce convergence.

(2) Stratigraphic convergence is the reduction of the vertical interval between two beds by the thinning of the intervening strata, as a result of the form of the area of deposition.

(3) In palaeontology, evolutionary convergence is said to have taken place when two groups of organisms develop similar features, even though not genetically related.

Convergent light. Normally a petrological microscope is used with a parallel light beam, but convergent light can be used for special purposes. By using convergent light, a section can be viewed with light passing through it at various angles at the same time. Use is made of convergent light to produce an ◊ *interference figure*.

31. Convolute bedding.

Convolute bedding. A structure in which the sedimentary laminae are contorted into a series of 'anticlines' separated by broad 'synclines'; distortion increases upwards, is confined to one bed, and is often abruptly truncated by overlying sediments. The distortions may be produced by high-velocity currents, expulsion of pore waters, or post-depositional sliding of sediments. (◊ Fig. 31.)

Conybeare, William (1787–1857). English geologist, later Dean of Llandaff, responsible with William Phillips (1773–1828), English geologist and printer, for the classic *Outlines of the Geology of England and Wales*. This summarised the total knowledge of rocks at that time, and postulated a stratigraphic system for them based on the fossil content of strata.

Coombe rock. ◊ *Combe rock*.

Cope, Edward Drinker (1840–97). Avid American palaeontologist with over ten thousand fossil vertebrate specimens in his collection when he died, covering more than four hundred species. He was in constant competition throughout his life with ◊ *Marsh* to name and classify new discoveries. His collection is now the property of the American Museum of Natural History.

Copper. A native metal, Cu, found in ◊ *hydrothermal* and ◊ *metasomatic* deposits, in the cavities of basic igneous rocks, and in the zones of oxidisation of copper veins. (◊ Appendix.)

Coprolites. Fossilised faecal pellets of fish, reptiles, birds, or mammals. They are generally ◊ *phosphatic* in character.

Coquina. A clastic ◊ *limestone* of cemented coarse shell debris, e.g. some of the shelly 'grits' of the English Mid-Jurassic.

Coral. ◊ *Coelenterata* (Anthozoa); *Reef*.

Corallian. A stratigraphic stage name for the European mid-Upper ◊ *Jurassic*, a synonym of Lusitanian.

Corallite. ◊ *Coelenterata* (Anthozoa) and Fig. 20.

Coral mud, Coral sand. Detrital deposits composed of clay-grade and sand-grade sized particles of calcareous material derived from ◊ *reefs* and generally deposited in the vicinity of such features. (◊ *Calci-lutite*; *Calcarenite*.)

Cordier, Pierre (1777–1861). French mining engineer, later professor, who tried to assess the composition of rocks by breaking them down for microscopic study and testing the separate particles.

Cordierite. ◊ *Cyclosilicates* and Appendix.

Cordillera. A string of islands in a ◊ *geosyncline*, a surface expression of a ◊ *geanticline*. They may divide the geosyncline into a number of separate basins.

Core of the Earth. That portion of the Earth below the Gutenberg ◊ *Discontinuity*, i.e, from a depth of 2,900 km. to the centre of the Earth. Because it will not transmit the S-waves of ◊ *earthquakes* the core must, in part at least, be liquid. However, there is considerable evidence that an inner core starting at 5,000 km. depth may be solid. The density of the core ranges from 9·5 to at least 14·5, and possibly higher. The most usual theory for the composition of the core is that it is an alloy of nickel (Ni) and iron (Fe), hence the use of the acronym NiFe. It is estimated that the temperature in the core is more than 2,700°C., with a pressure of $3·5 \times 10^6$ bars. (◊ *Mantle*; *Earth model*.)

Cornstones. Concretionary ◊ *limestones*, generally formed under arid conditions; typical of the Devonian and Permo-Trias of Great Britain.

Cornute. A synonym of trochoid (◊ *Coelenterata* and Fig. 23).

Corona structure. The development of a concentric zone or zones of one or more minerals, commonly consisting of radially arranged crystals, surrounding another mineral. The secondary mineral may be a ⬦ *reaction rim* (an alteration or corrosion product of the primary mineral) or an overgrowth of a later mineral on the primary mineral (e.g. the development of ⬦ *idocrase* layers on grossularite ⬦ *garnets*). 'Kelyphitic border' is in part a synonym. (⬦ *Epitaxy*; *Bowen's reaction series*.)

Corrasion. Vertical erosion by a river leading to downcutting.

Corrie. A ⬦ *cirque*.

Corsite. An ⬦ *orbicular* ⬦ *diorite*, containing large layered spherical arrangements of crystals.

Cortlandite. An ⬦ *ultrabasic rock*, hornblende peridotite.

Corundum. Al_2O_3, a mineral valued as an abrasive and gemstone (ruby, sapphire, etc.), found widely in metamorphosed shales and in certain kinds of metamorphic limestone veins and in some ⬦ *undersaturated* igneous rocks. (⬦ Appendix.)

Coset. A group of units of ⬦ *cross-bedding* which show a uniform direction of current flow.

Cosmogeny. The study of the age and origin of the Universe, and more especially the age and origin of the Solar System including the Earth.

Cotteau, Gustave (1818–94). French lawyer who was twice elected President of the Geological Society of France, and is remembered particularly for his authoritative account of the fossil Echinoids.

Country rock. The rock bodies which enclose an intrusive mass of igneous rock, or series of mineral veins, or replacement bodies. (⬦ *Envelope* (1).)

Couvinian. A synonym of ⬦ *Eifelian* (Devonian System).

Covellite. A copper ⬦ *ore mineral*, CuS, found in the zones of ⬦ *secondary enrichment* of copper veins. (⬦ Appendix.)

Crag and tail. A hill with a steep face on one side and a comparatively gentle downward slope on the other, formed where a highly resistant rock mass (the 'crag') on a valley floor has obstructed the glacier movement and afforded some protection from erosion to the softer sediments (the 'tail') immediately behind the resistant rock. Alternatively, the resistant mass may acquire a 'tail' of glacial debris on its 'downstream' side. The 'crag' portion of these features is commonly a volcanic ⬦ *plug* or neck.

Cranidium (Arthropoda). ⬦ Fig. 3.

Crater. In general, a depression which is more or less circular in plan. Two main types exist: (a) Impact craters, formed by bodies such as meteorites striking the Earth's surface; (b) Explosion craters, such as those produced by ◊ *volcanoes*. (◊ *Caldera*.)

Craton. ◊ *Kraton*.

Creep. A slow or gradual movement; applied to (a) soil and superficial accumulations moving under gravity (◊ *Gravity transport*); (b) progressive deformation of a rock by long-applied stress just below the elastic limit. When the stress is removed part of the deformation remains and part is lost by elastic recovery.

Crest line. The line joining the highest points on the same bed in an anticlinal ◊ *fold*.

Crest plane. The plane connecting the ◊ *crest lines* of successive beds in an anticlinal ◊ *fold*.

Cretaceous (Greek *creta*, 'chalk'). The Cretaceous Period has a duration of approximately 72 m.y., from 136 to 64 m.y. The name was adopted by the Belgian d'Halloy in 1822 and introduced into England by Fitton. The division of the period into lower and upper is at about 100 m.y., each of these main divisions being further subdivided into six stages. The lower limit of the period is defined as the base of the Craspedites zone; the upper limit is the top of the Maastrichtian, although some authorities include the Danian, in whole or in part, as the uppermost stage.

The Lower Cretaceous sediments continue the pattern of ◊ *Jurassic* sedimentation with the local development, as in Britain, of lacustrine, deltaic and estuarine facies. At the beginning of Upper Cretaceous times a widespread marine transgression occurred, the Cenomanian transgression, which, at its maximum extent, produced the greatest proportion of sea relative to land on the Earth's surface since Palaeozoic times.

In northern Europe and part of the mid-western United States the Upper Cretaceous is represented by the unique white limestone known as the Chalk. (For chalk as a rock, ◊ *Limestone*.) In ◊ *Tethys*, the early stages of the Alpine Orogeny occurred and in the trough so produced substantial thicknesses of marine sediment were deposited.

The end of Cretaceous times marks the end of the ◊ *Mesozoic* era.

The fauna of the Cretaceous, to begin with, is closely related to that of the Jurassic. Ammonites (◊ *Mollusca*) were abundant, especially in the lower part, but became rarer in the later beds, and finally became extinct at the end of the period. They are used as zone fossils wherever possible. Echinoderms (◊ *Echinodermata*) and lamellibranchs (◊ *Mollusca*) were locally important and are used as zonal indices in the absence of ammonites. Belemnites (◊ *Mollusca*) also became extinct during the period. ◊ *Brachiopoda* flourished but suffered a marked reduction in numbers at the end of the period. On the whole, corals (◊ *Coelenterata*) were distinctly less abundant than in previous eras.

On land, the dinosaurs continued to be dominant, but became extinct at the end of the period; Cretaceous mammals are known but are insignificant in size and numbers. Advanced flowering plants (Angiosperms) became important in the Cretaceous.

Crevasse. ◊ *Glaciers and glaciation.*

Crinanite. ◊ *Teschenite.*

Crinoidea. ◊ *Echinodermata.*

Cristobalite. ◊ *Silica group of minerals.*

Crocidolite. A common variety of ◊ *asbestos.*

Croixian. A stratigraphic stage name for the North American Upper ◊ *Cambrian.*

Cromerian. A stratigraphic stage name for the British and European basal Middle Pleistocene (interglacial between the Weybourne and the Lowestoft Till) (◊ *Tertiary System*).

Cross-bedding, Cross-stratification. A series of inclined bedding planes having some relationship to the direction of current flow, the angle of rest of the sediment and the rate of supply of sediment: hence the often used term 'current-bedding'. (◊ Fig. 32A.) Cosets are groups of units of cross-bedding which show a uniform direction of current flow. The boundaries of sets and cosets may be plane, curved, or irregular and these may be either erosional or non-erosional in origin. It is convenient to divide them into large-scale and small-scale units, the boundary being drawn at 10 cm. in height.

Although cross-bedding is most commonly found in ◊ *arenaceous* sediments and oolitic ◊ *limestones*, some types of clastic lime stone also display these features.

LARGE-SCALE. The formation of sets of planar stratification can most easily be explained by the growth of banks migrating down-current, as in a large ripple mark. Large ◊ *deltas* are much more complex bodies. Another common type of large-scale planar stratification is dune-bedding, sets of which show non-erosional contacts. An isolated set of trough cross-bedding is often termed a channel or ◊ *washout* and may roughly truncate evenly bedded or cross-bedded strata. Such structures are probably produced by the infilling of stream channels and repetition of the process produces a coset of beds.

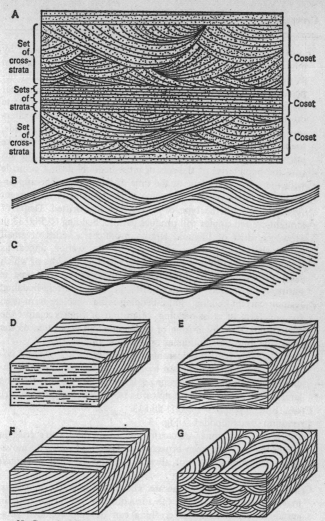

32. Cross-bedding. (A) Sets of strata, sets of cross-strata, cosets of strata; (B, C) Sections of ripple-drift bedding produced by the migration of straight-crested ripples; (D, E, F, G) Patterns of cross-stratification (as seen in different sections) produced by variation in ripple crest pattern and rate of supply of sediment: (D+F) Straight-crested ripples; (E+G) Sinuous-crested ripples; (D+E) High rate of sediment supply; (F+G) Low rate of sediment supply.

SMALL-SCALE. Small-scale cross-bedding is produced by migration of various types of ◊ *ripple marks* – a process known as ripple-drift. (◊ Fig. 32B–G.) Such drift is accomplished by deposition on the steeper ('lee') side of the ripples exceeding that on the 'stoss' (gentler sloping) side. Thus the ripple crests migrate down-current (◊ Fig. 32B, drawn in a plane parallel to the direction of current flow). If however erosion occurs on the stoss side, the sets of cross-bedding produced on the lee sides of progressive migrating ripples will have abrupt contacts (◊ Fig. 32C). In this case the boundaries of the sets of cross-bedding do not coincide with the horizontal plane at the time of deposition. The appearance of sections viewed in a plane at right-angles to the current flow depends on the type of ripple that has drifted. Drifts of straight-crested ripples will produce parallel bedding (◊ Fig. 32D). However, drift of linguoid ripples produces a series of small trough-shaped sets; these, when seen in plan section, appear as a long series of crescent-shaped ridges, the concave sides of which face down-current. (◊ Fig. 32G). Ripple-drift of this type is variously described as ripple-drift bedding, current-bedding, festoon current-bedding. False-bedding is an obsolete term used for any form of cross-bedding. Migration of ripples against the current direction, by erosion of the leeside and deposition on the stoss side, produces antidunes or antidune ripples.

Cross-bedding is of considerable importance as a ◊ *way-up criterion*. Two main criteria can be used: (a) the truncation of individual cross-laminae by younger laminae; (b) the fact that the laminae are always concave upwards. (◊ *Sedimentary structures*).

Crura (Brachiopoda). ◊ Figs. 11 and 13.

Cruralium (Brachiopoda). ◊ Fig. 13.

Crush belt (Crush zone). A narrow region of the Earth's crust in which the rock has been broken and pulverised owing to movement. Sometimes, but not always, it coincides with a ◊ *fault* zone.

Crush breccia, Crush conglomerate. The coarse-grade (◊ *rudaceous*) material developed during the formation of a ◊ *crush belt*. In certain circumstances the crushing can develop rounded fragments; the product is then termed a crush conglomerate. (◊ *Mylonite*; *Cataclastic rock*; *Fault breccia*.)

Crustacea. ◊ *Arthropoda*.

Crust of the Earth. That portion of the Earth lying above the Mohorovičić ◊ *Discontinuity*. It is divided into two shells, a lower, continuous, one – the sima (acronym of *si*lica and *ma*gnesia) – and an upper, discontinuous, layer – the sial (acronym of *si*lica and *al*uminium). The sial is apparently confined to the continental masses.

	Sial	Sima
Thickness	10–12 km.	15–20 km.
Density	2·7	2·95
Velocity of	P 5·57 km./sec.	6·5 km./sec.
earthquake	S 3·36 km./sec.	3·74 km./sec.
waves		
Composition	Granite	Basalt

The crust as a whole is thickest beneath the mountains and thinnest under the oceans. It will be noted that the thin skin of sedimentary cover has an insignificant thickness. (◊ Fig. 33.)

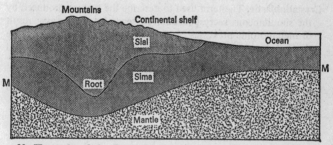

33. The crust of the Earth. M–M, the Mohorovičić Discontinuity.

Cryolite. An uncommon mineral, Na_3AlF_6, found in ◊ *pegmatite* veins in Greenland and used in the smelting of aluminium ores. (◊ Appendix.)

Crypto- (prefix). Hidden.

Cryptocrystalline. A term used to describe a very finely ◊ *crystalline* aggregate in which the crystals are so small as to be indistinguishable except under powerful magnification.

Crystal. A three-dimensional body, the bounding surfaces of which are arranged symmetrically and, for a given substance, with constant angular relationships, arising from the inherent regular atomic structure of the substance. (◊ *Crystallinity*; *Atomic structure*; *Symmetry*.)

Crystalline rock. A term used loosely to imply an igneous or metamorphic rock, as opposed to a sedimentary rock, e.g. crystalline limestone is equivalent to marble.

Crystallinity. The extent to which the ◊ *atomic structure* has controlled the outward form of a substance. Taking silica (SiO_2) as an example, it may develop in four ways: (1) as perfect crystals of quartz, i.e. showing crystal faces – crystallised; (2) as irregular grains of quartz with a fully developed atomic arrangement but not displaying crystal faces – crystalline; (3) as a very finely crystalline aggregate, chalcedony, in which the individual units are so small as to be indistinguishable except under powerful magnification – cryptocrystalline; (4) as the material opal, in which there is no regular arrangement of atoms – amorphous. (◊ *Form*; *Habit*.)

Crystallite. A microscopic, often skeletal, crystal unit which is the initial form of crystalline material developing in a volcanic ◊ *glass*. (◊ *Devitrification*; *Microlite*.)

Crystalloblastic. The term used to describe the texture produced by the simultaneous recrystallisation of several minerals as a result of metamorphic processes. A characteristic feature of the texture thus produced is that any of the minerals involved may be found as inclusions in any of the other minerals.

Crystal structure. Equivalent to ◊ *atomic structure*.

Crystal system. A group of ◊ *symmetry* classes which are associated because of their common features. The following is a list of systems:

System	Number of classes	Characteristic symmetry elements*
Cubic or Isometric	5	4 iii-fold axes
Tetragonal	7	unique iv-fold axis or a iv-fold inversion axis
Hexagonal	7	unique vi-fold axis or a vi-fold inversion axis
Trigonal	5	unique iii-fold axis or a iii-fold inversion axis
Orthorhombic	3	one or three ii-fold axes of symmetry, and if planes of symmetry are present, there are at least two

* Characteristic symmetry elements is used here to mean symmetry element(s) whose presence is diagnostic of a particular system. Other associated symmetry elements define the class within the system.

| Triclinic | 2 | no axes or planes of symmetry |
| Monoclinic | 3 | unique ii-fold axis or a ii-fold inversion axis, with a maximum of one plane of symmetry |

$$\overline{32}*$$

* Of the thirty-two crystal classes, at least two are doubtfully represented by known crystals, and several others have few mineralogical examples.

Within the limits of a crystal system, all crystals may be referred to a common pattern of referent axes (◊ *Axes, crystallographic*).
The trigonal and hexagonal systems are united by some crystallographers.
It is estimated that 50% of the known crystals belong to the ◊ *monoclinic system*, 25% to the ◊ *orthorhombic system,* and 15% to the ◊ *triclinic system*. Of the remaining four systems the ◊ *cubic* is most abundant followed in order by the ◊ *tetragonal,* ◊ *trigonal* and ◊ *hexagonal systems*. Within each crystallographic system the vast majority of examples occur in the ◊ *holohedral class*.

Cubic system. A ◊ *crystal system* divided into five classes:

	Inter-national symbol	Centre	Planes of sym-metry	Axes of rotation symmetry	Axes of rotary inver-sion	Example
1	4/m3m (m3m)	C	9	3 iv, 4 iii, 6 ii	—	Rock salt Garnet Fluorite
2	43m	—	6	4 iii	3 iv*	Zinc blende
3	432 (43)	—	—	3 iv, 4 iii, 6 ii	—	No known crystal
4	2/m3 (m3)	C	3	3 ii, 4 iii	—	Pyrite
5	23	—	—	3 ii, 4 iii	—	Cobaltite (CoAsS)

* The 3 iv-fold rotary inversion axes of this class are equivalent to 3 ii-fold rotation axes.

All cubic crystals are referred to three axes mutually at right angles, employing equal parameters on all three axes. (For international symbols, ◊ *Symmetry*; ◊ *Axes, crystallographic*.)

Isometric is a synonym of cubic.

Cuesta (Spanish, 'flank, slope'). A geomorphological unit consisting of a gently inclined surface parallel to the ◊ *dip* of the bedding planes, and an escarpment or scarp face which is steeply inclined in the opposite direction to the dip slope and cutting across the bedding planes. Typical examples of cuestas are to be found in the North and South Downs of England. Some authors have used the term 'escarpment' as synonymous with cuesta. 'Wold' is an obsolete synonym for cuesta preserved in English place names such as Cotswolds. (◊ *Hog's back*.)

Cuisian. A stratigraphic stage name equivalent to ◊ *Ypresian* (Tertiary System).

Culm. A miner's term for carbonaceous shale and fissile varieties of anthracite (◊ *Coal*). Applied stratigraphically to Carboniferous rocks in Devon and Cornwall.

Culmination. The result of the coincidence of two ◊ *anticlines* belonging to different sets of ◊ *folds*, in cross-folding.

Cummingtonite. A mineral of the ◊ *amphibole* group (monoclinic).

Cumulates (Latin, *cumulus*, 'heap'). ◊ *Layered igneous* masses commonly contain rocks which appear to have formed by 'accumulation' of 'primary precipitate crystals'. Wager, Brown, and Wadsworth (*Journal of Petrology*, Vol. 1, Pt 1, 1960) have suggested that the term 'primary precipitate crystal' should be replaced by the term 'cumulus crystal', and that the resulting rocks should be called cumulates. Their definition may be quoted as follows: 'The term cumulus crystal means a unit of the pile of crystals as originally precipitated from the magma before any modification by later crystallisation. The liquid in the interstices of the cumulus may then be called intercumulus liquid and the crystalline material occupying this position ... may be called intercumulus material.'

According to the nature of the intercumulus material and its relationship to the cumulus crystals, Wager, Brown, and Wadsworth have suggested three major types of igneous cumulates: (1) Orthocumulates – the intercumulus liquid has crystallised out into one or more minerals which enclose the original cumulate crystals. (2) Adcumulates – the cumulus crystals are enlarged by crystallisation of material of the same composition, and the amount of residual trapped liquid (pore material) is gradually reduced. Adcumulus crystals are not infrequently found set ◊ *poikilitically* in large crystals of another mineral or minerals; this type is referred to as a heteradcumulate. If the accumulate crystals form branching (arborescent) or parallel growth together with the development of a heteradcumulate texture the term 'harrisitic cumulate' or 'crescumulate' has been proposed. (3) Mesocumulates – intermediate in character between ortho-cumulates and adcumulates. Some overgrowth of material occurs on the cumulus crystals but some pore material crystallises as in orthocumulus types.

L. R. Wager and G. M. Brown, *Layered Igneous Rocks* 1968.

Cumulophyric. A synonym of ◊ *glomeroporphyritic*.

Cupola. A small dome-like protuberance projecting from the main body of a larger igneous intrusion. It is possible that some bodies of igneous rocks described as ◊ *stocks* and ◊ *bosses* may in fact be cupolas. (◊ *Batholith* and Fig. 8.)

Cuprite. An important copper ore, Cu_2O, found in the zones of weathering of copper veins. (◊ Appendix.)

Curie point. ◊ *Palaeomagnetism*.

Current-bedding. ◊ *Cross-bedding*.

Cursorial. A term applied to animals with one pair of limbs adapted for running, e.g. ostriches, some dinosaurs, man. (Cf. ◊ *Un-guligrade*.)

Cuvette. A basin of deposition which is not ◊ *tectonic* in origin; i.e. often a lower region between mountain ranges. The Old Red Sandstone in Scotland was deposited in a series of such basins. (◊ *Devonian*.)

Cuvier, Georges (1769–1832). A Frenchman who is sometimes referred to as the founder of vertebrate palaeontology. By a process of correlation, he produced drawings of entire vertebrates from the bone fragments available, and established that fossils came from now-extinct forms allied to living creatures. He suggested that fossils could be used to establish the age of strata, and believed that the extinction of species was governed by the occurrence of sudden catastrophic change. (Cf. ◊ *Lamarck*.)

Cwm. Welsh word for 'corrie', sometimes Anglicised to 'coombe' or 'combe'. (◊ *Cirque*.)

Cycle of erosion. A concept in which the various stages in the erosion of a region are regarded as parts of a 'cyclic' process. Ideally the cycle begins with uplift of a land surface, followed by initiation and development of relief, and finally degeneration of relief and ultimate planation. Further uplifts cause the cycle to recommence (◊ *Rejuvenation*). It is very likely that complete planation is never achieved, uplift occurring at almost any stage in the cycle. In certain circumstances traces of an earlier cycle may be preserved, even after several later cycles of erosion have operated.

The Normal cycle of erosion is developed in temperate climates by the action of rivers. The end stage of planation is termed a peneplain (after W. M. Davis, and hence 'Davisian Cycle'). In arid and semi-arid climates the final stage is the pediplain (after W. Penck and L. C. King) (◊ *Pediment*). This pediplain is produced largely by intermittent sheet flood action. In some areas it is possible to recognise features of both peneplanation and pediplanation, suggesting that changes in climatic regime have occurred.

Although cycles of ◊ *marine erosion* and glacial erosion have been described, and to a limited extent the idea is useful, they are not comparable with the cycles outlined above (◊◊ *Karst scenery*). All the processes operating in the various cycles of erosion tend to reduce the relief of the continent and inevitably result in ◊ *isostatic* readjustment (rejuvenation).

Cyclic sedimentation. A term applied to a sequence of sediments which change their character progressively from one extreme type to another followed by a return to the original type. A typical example would be:

| Decreasing salinity | ↑ | { Fresh water
  Brackish water |

| Increasing salinity | ↑ | { Marine
  Brackish water
  Fresh water |

representing a marine incursion into an otherwise freshwater environment. The essential features of the cyclic unit is the steady sequence of changes, starting and finishing with similar types. This may be symbolically expressed

a
b
c
d
c
b
a

Cyclic units tend to be large in thickness and extent. ($\diamondsuit$ *Cyclothem*; *Rhythmic sedimentation*.)

Cyclosilicates. This group consists of those $\diamondsuit$ *silicates* in which the SiO_4 units are linked so as to form three-, four-, or six-membered rings. In each SiO_4 unit two oxygens are shared with adjoining units, giving rings with the formulae Si_3O_9, Si_4O_{12}, Si_6O_{18} respectively.

THREE-MEMBERED RINGS. There are no common minerals belonging to this group, the very rare mineral benitoite ($BaTiSi_3O_9$) being the best known.

FOUR-MEMBERED RINGS. Apart from some $\diamondsuit$ *zeolites*, the best-known example is the triclinic mineral $\diamondsuit$ *axinite* – $Ca_2(MnFe'')Al_2(BO_3)Si_4O_{12}(OH)$ – a product of $\diamondsuit$ *pneumatolysis*, especially of calcareous rocks, involving boron.

SIX-MEMBERED RINGS. (1) Beryl. This is the simplest six-membered ring, with the formula $Be_3Al_2Si_6O_{18}$. It is hexagonal holohedral (6/mmm). The mineral has a hardness of 8 and is used as a gemstone under the names aquamarine (blue) and emerald (green). Beryl is the only common beryllium-bearing mineral and is much sought after as a source of the metal. It occurs mainly in granites and granite pegmatites. ($\diamondsuit$ Appendix.)
(2) Cordierite is closely related to beryl, despite the fact that it is orthorhombic, the composition being $Al_3(Mg,Fe'')_2(Si_5Al)O_{18}$. It occurs as pseudo-hexagonal triplets and is common in thermally metamorphosed argillaceous rocks. ($\diamondsuit$ Appendix.)
(3) Tourmaline – hexagonal trigonal (3m) – perhaps the best-known six-membered ring silicate. The basic structural units are an Si_6O_{18} ring and three BO_3 units, together with six AlO_4 and four (OH,F) units. The cations linking these may be almost any combination of the common cations (K, Na, Mg, Mn, Li, Fe", Fe''', Al). As a result of this tremendous range of possible combinations, many varietal names have become established – the following is not an exhaustive list:

Black	Schorl
White	Achroite (K-rich)
Pink	Rubellite (Li-rich)
Blue	Indicolite (Na-rich)
Green	Elbaite
Brown	Dravite

Many tourmalines display colour zoning and the mineral is $\diamondsuit$ *piezoelectric* and $\diamondsuit$ *pyroelectric*. It occurs mainly as a result of $\diamondsuit$

pneumatolysis associated with granites, and often occurs in granite pegmatites. It is a common detrital mineral and also occurs frequently in metamorphic rocks. (◊ Appendix.)

Cyclothem. A term applied indiscriminately to the 'repeat' unit of either ◊ *cyclic sedimentation* or ◊ *rhythmic sedimentation*. In the examples quoted, the cyclothem for cyclic sedimentation would be the whole succession

> a
> b
> c
> d
> c
> b
> a

whereas for rhythmic sedimentation the cyclothem is the unit

> d
> c
> b
> a

from the sequence

> d
> c
> b
> a
> ——
> d
> c
> b
> a

The cyclothem for a given set of deposits is probably an idealised concept, since detailed examination of the actual units will usually show gaps in or repetitions of parts of the cyclothem, e.g. one cyclothem might be

	a another	a or a
	b	c	b
	d	d	c
			b
			c
			d

The cyclic unit may be regarded as a symmetrical cyclothem, while the rhythmic unit is an asymmetrical cyclothem. The meaning of the terms 'cyclic' and 'rhythmic' has unfortunately become clouded by imprecise usage, which has led to the view that rhythms, as here defined, are merely a kind of cycle or vice-versa. In view of the fact that rhythms and cycles probably arise in different tectonic environments, the use of two distinct terms seems justified.

A cycle of cyclothems is called a megacyclothem and a sequence of megacyclothems is called hypercyclothem.

It has been claimed that in any sequence of varied sediments which is not a simple alternation of two types, a rhythm or cycle can be established, providing that one works on a suitable scale.

Cystoidea. ◊ *Echinodermata.*

Dacian. A stratigraphic stage name for the base of the European Upper Pliocene (◊ *Tertiary System*).

Dacite. ◊ *Rhyolite*.

Dalradian. The youngest stratigraphic division of the ◊ *Precambrian* in Scotland and Ireland. The upper part contains a ◊ *Cambrian* fauna.

Dana, James Dwight (1813–95). Distinguished and influential American geologist who produced the first full account of the geology of that country. He was the first to postulate clearly the idea of horizontal compression as an orogenic force caused by the cooling and contraction of the core of the earth. He is best known as the author of *The System of Mineralogy*, which remained for eighty years the most comprehensive work on mineralogy available – and is still a valuable reference book today.

Danian. A stratigraphic stage name for the base of the European Palaeocene (◊ *Tertiary System*), or, in some authors, the top of the ◊ *Cretaceous*.

Darcy. A measure of the ◊ *permeability* of a rock. One darcy (D) equals a permeability such that one millilitre of fluid, having a viscosity of one centipoise, flows in one second under a pressure differential of one atmosphere through a porous material having a cross-sectional area of one square centimetre and a length of one centimetre. The working unit is the millidarcy (mD), one thousandth of a darcy. Darcy's Law relates to the flow of fluids, especially gas, oil, and water, in underground rocks. (The word comes from Henri Darcy, a nineteenth-century French water engineer.)

Daubrée, Gabriel (1814–96). French mining engineer and professor who produced a comprehensive survey of the origins, distribution and properties of surface and underground waters. A pioneer of experimental geology, he was particularly interested in the effect of the heating of water at great depth in the origin of metamorphic rocks, and distinguished the terms regional, contact and dynamic metamorphism.

Daughter element. An element produced by the radioactive decay of a pre-existing element. (◊ *Radioactive dating*.)

Decken structure (German, 'cover'). A recumbent ◊ *fold*; equivalent to 'nappe structure'.

Décollement (literally, an unglueing or unsticking). The plane of dislocation caused by an upper series of rocks folding and in the process sliding over a lower series of rocks, which may be unfolded or only slightly folded. It has been likened to the crumpling of a tablecloth when pushed over a polished table. For a successful and widespread décollement to occur, a bed or beds which can act as a lubricant is essential – the commonest types are salt, anhydrite, and certain types of shale. The classic example of décollement occurs in the Jura Mountains. (◊ Fig. 34.)

34. A décollement. D–D, plane of décollement.

Decrepitation. ◊ *Geological thermometry*.

Decussate texture. A random arrangement of lath-like or columnar crystals making up a rock.

Dedolomitisation. A process whereby a ◊ *dolomite*-bearing rock is converted to a ◊ *calcite*-bearing rock. Two ways have been described:

(1) By ◊ *metamorphism* – the magnesium ions in dolomite tend to react more readily than the calcium ions with impurities such as silica or clay minerals, e.g.

$$2CaMg(CO_3)_2 + SiO_2 = 2CaCO_3 + Mg_2SiO_4 + 2CO_2$$
Dolomite   Silica   Calcite   Olivine

(2) By ◊ *diagenesis*. Recently, accounts have been given of limestones which contain irregular mosaic ◊ *pseudomorphs* of calcite after dolomite crystals. The sequence of changes seems to have been: original calcite limestones → development of dolomite crystals → removal of magnesium, leaving calcite pseudomorphs. The precise mechanism for this change is not fully understood, but presumably percolating waters are somehow responsible. (◊ *Dolomitisation*.)

Deep-sea deposits. ◊ *Abyssal deposits*.

Deerparkian. A stratigraphic stage name for the North American mid Lower ◊ *Devonian*.

Deflation. Wind transport of loose surface debris.

Deformation. A structural term used to describe any change in attitude, shape, or volume of a bed or layer, after its formation. Recently the term has come into more widespread use to cover such things as folding, faulting, development of schistosity, and nearly all other features produced by ◊ *tectonic* forces. (◊ *Stress*; *Strain*; *Pressure*.)

Deglaciation. The stage of retreat of a widespread glaciation.

Degradation. The general lowering of land level by processes of ◊ *denudation*.

Delta. A deposit of sediments formed at the mouth of a river where it enters a lake or the sea. It is normally built up only where there is no tidal or current action capable of removing the sediment as fast as it is deposited, and hence the delta builds forward from the coastline. This process of building-up is complex, and leads to the formation of a number of separate channels (distributaries), isolated lagoons, ◊ *levées*, marshy ground, and a network of small creeks. Most deltas are complicated and multiple, but in a simple delta three main types of bed may be distinguished: (1) The bottomset beds – forming on the sea-bottom beyond the seaward face of the delta. (2) The foreset beds – building outward from the seaward face of the delta over the bottomset beds. (3) The topset beds – which are deposited above the already deposited foreset beds. This triple structure may be seen on a small scale in ◊ *cross-bedding*.

Deltaic sediments can generally be recognised by this type of sedimentary feature. Sediments found are, in general, sands of various sorts, clay material and silts, together with a certain amount of organic debris. A good deal of deposition takes place by the ◊ *flocculation* of colloidal material in the river water when it comes into contact with the sea. Occasionally conglomerates are deposited during extreme floods.

Delthyrium (Brachiopoda). ◊ Fig. 11.

Deltidial plates (Brachiopoda). ◊ Fig. 11.

Demantoid. ◊ *Garnet*.

Dendritic (lit. 'tree-like'). A term used of any form which is branching, ramifying or dichotomising, thus giving the appearance of a tree in silhouette. It may be applied to drainage (◊ *Drainage patterns*), certain mineral forms, organic borings, etc.

Dendritic minerals include branching aggregates of crystals, for example metallic elements such as silver, gold, copper, etc., but more generally they occur in the form of film-deposits in cracks, joints, crevices, etc., deposited by percolating water.

The term is also applied to branching systems of arms, tendrils, feelers, stipes, etc. in organisms, e.g. crinoid arms (◊ *Echinodermata*) and graptolites (◊ *Chordata*).

Dendrochronology. The measurement of time intervals (◊ *Geochronology*) by counting the annular rings of trees; applicable only to the last 3–4,000 years.

Dendroidea. ◊ *Chordata* (Graptoloidea).

Density current. A turbidity current (◊ *Turbidite*).

Dental sockets (Mollusca). ◊ Fig. 101.

Denudation. The sum total of the processes which result in the general lowering of the land surface. It is normally taken to consist of the processes of ◊ *weathering*, ◊ *transportation*, and ◊ *erosion*.

Depression. The result of the coincidence of an ◊ *anticline* and a ◊ *syncline*. (◊ *Fold*.)

Derived fossil. Any organic trace found in one sedimentary layer which was originally preserved in an older formation, from which it has since been removed by the natural processes of denudation and redeposition.

Deroofing. (1) The uncovering of an intrusive ◊ *plutonic* mass of igneous rock by the normal processes of ◊ *denudation*.

(2) The foundering of the roof rocks above an intrusive magma body which allows the molten rock to flow out at the Earth's surface. (◊ *Cauldron subsidence*.)

Desert. A term loosely applied to any arid region, but more accurately defined as a region which receives less than 25 cm. of rain in a year. It is not necessarily an area of hot climate, since cold deserts also occur. Regions having between 25 and 50 cm. of rain per year are often referred to as semi-desert. There are more precise formulae for determining desert climates, but these are not particularly useful to the geologist.

Desiccation cracks. Cracks, usually in mud, produced by the rapid drying-up of the surface. (◊ *Sedimentary structures*.)

Desmarest, Nicolas (1725–1815). Frenchman, often described as the father of volcanic geology, who first put forward the theory that basalt was a product of vulcanicity, and produced a classification of volcanic rocks according to age and alteration.

Desmoinesian. A stratigraphic stage name for the North American Middle ◊ *Pennsylvanian* (Carboniferous System).

Detrital. A term applied to any particles of minerals, or, more rarely, rocks, which have been derived from pre-existing rock by processes of ◊ *weathering* and/or ◊ *erosion*. Detrital minerals are usually those which are chemically stable and resistant to mechanical abrasion, and without cleavage. Some detrital minerals are economically important: most of the world's tin supplies come from detrital ◊ *cassiterite*, while gold, diamonds, and titanium (in the form of ◊ *ilmenite* and ◊ *rutile*) are also found in this form. (◊◊ *Clastic rocks*.)

Deuteric (Synantectic). The term applied to the alterations in a body of igneous rock produced by the action of vapours and volatiles, which are derived from the magma itself during the later stages of consolidation. Amongst the processes which can be included here are tourmalinisation, kaolinisation, serpentinisation, and greisening (◊ *Pneumatolysis*; *Hydrothermal processes*).

Devitrification. The development of crystals, initially on a very small scale, in a glassy igneous rock such as obsidian (◊ *Rhyolite*) or ◊ *pitchstone*. Many of the initial crystal growths have special names, such as globularites, margarites, belonites, etc. One of the commonest ways in which devitrification appears is in the formation of radial masses of ◊ *crystallites*, in the form of ◊ *spherulites*, e.g. spherulitic obsidian. Devitrification may ultimately spread throughout a glassy rock leaving it completely crystalline. These completely devitrified rocks are sometimes called ◊ *variolites*.

Devonian. A period named from the county of Devon in south-west England. It extends from 395 to 345 m.y., having a duration of 50 m.y. (For the lower limit of the period, ◊ *Silurian*.) The upper limit of the marine Devonian is the top of the Famennian; the upper limit of the continental facies of the Devonian is variable, as the overlying Lower Carboniferous is ◊ *diachronous* upon it. Higher zones of the continental Devonian which occur in Greenland are not represented in Great Britain.

The Devonian includes the oldest widespread ◊ *phanerozoic* continental deposits. These interdigitate with the marine deposits in a number of areas, but even so their exact equivalence cannot always be determined. The continental deposits are referred to as the Old Red Sandstone.

Although the marine Devonian was first defined from the type area of Devon, this area has not proved satisfactory for correlation, owing to the absence of well-preserved fossils. For this reason the stages established in the fossiliferous marine deposits of the Ardennes, Belgium, are now commonly referred to.

The marine Devonian is zoned on early ammonoid cephalopods (◊ *Mollusca*), whereas the Old Red Sandstone is zoned largely on brackish or freshwater fish. No new fossil groups are represented in the marine Devonian, but some of the groups common in the Lower Palaeozoic sediments became rare or extinct. Graptolites (◊ *Chordata*) died out early in the period, and trilobites (◊ *Arthropoda*) were generally rare. Corals (◊ *Coelenterata*) were abundant especially in the Middle Devonian, where reefs occurred. Ammonoid cephalopods became relatively common, whereas the nautiloids (◊ *Mollusca*) were no longer important. Crinoids (◊ *Echinodermata*) flourished as did the ◊ *Brachiopoda*, while the first undoubted marine fish occurred.

In the Old Red Sandstone fish, plants, and freshwater molluscs (◊ *Mollusca*) are the only common forms found fossil. The fish underwent considerable development, starting with primitive, armoured, jawless forms which developed into advanced jawed forms, and which, at the top of the Old Red Sandstone, developed lungs and were on the threshold of an amphibious mode of life. Plants developed initially as unspecialised swamp forms rarely exceeding 60 cm. in height, which later evolved to give rise to the large tree-ferns which appeared at the close of the period. The most abundant of the freshwater molluscs were lamellibranchs similar in form to the modern freshwater mussels.

The climax of the ◊ *Caledonian* orogeny occurred during the Devonian (and was accompanied by widespread volcanicity).

Dextral. A term applied to tear (strike) ◊ *faults* to describe the apparent direction of relative movement of each side, in this case to the right.

Dextral coiling. ◊ *Mollusca* (Gastropoda) and Fig. 100.

Dia- (prefix). Through.

Diabase. A term utilised in several different ways in different countries, so that confusion prevails. In Britain, it was formerly used to describe slightly metamorphosed medium-grained basic igneous rocks, in which the ◊ *pyroxene* has been altered to ◊ *amphibole*. This usage derives from the former Continental term for a pre-Tertiary medium-grained basic igneous rock, ◊ *dolerite* being the name used for equivalent Tertiary and Recent rocks. In the United States, the term is a precise synonym of the British usage of the term dolerite, with no restriction as to age. It is perhaps not too late to suggest abandoning the term by reason of its ambiguity.

Diachronous (lit. 'across time'). A term applied to lithological units (e.g. a bed of sandstone, a reef limestone) which appear to be continuous beds but which in fact represent the development of the same ◊ *facies* at different places at different times, i.e. the lithological unit transgresses the absolute time planes. Sometimes fossils are described as being diachronous, but this is not a proper usage of the word. (◊ Fig. 35.)

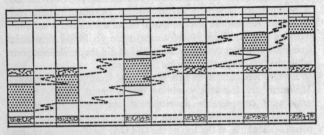

 Diachronous bed

▭▭▭ ▨▨▨ ▨▨▨ Marker bands

35. A diachronous bed. The marker bands (◊ *Marker beds*) represent a particular ◊ *facies* occurring consistently right across the area. The diachronous bed displays a lithological facies developed at different times in different parts of the area; e.g. at the left of the diagram it is between the lower and middle marker bands and on the right it is between the middle and upper marker bands, thus showing that the lithological bed transgresses the time plane.

Diadochy. The replacement of one ion in a lattice by another. (◊ *Geochemistry*.)

Diagenesis. Those processes affecting a sediment while it is at or near the Earth's surface, i.e. at low temperature and pressure. Changes which take place as a result of Earth movements and increased pressure are ◊ *metamorphic* in character and are excluded from the definition; however it is clear that diagenetic changes grade into metamorphic ones. Sub-aerial weathering is also specifically excluded.

Lithifaction is the term used to describe those changes which result in the formation of a massive rock from a loose sediment. Two important processes in diagenesis which appear to operate more or less continuously up to the completion of the lithifaction process are compaction (consolidation is approximately synonymous) and cementation. Compaction involves the close-packing of the individual grains mainly by the elimination of pore-space and expulsion of entrapped water; this is normally brought about by the weight of the overlying sediments. Cementation is the process by which the individual particles of a sediment are held together by a secondarily developed material. This may either be a substance introduced by ground water percolating through the pores of the rock, or be derived from solution of part of the mineral matter of the rock followed by redeposition. (✧ *Arenaceous rocks.*)

Many diagenetic changes are controlled by the eH (oxidising potential) and pH (acidity/alkalinity) of the water. Diagenesis is recognised as taking place in three stages, pre-burial, early burial, and late burial, during which the following changes occur:

PRE-BURIAL CHANGES. (1) Disruption of sedimentary structures and mixing of layers by the action of burrowing organisms. (2) Adjustment of part of the sedimentary material to the eH and pH of the water of the basin of deposition. (3) Adjustment of clay minerals to changes in concentration of the major cations present in the surrounding waters. (4) Development of certain ✧ *authigenic* materials, e.g. glauconite, phosphorite, pyrite; possibly certain iron hydroxides.

EARLY BURIAL CHANGES. (1) Continuation of processes 2 and 3 above to a greater extent. (2) Oxidation of carbonaceous material; especially important under normal conditions. At this stage there is a maximum development of sulphides if the oxidising-reducing conditions are right. (3) Continuation of compaction (de-watering). (4) Probable initiation of development of ✧ *concretions.* (5) Start of primary cementation. (6) Possibly some conversion of primary aragonite to calcite, and development of dolomite.

LATE BURIAL CHANGES. (1) eH and pH cease to be of great importance. (2) Completion of de-watering of clays and an increase in the 'grade' of the clay minerals. Authigenic micas and chlorites may develop. (3) Completion of development of concretions and cementation processes. (4) Replacement of aragonite by calcite; ✧ *dolomitisation* is an important factor.

For the diagenesis of carbonaceous sediments, ✧ *Coal*; coalification is essentially a diagenetic process.

J. H. Taylor, *Advancement of Science*, Vol. 20, No. 87, January 1964.

Diallage. An obsolete synonym for the mineral ✧ *augite*, when displaying ✧ *schiller*.

Diallagite. A rock consisting essentially of ✧ *diallage*.

Dialogite. An obsolete synonym for the mineral ◊ *rhodochrosite*.

Diamond. A cubic mineral variety of crystalline carbon, C, which may be of ◊ *gem* quality. It is found in ◊ *ultrabasic* and alluvial deposits and is used for cutting and as an abrasive (◊ *Bort*). (◊ Appendix.)

Diaphthoresis. Retrograde ◊ *metamorphism.*

Diapir. An intrusion which domes the overlying cover after piercing lower layers. It may develop a narrow neck at depth, giving a balloon-like form. Igneous rocks may form diapirs. ◊ *Salt domes* are another diapiric structure, and these sometimes give rise to diapiric ◊ *folds*, i.e. folds which develop forcibly upwards and pierce the overlying strata. Occasionally diapiric bodies pierce through to the surface.

Diastem. A term often used for an ◊ *unconformity* produced by a period of non-deposition which is localised and/or short in duration.

Diastrophism. The large-scale deformation of the Earth's ◊ *crust* which produces continents, ocean basins, mountain ranges, etc.

Diatomite (Diatomaceous earth). A siliceous sediment made up more or less entirely of the 'skeletal' remains of the microscopic plants called diatoms. It is exceedingly fine-grained, incoherent, and highly absorbent. It is used in filters, as an absorbent for nitroglycerine, and as an ultrafine abrasive. Kieselguhr is a synonym.

Diatreme. A volcanic vent piercing sedimentary strata, usually the result of an explosive eruption.

Dichotomous. Regularly branching into two. (Cf. ◊ *Anastomosing*.)

Dichroism. A variety of ◊ *pleochroism*, in which two colours or shades are displayed. A dichroiscope is a simple device for observing pleochroism.

Dichroite. Obsolete synonym for ◊ *Cordierite.*

Dickite. A ◊ *clay mineral* of the kaolinite group.

Differential erosion/weathering. These features develop in rocks which have varying resistance to the agencies of erosion and/or weathering, so that parts of the rock are removed at a greater rate than others. A typical example is the removal of soft beds from between harder beds in a series of sedimentary rocks. The term may be applied to any size of feature, from small-scale 'etching' to the regional development of hills and valleys controlled by hard and soft rocks.

Differentiation. (1) The separation of an igneous magma into two or more fractions, which can then consolidate as different rock types (◊ *Magmatic differentiation*).

(2) The migration and concentration of elements during metamorphism so as to produce an inhomogeneous rock from an originally homogeneous one (metamorphic differentiation). The process can give rise to banding (◊ *Gneiss*), more or less spherical zoned bodies, and occasionally segregation veins.

Dike. American spelling of ◊ *dyke.*

Diluvium. An obsolete term formerly used for what are now referred to as superficial deposits, supposedly the deposits from the Noachian deluge. It was subsequently used for any flood deposit and then for any superficial deposit.

Dimorphism. (1) (Mineralogy). ◊ *Polymorphism.*
(2) (Palaeontology). The occurrence of certain species of fossil in two distinct forms, e.g. the megaspheric and microspheric forms of some Foraminifera. (◊ *Protozoa.*)

Dinantian. A stratigraphic stage name for the European Lower ◊ *Carboniferous.* (◊ *Mississippian.*)

Diopside. A mineral of the ◊ *pyroxene* group, $CaMgSi_2O_6$, found in igneous rocks and metamorphosed impure dolomites. (◊ Appendix.)

Diorite. A coarse-grained plutonic intermediate igneous rock, consisting essentially of intermediate plagioclase ◊ *feldspar* (oligoclase to andesine), and one or more of the ◊ *ferromagnesian minerals* – ◊ *biotite,* ◊ *hornblende,* ◊ *augite*; quartz may be present in small amounts, up to 10%, and alkali feldspar may also occur, up to one third of the total feldspar. ◊ *Sphene* (titanite) is a common accessory.

The problem of defining the limits of the term diorite has been discussed by many authors. It has been complicated by variable use of the terms tonalite (here taken to be any diorite, as defined above, containing accessory quartz), granodiorite, and, to some extent, adamellite. We take the view here that the term diorite should be applied to any rock satisfying the definition set out above.

With increasing alkali feldspar, diorites grade into ◊ *monzonites,* and with increasing quartz they grade into ◊ *granodiorites* (quartz-bearing types usually contain oligoclase rather than andesine). When the plagioclase becomes more basic than andesine, e.g. labradorite, the rock becomes a ◊ *gabbro.* Of the finer-grained equivalents of diorite, some are microdiorites and some are ◊ *lamprophyres.*

Diorites containing more than 70% ◊ *ferromagnesian minerals* are called meladiorites (this term is taken to include the ◊ *appin-*

ites in part). Increasing the ferromagnesian content to 100% produces a ◊ *perknite*. A decrease in ferromagnesian minerals to 20% or less gives rise to a leucodiorite, and forms containing only very small amounts of ferromagnesian minerals, 5% or less, are oligoclasites or andesinites.

Diorites tend to be equigranular, i.e. ◊ *porphyritic* types are rare. The well-known orbicular diorites (corsite or napoleonite) contain large (2–15 cm. in diameter) layered spherical arrangements of crystals. Pyroxene diorites normally contain ◊ *augite*, but ◊ *hypersthene* diorites have been described. Hornblende diorite is probably the commonest type. Mica diorites usually contain some quartz.

The most important difference between the composition of diorites and granodiorites is the markedly lower SiO_2 content of diorites, while they differ from gabbros in having higher SiO_2 and much lower Mg and Ca contents. Alkalis are generally higher than in normal gabbros.

Undersaturated diorites are exceedingly rare; plumasite is an oligoclase-corundum rock in this group, while dungannonite is an andesine-corundum rock with accessory nepheline. (This latter rock is associated with nepheline syenites.)

Microdiorites are not very common dyke rocks. Porphyritic varieties (ambiguously referred to sometimes as diorite porphyries) correspond to the old term 'porphyrite'. The name malchite has been used for non-porphyritic microdiorites, although it is sometimes applied to lamprophyres. Markfieldite has been described as a microdiorite, but the content of alkali feldspar suggests monzonitic or even syenitic affinities. A number of microdiorites show well-marked micrographic textures. (For ◊ *melanocratic* microdiorites, ◊ *Lamprophyres*.)

Diorites are the plutonic equivalents of ◊ *andesites*. They are rather uncommon rocks, rarely forming large independent masses, and are usually found either as rather small ◊ *plugs* or ◊ *bosses* (rarely large ◊ *dykes* or ◊ *sills*) or as marginal or satellite masses to granodiorites, gabbros, or, more rarely, syenites.

Dip. The true dip of a plane is the angle that it makes with a horizontal plane – the angle being measured in a direction perpendicular to the ◊ *strike* of the plane. Apparent dip is the angle measured in any other direction. Given the strike and an apparent dip, or two apparent dips, it is possible to obtain the true dip. (◊ Fig. 36.) In geological work, the dip of a ◊ *bedding plane* or other planar structure cannot usually be measured with an accuracy greater than $\pm 1^0$, and in many cases dips are recorded which are generalised in amount and direction. A clinometer is used for measuring dip.

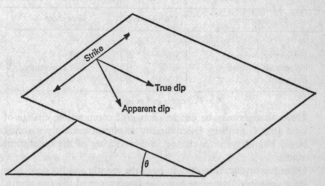

36. Dip and strike.

Regional dip is a generalised statement of the direction of dip over a region, ignoring minor variation of amount or direction.

The term is also used as an adjective implying a direction, e.g. a dip fault is a fault whose strike is parallel to the direction of dip of the beds affected; cf. hade (◊ *Fault*). For the dip-slip movement ◊ *Fault*.

Dip slope. A topographic surface which dips in the same direction as the underlying beds and is often more or less parallel to them. The term is often applied to the back-slope of a ◊ *cuesta*.

Dirt beds. Colloquial name for a fossil ◊ *soil*.

Disconformity. An imprecise term often used as a synonym for ◊ *Unconformity*.

Discontinuity. Seismologists have located within the Earth two major layers which separate zones within the Earth having markedly different properties. The outer one – the Mohorovičić Discontinuity – the 'Moho' – separates the ◊ *crust* from the ◊ *mantle*, its average depth being 35 km. Separating the mantle from the ◊ *core* is the Gutenberg Discontinuity (sometimes Weichert-Gutenberg Discontinuity) at about 2,900 km.

The following table represents conditions immediately adjacent to the discontinuities, not average uniform conditions (for P-waves and S-waves, ◊ *Earthquake*):

Velocity of P-waves	6·5	Crust	16·6	Mantle
Velocity of S-waves	3·74		7·4	
Density	2·95		5·7	
Above				
Discontinuity	Mohorovičić		Gutenberg	
Below				
Velocity of P-waves	3·76	Mantle	8·1	Core
Velocity of S-waves	4·36		Not transmitted	
Density	3·3–3·5		9·5	

Velocities in km./sec.

These changes may be due to change of composition, change of state (the Gutenberg Discontinuity is almost certainly a solid/liquid boundary) or a change in the packing of the constituent atoms.

Other minor discontinuities in both the mantle and the core have been described, but none have the importance of the two above.

Discordant. A term used to describe an igneous rock showing a cross-cutting relationship to bedding or foliation. (Cf. ◊ *Dyke*; ◊ *Concordant*.)

Dismicrite. ◊ *Limestone*.

Dispersion. Violet light is refracted slightly more than red light and hence the refractive index of a transparent material differs for red and violet lights. The difference between these refractive indices is termed the dispersion. Dispersion is normally detected during the examination of the ◊ *interference figure*. It should be noted that dispersion is not a function of the crystal symmetry of minerals and even non-crystalline substances display dispersion.

Dissemination. A term usually employed to describe an ore deposit consisting of fine particles of the ore mineral dispersed through the enclosing rock. (Cf. ◊ *Impregnation*.)

Dissepiment, Dissepimentarium. (Chordata) ◊ Fig. 17; (Coelenterata) ◊ Fig. 21.

Disthene. Kyanite (◊ *Aluminium silicates*).

Ditroite. ◊ *Alkali syenite*.

Dittonian. A stratigraphic stage name for the British Lower ◊ *Devonian* (Old Red Sandstone facies).

Divaricator muscles (Brachiopoda). ◊ Fig. 11.

Divide. ◊ *Watershed*.

Djulfian. A stratigraphic stage name for the top of the European Upper ◊ *Permian*.

Do- (prefix). With one factor dominating another (petrology).

Dogger. (1) A large spherical or oblately spherical calcareous ◊ *concretion*. The term is usually applied to large masses, as opposed to the smaller ◊ *nodules*. No formal size limits having been proposed; it is here suggested that the term dogger should be applied to concretionary masses with a diameter greater than 256 mm. (cf. 'boulder' on Wentworth size scale – ◊ *Particle size*).

(2) An obsolete name for the Middle ◊ *Jurassic* in Europe.

Dog-tooth spar. ◊ *Calcite* crystallised in acute scalenohedral forms.

Dolerite. A medium-grained basic hypabyssal igneous rock, mineralogically and chemically the same as ◊ *gabbro* and ◊ *basalt*. The systematic name microgabbro is sometimes used (cf. ◊ *Diabase*). Porphyritic dolerites are uncommon, but ◊ *ophitic* textures are of frequent occurrence. Quartz dolerites are related to ◊ *tholeiites*. Hypersthene dolerites (micro-norites) are unusual, although some dolerites contain two or even three pyroxenes, which makes naming difficult.

For dolerites containing alkali feldspars and/or feldspathoids, ◊ *Alkali gabbros*. Picrite-dolerite or picrodolerite is a plagioclase-poor variety, a mela-dolerite.

Dolerites occur mainly as ◊ *dykes* (often in swarms), ◊ *sills*, and small ◊ *plugs* which are often old volcanic necks. It should be noted that a fine-grained basic rock occurring as a dyke or sill should be termed basalt, regardless of its mode of occurrence.

Doline. A synonym of swallow-hole, down which surface waters proceed underground in limestone country. (◊ *Karst scenery*.)

Dolomieu, Guy S. Tancrède de (1750–1801). French army officer, later Professor of Mineralogy, who studied the origin and composition of igneous rocks by comparing them with volcanic products. He believed that sulphur was responsible both for volcanic eruption and for the viscosity in rocks which resulted in their glassy or crystalline structure. He carried out a chemical analysis of the limestones in the South Tyrol and proved a high magnesium content, since which time both this region and limestones of this type elsewhere have carried his name – Dolomites.

Dolomite. (1) (mineral). ◊ *Carbonates*; Appendix.

(2) (rock). Many limestones contain small amounts of dolomite (◊ *Magnesian limestone*) and the term dolomite as a rock name is usually confined to rocks with more than 15% magnesium carbonate. Dolomites may occur as ◊ *evaporites*. Calcite limestones are readily dolomitised, and in modern sediments dolomite

seems to form as a result of penecontemporaneous ◊ *dolomitisation* by hypersaline water (brine).

Dolomitisation. The process by which an original calcium carbonate rock is converted into the double calcium magnesium carbonate, either wholly or in part.

Dolomitisation often destroys contained fossils, original sedimentary structure etc., but occasionally there will be alteration of only part of the rock, e.g. in some limestones only the smallest grains are converted to dolomite, the larger crystals and shell fragments remaining unaltered or merely developing a thin dolomitic skin. On the other hand, in some rocks all fossil remains are dolomitised and only the matrix is unaltered. Dolomitisation may take place at any time during or after the deposition of the calcareous sediment. It appears to be in many cases the result of a reaction between hypersaline marine brines and calcite, e.g. in the deeper parts of coral reefs dolomitisation has proceeded to a remarkable extent. (◊ *Dedolomitisation*.)

Dolostone. A term suggested for a rock consisting entirely of the mineral ◊ *dolomite*, to avoid the confusion caused by using the term 'dolomite' for the rock as well.

Dome. ◊ *Fold* and Fig. 64.

Domerian. A stratigraphic stage name for the European Upper ◊ *Charmouthian* (Jurassic System).

Domite. A ◊ *biotite* ◊ *trachyte*, named from its occurrence in the Puy-de-Dôme, France.

Dordonian. A synonym of ◊ *Maestrichtian* (Cretaceous System).

Double refraction. ◊ *Birefringence*.

Doublure (Arthropoda). ◊ Fig. 3.

Downtonian. A stratigraphic stage name for the top of the European ◊ *Silurian* (basal Devonian of some authors).

Dragonian. A stratigraphic stage name for the North American equivalent of the mid ◊ *Montian* (Tertiary system).

Drainage pattern. The arrangement and disposition of streams which a drainage system etches into the land surface and which may reflect the sum total of factors which influence the number, size, and frequency of streams in a particular area. Patterns of streams are influenced by: (1) Initial slope. (2) Lithology and lithological variation. (3) Structure – in its broadest sense – including earth movements and vulcanicity. (4) The geological and geomorphological history of the area. (5) The climate and rainfall regime of the area.

There are three main types of pattern (◊ Fig. 37):

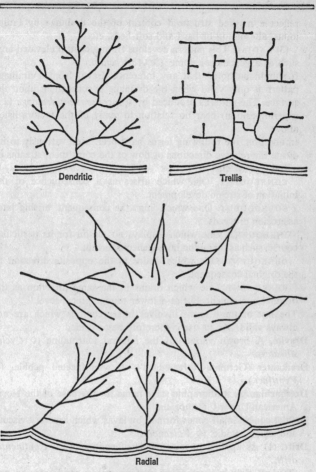

Dendritic

Trellis

Radial

37. Drainage patterns.

(1) DENDRITIC. This is by far the most common pattern and is characterised by irregular branching of tributary streams in a similar manner to that in which a tree branches (Greek *dendros*, 'tree').

(2) TRELLIS. A network of parallel or sub-parallel streams developed along strike and dip directions. Such patterns generally

133

reflect a marked structural control of the drainage by faults, joints, alternation of hard and soft beds, etc.

(3) RADIAL. This pattern develops from a central elevated area such as a volcano or dome. (◊ *Abstraction.*)

It should be noted that any inference drawn from a drainage pattern is only valid when the drainage is consequent upon the surface. The patterns produced by ◊ *inconsequent drainage* (e.g. superimposition) bear no relation to many of the factors listed above.

In addition, the following terms have been used variously to indicate the relative directions of flow of the constituent streams of a drainage system:

CONSEQUENT. One which arises as a consequence of the initiation of stream development.

SUBSEQUENT. One which joins the consequent, arising later as erosion proceeds.

INSEQUENT. One which displays no reason for its particular course, such as that upon homogeneous terrain.

OBSEQUENT. One which drains in the opposite direction to the original consequent.

RESEQUENT. One which drains in the same direction as the original consequent, but at a lower topographical level.

The last of these terms involves interpretations which are not always valid, and its use is therefore best avoided.

Dravite. A brown variety of the mineral tourmaline (◊ *Cyclosilicates*).

Dreikanter (German, 'three-edged'). A wind-faceted pebble. (◊ *Ventifact.*)

Dresbachian. A stratigraphic stage name for the base of the North American Upper ◊ *Cambrian.*

Driblet cones. Small cones formed by lavas which are very viscous and usually acidic. (◊ *Volcanoes.*)

Drift. (1) As applied to movements of continents, ◊ *Continental drift.*

(2) Any slow movement of surface water, sand, etc. or any deposit which has been transported (e.g. ◊ *loess*) by wind.

(3) Specifically, all the glacial and fluvio-glacial deposits left after the retreat of ◊ *glaciers* and ice sheets. Fluvio-glacial drift is water-lain, which distinguishes it from the unstratified condition of direct glacial deposits, i.e. those either laid down beneath the ice or dropped from the surface as the ice melted. The terms boulder clay and till are also applied to this unsorted

material, which ranges from ◊ *rock flour* to boulders and rocks of great size, in varying composition according to the nature of the bedrock. Drift thickness varies considerably, and can be more than a hundred metres. (◊◊ *Erratic*; *Moraine*; *Drumlins*; *Esker*.)
(4) A term used to describe an edition of a geological map upon which drift (see 3 above) and other superficial deposits (alluvium, river terraces) are shown. (Cf. ◊ *Solid*.)

Dripstone. An alternative name for stalagmite (◊ *Calc tufa*).

Drumlins. Elongated whale-backed mounds of ◊ *boulder clay*, sometimes reaching a kilometre or two in length, and ranging from 6 to 60 m. in height. They commonly occur in swarms in previously glaciated areas of low relief; the most notable examples are in County Down, N. Ireland, and in Wisconsin and central New York State, U.S.A. Characteristically they are highest and steepest at the end which faced the advancing ice, then taper away (cf. ◊ *Roches moutonnées*). They were probably formed by a rhythmic moulding action of the ice-sheet on newly deposited ground ◊ *moraine*, although no explanation of their origin and occurrence has proved completely satisfactory.

Druse (Drusy). A cavity in a rock or mineral vein into which euhedral crystals of the minerals forming the rock or vein project. (Cf. ◊ *Amygdale*; ◊ *Geode*; ◊ *Vugh*. ◊◊ *Miarolitic rocks*.)

Dry valley. A valley in which there is no existing river. There are two sorts of dry valley: (a) those cut by rivers which have subsequently been diverted or otherwise lost, and (b) those developed by other erosive agencies (e.g. ice or wind) which have not yet become integrated into the local system of drainage. A river may be lost to a valley (1) because of natural or artificial damming of the valley; (2) by ◊ *river capture*; (3) by loss of water to an underground drainage system, as in limestone areas (◊ *Karst scenery*); (4) because of a decrease in the rainfall of the area, resulting in a major lowering of the ◊ *water table*; (5) by glacial breaching, in which ice develops a new course for a river, causing it to abandon an old one.

Duchesnian. A stratigraphic stage name, the North American equivalent of the European ◊ *Ludian* (Tertiary System).

Dune-bedding. A type of large-scale ◊ *cross-bedding*.

Dunes. In areas of loose, unconsolidated, surface deposits, e.g. sandy deserts, a well-established prevailing wind will tend to heap up the sand into regular accumulations which are known as dunes. Four major types are recognised:

   (1) BARCHANS or CRESCENT DUNES. Crescent-shaped and

moving forward continually, the horns of the crescent pointing downwind. They have some of the characteristics of giant ripple marks, and in cross-section display dune-bedding. (◊ *Cross-bedding*.)

(2) SEIFS or LONGITUDINAL DUNES. Elongated in a downwind direction, they appear to have been derived originally from barchans which have become partially anchored. Observation shows that quite slight features, such as a patch of vegetation, may hold one end of a barchan, allowing the free end to rotate and become a seif.

(3) TRANSVERSE DUNES. Elongated dunes which form at right angles to the prevailing wind. They appear to arise by the coalescence of a large number of barchans, and there is some evidence that they develop when more sand is available than can be accommodated in a normal barchan 'fleet'.

(4) WHALEBACK DUNES. Very large longitudinal dunes with flat tops on which barchans or seifs may occur. Two suggestions have been made as to their origin – that they formed by the coalescence of several seifs, or, because there is commonly a rock floor exposed between whalebacks, that they are erosional features.

Fossil sand dunes are known from the Permo-Trias and other horizons; they are firm indicators of an arid, continental environment.

Dungannonite. An andesine-corundum ◊ *diorite* with accessory nepheline.

Dunite. An ◊ *ultrabasic*, ◊ *monomineralic rock* consisting of more or less pure ◊ *olivine*.

Durain. ◊ *Coal*.

Duricrust. A near synonym of ◊ *hard-pan*.

Durite. ◊ *Coal*.

Dutton, Clarence Edward (1841–1912). American geologist who developed and named the theory of ◊ *isostasy* to explain ◊ *Hall*'s concept of raised accumulation of sediments.

Dwyka. A stratigraphic stage name for the lower ◊ *Karroo* of South Africa.

Dyke. A sheet-like body of igneous rock which is discordant, i.e. cuts across the bedding or structural planes of the host rock (cf. ◊ *Sill*). Dykes may be ◊ *composite* or ◊ *multiple*, and they may occur in association with almost any larger igneous body. They occasionally occur in large swarms, which tend to be either parallel or radial in pattern. (◊ Fig. 38.) Dykes sometimes extend for a considerable distance, e.g. a dyke belonging to the Mull swarm extends to North Yorkshire, a distance of 260 miles.

Ring dykes have a circular outcrop, and are either vertical, or dip steeply away from a centre. They are typically associated

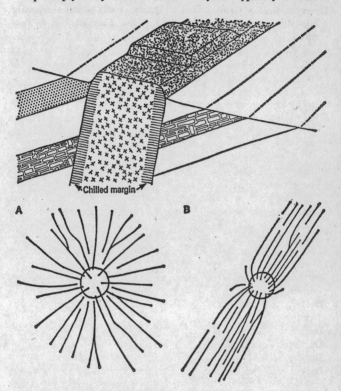

38. A dyke, showing the chilled margin which is formed when the hot dyke comes into contact with the cold country rock. (A) A radial dyke swarm; (B) A parallel dyke swarm.

with the Brito-Icelandic-Greenland Tertiary igneous Province. (Cf. ◊ *Cone-sheet*; ◊ *Ring structure*.)

Dyke rocks are usually medium-grained, but small dykes which have cooled more quickly are fine-grained, while very large dykes, particularly some of the ring dykes, are coarse-grained enough to be described as plutonic. It has been calculated that some of the Tertiary dyke swarms have caused a local stretching of the crust by as much as 65 km. (40 miles).

Basic dykes are by far the commonest type.

Dyscrystalline. A term applied to igneous rocks which are poorly crystalline and poorly crystallised. (Cf. ◊ *Eucrystalline*; ◊ *Texture*.)

E- (prefix). Without.

Earth constants.

DIMENSIONS	Km.	Miles
Equatorial radius	6,378·3	3,963·4
Polar radius	6,356·9	3,956·9
Mean radius	6,371·0	3,956·4

	Millions of	
AREA	Sq. km.	Sq. miles
Land (29·22%)	149	57·5
Oceans (70·78%)	361	139·4
Total Surface Area	510	196·9

RELIEF	Metres	Feet	
Land:			
Greatest height	8,848	29,028	Mt Everest
Average height	840	2,757	
Oceans:			
Greatest depth	11,035	36,204	Marianas Trench
Average depth	3,808	12,460	

MASS, VOLUME, DENSITY

Mass	$5·976 \times 10^{27}$ grams
Volume	$1·08 \times 10^{27}$ cc
Density	5·517 grams/cc.

GRAVITY

Acceleration due to gravity	981·183 cm./sec.^{2}
Heat loss	$1·25 \times 10^{-6}$ calories cm.^{2}/sec.
Moment of inertia	0·334 mr^2

when m = mass, r = radius, of the Earth

Earth flow. The rapid movement of water-laden soil material down a slope. ($\diamond$ *Gravity transport*.)

Earth model. The term applied to any attempt at a synthesis of data relating to the Earth as a whole, so as to produce a consistent representation of that data. For example, the density of the Earth as a whole is 5·5 grams/cc.; the average density of the crustal rocks is 2·7 grams/cc.; hence it may be deduced that there is a core to the Earth having a density considerably in excess of 5·5. This is supported by the data from P- and S-waves (⟡ *Earthquake*) and by a consideration of the moment of inertia. The moment of inertia, 'I', of a uniform sphere is 0·4 [mass, (m) x square of radius, $(r)^2$]; 'I' for the Earth is measured as ·334mr², which is also consistent with the existence of a dense core.

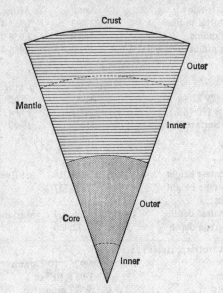

39. A simple Earth model.

Amongst other factors which most detailed models take into account are magnetic and gravitational variations, the spin of the Earth and its orbit around the Sun, the composition of the Earth and possible astronomical events. (⟡⟡ *Core*; *Crust*; *Mantle*; *Discontinuity*.) A model which has received wide acceptance is shown in diagrammatic form in Fig. 39.

Earth pillars. Columns of clay capped by large boulders, which develop from spurs arising from valley slopes. As rain erodes the surface, the boulder protects the earth beneath it and is left perched on a pillar of clay. Where the slope is sheltered from oblique rain, particularly tall pillars may be found – heights of

60 m. have been recorded in the Tyrol. The material in which the formation of earth pillars is most commonly observed is boulder clay (◊ *Drift* (3)).

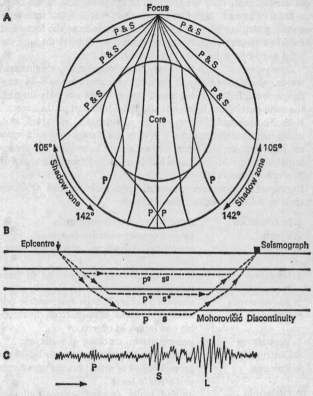

40. (A) Paths of earthquake waves through the mantle and core; (B) Paths of earthquake waves in the crust (pᵍ, sᵍ, p*, s* differentiate conventionally between different sets of p and s waves, p and s being the fastest, pᵍ and sᵍ being the slowest); (C) Seismograph trace.

Earthquake. A series of shock waves generated at a point (the focus) within the Earth's ◊ *crust* or ◊ *mantle*. The point on the surface of the Earth above the focus is called the epicentre. Three main types of wave motion are generated by an earthquake (◊ Fig. 40):

(1) L-WAVES: low-frequency, long-wavelength, transverse vibrations, which develop in the immediate neighbourhood of the epicentre and are responsible for most of the destructive force of earthquakes. They are confined to the outer skin of the crust.

(2) S-WAVES: high-frequency, short-wavelength, transverse waves, which are propagated in all directions from the focus and travel at varying velocities (proportional to density) through the solid parts of the Earth's crust, mantle, and core.

(3) P-WAVES: high-frequency, short-wavelength, longitudinal waves, which have many of the same characteristics as S-waves – the major difference being that P-waves travel not only through the solid part of the Earth but also the liquid part of the core.

Both P-waves and S-waves can be reflected and refracted, and, under certain circumstances, a P-wave can change into an S-wave on reflection, or vice-versa. Much of our knowledge of the internal structure of the Earth has been gained from the study of P- and S-waves.

Earthquake shocks are measured on a more or less arbitrary scale called the Modified Mercalli Scale:

1. Instrumental	detected only by seismographs
2. Feeble	noticed only by sensitive people
3. Slight	resembling vibrations caused by heavy traffic
4. Moderate	felt by people walking; rocking of free-standing objects
5. Rather strong	sleepers awakened; bells ring; widely felt
6. Strong	trees sway; some damage from overturning and falling of objects
7. Very strong	general alarm; cracking of walls, etc.
8. Destructive	chimneys fall; some damage to buildings
9. Ruinous	ground begins to crack; houses to collapse and pipes to break
10. Disastrous	ground badly cracked; many buildings destroyed; some landslides
11. Very disastrous	few buildings stand; bridges, railways destroyed; water, gas, electricity, telephones, etc. out of action
12. Catastrophic	total destruction; objects thrown into air; much heaving, shaking, and distortion of surface

Of the many thousands of earthquakes each year, only a few are destructive or even noticed by man. The great earthquake regions of the Earth are the west coast of North, Central, and South America, Japan, the Philippines, South-east Asia, New Zealand, India, the Middle East, and the Mediterranean. (Cf. ◊ *Island arc* distribution.) Submarine earthquakes occur along the lines of the ocean basin rifts, e.g. the Mid-Atlantic Ridge (◊ *Mid-oceanic ridge*). It will be noticed that the present earthquake regions are associated with the younger fold-mountain regions, and the present earthquake activity is a phase of the end of the Alpine ◊ *orogeny*.

(◊ *Isoseismic line*; *Plate tectonics*.)

Earth science. A name acquiring currency as an all-embracing term for geology, geography, geophysics and geochemistry (in their widest senses), geodetics, climatology and meteorology, oceanography, and the astronomical aspects of the Earth–Moon system.

Earth's crust. ◊ *Crust of the Earth*.

Eburonian. A stratigraphical stage name for the European mid Lower Pleistocene (◊ *Tertiary System*).

Ecca. A stratigraphic stage name for the base of the Middle ◊ *Karroo* in South Africa.

Echinodermata. A group of entirely marine animals having a skeleton or test made up of calcareous plates or spicules. The skeleton differs from that found in all other invertebrates in that it is secreted by the middle rather than the outer body layer; thus the test is enveloped in soft tissue. Each plate is a single crystal of calcite. The echinoderms are divided into the following classes:

(1) Cystoidea/Carpoidea ⎤
(2) Blastoidea   ⎥   Pelmatozoa or fixed forms
(3) Crinoidea   ⎥
(4) Edrioasteroidea   ⎦

(5) Echinoidea   (the Sea Urchins)   ⎤
(6) Holothuroidea   (the Sea Cucumbers) ⎥ Eleutherozoa or free-
(7) Asteroidea   (the Starfish)   ⎥   living forms
(8) Ophiuroidea   (the Brittle Stars) ⎦

(1) CYSTOIDEA (◊ Fig. 41). Presumed to be the most primitive group of the Echinodermata, the test consists of irregularly arranged plates with little trace of the typical pentameral symmetry, while considerable variation is seen in the character of the stalk and number of arms present. The cystoids range from the Upper ◊ *Cambrian* to ◊ *Carboniferous*, and are rare forms

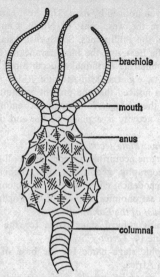

41. A generalised cystoid.

of no geological importance. Some authorities consider that those cystoids which do not show radial symmetry constitute the class Carpoidea, whose systematic position is uncertain.

(2) BLASTOIDEA (◊ Fig. 42). Presumed to have evolved from

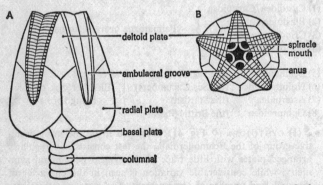

42. A blastoid (Pentremites). (A) General view of calyx; (B) Oral view.

the cystoids, they consist typically of a stalk surmounted by a calyx or cup, of which the arms form an integral part. Pentameral symmetry is common. The number of plates in the calyx is fixed at thirteen, arranged in three rows or circlets. The range of the group is from ◊ *Ordovician* to ◊ *Permian*; they have no use as zonal indices. They are occasionally found in reefs in the Lower ◊ *Carboniferous*.

(3) CRINOIDEA (◊ Fig. 43). Fixed or free-living marine Echinodermata consisting of a stem (in fixed forms only) and a calyx or cup, bearing five pinnate arms, the whole structure being made up of separate plates and ossicles. The stem, which has 'rootlets' at the base, is made up of individual ossicles which are usually round, oval, pentagonal, or star-shaped, and constitute the majority of crinoid remains. The calyx consists of two or three rows of plates, the uppermost row bearing the articulation for the arms. The arms, which may or may not branch, bear small alternate pinnules which give them a feathery appearance. They are responsible for collecting small food particles which are passed down to the mouth, situated at the top of the calyx. Surmounting the calyx and surrounded by the arms there is a domed or elongated structure of small irregular plates bearing the anus at the apex. The crinoids are divided into four subdivisions, based on the character of the cup or calyx. The following terms are used to describe certain features of the group:

Biserial   Arms consisting of a double axial row of ossicles.
Uniserial   Arms consisting of a single axial row of ossicles.
Dicyclic   A cup consisting of three circlets of plates.
Monocyclic   A cup consisting of two circlets of plates.

Their range is from ◊ *Cambrian* (Lower Ordovician) to Recent. They are important rock-formers, and have also been used as zonal indices in the Upper ◊ *Cretaceous*.

(4) EDRIOASTEROIDEA. An enigmatic group of extinct Echinodermata, consisting of a spherical test, over the surface of which ramify five food grooves. In many aspects they resemble the cystoids, from which they may have evolved. They range from Lower ◊ *Cambrian* to ◊ *Carboniferous*.

(5) ECHINOIDEA (◊ Figs. 44 and 45). A group of Echinodermata having a characteristically spinose, spherical and radially symmetrical test, which in the more advanced forms may become elongated and bilaterally symmetrical. The test is typically composed of twenty vertical rows of plates arranged in pairs, of

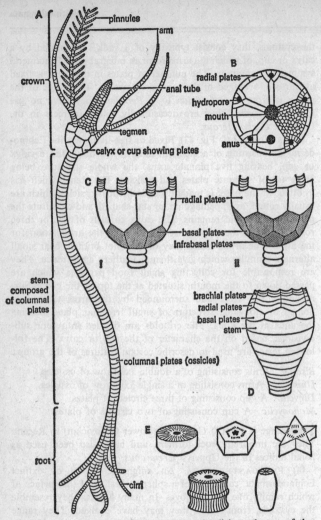

43. Crinoids. (A) Generalised crinoid (for simplicity only two of the five arm systems are shown, and only one branch of one arm is shown with pinnules); (B) Oral view; (C) i, arrangement of plates in a calyx of a monocyclic crinoid; ii, arrangement of plates in a calyx of a dicyclic crinoid; (D) *Apiocrinus*, showing the expanded stem forming a transition towards the calyx, and the incorporation of brachial plates into the calyx; (E) Variation in form of the stem columnals.

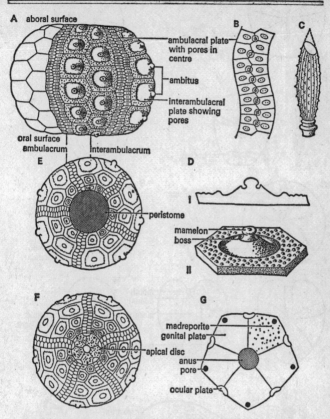

44. Regular echinoids. (A) Equatorial view; (B) Detail of ambulacral plates, each plate displaying one pair of pores (pore pair or biserial pores); (C) Detail of spine showing calcite cleavage; (D) i, cross-section through ii; ii, detail of interambulacral plate; (E) Oral view; (F) Aboral view; (G) Detail of apical disc.

The mouth lies in the centre of the peristomial membrane and the anus within the periproctal membrane.

which five rows of plates are perforate (ambulacral areas), alternating with five imperforate (interambulacral areas) rows. The plates of the ambulacral area each bear pores from which emerge the tube feet, connected to an internal water-vascular

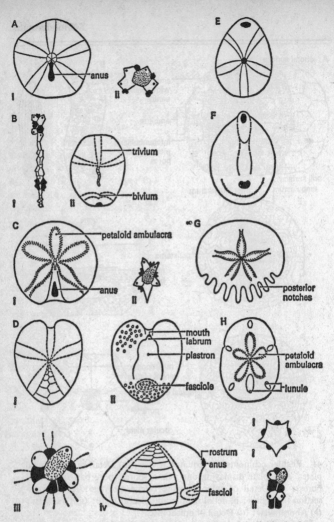

45. Irregular echinoids. (In general no details of individual plates are shown and ambulacra and interambulacra are shown conventionally. Ocular plates are shaded black in details of the apical disc.) (A) *Pygaster*: i, aboral view – anus retreating from apical disc; ii, apical disc showing loss of genital plates; (B) *Collyrites*: i, elongated dissociated apical system; ii, the bivium and trivium; (c) *Clypeus*: i, aboral view – anus in posterior groove; ii, apical disc; (D) *Micraster*: i, aboral view; ii, oral view; iii, apical disc; iv, side

system, which are responsible for locomotion and respiration. On the upper (aboral) surface of the test is situated the apical system, which consists of: (a) the ocular plates, which produce the ambulacral plates; (b) the genital plates, which contain the genital openings; the anus is situated in the centre of the apical system in the radially symmetrical forms, but outside the apical system, in a posterior position, in the bilaterally symmetrical forms. On the lower (oral) surface is situated the peristome, which consists typically of a circular leathery membrane in the centre of which is the mouth. In the advanced burrowing Echinodermata, the peristome area is reduced and the mouth is protected by a lip or labrum (◊ Fig. 46D). The spines of echinoids are usually large in the radially symmetrical forms, and small in the bilaterally symmetrical forms; their normal function is protective, although in the burrowing echinoids they may become flattened and used for locomotion.

Modern classification divides the echinoids into two groups, based on the characters of the peristome area and the nature of the ambulacral plates. An earlier, but simpler, classification divides them into Regular (radially symmetrical forms) and Irregular (bilaterally symmetrical forms). These two classifications do not coincide. 'Petaloid' echinoids have ambulacral areas with an elongate ovate (leaf-like) form on the aboral surface. Echinoids range from ◊ *Ordovician* to Recent, and have been used as zonal indices in the ◊ *Cretaceous*.

(6) HOLOTHUROIDEA. A group of Echinodermata in which the plates have been reduced and are present only as small 'ornamental' spicules situated in a leathery integument (outer layer). The spicules may be needle, anchor, star, or wheel shaped. The range is from ◊ *Cambrian* (Ordovician) to Recent. Holothuroid spicules occur rarely in residues prepared for microfossils and have no geological use.

(7) ASTEROIDEA (◊ Fig. 46). Stellate Echinodermata in which the arms of the animal merge imperceptibly into the central portion of the animal. The number of arms is characteristically five, but may be more, in multiples of five. The under-surface of the arm bears tube feet, which in the asteroids are used for

view; (E) *Conulus*, oral view; (F) *Echinocorys*, oral view; (G) *Heliophora*, aboral view; (H) *Encope*, aboral view; (I) Apical discs: i, *Clypeaster*, showing fused genital plates; ii, *Holaster*, showing elongation.

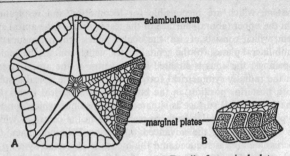

46. An asteroid. (A) Oral view; (B) Detail of marginal plate.

opening molluscs which is the usual food for this group. The skeleton consists of ossicles (plates) which are usually granular, calcareous, of regular shape, and may be associated in rows. Their range is from ◊ *Ordovician* to Recent; the group is of no geological importance.

(8) OPHIUROIDEA (◊ Fig. 47). Stellate Echinodermata with five or more (in multiples of five) flexible arms separate from and radiating from a central circular disc. The skeleton consists of

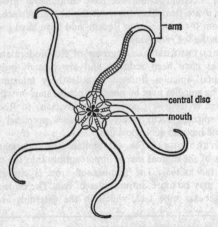

47. An ophiuroid, oral view, showing detail in one arm only.

ossicles which are plate-like in the central disc and rounded in the arms, where they form a central articular rib. The arms do not bear tube feet. Their range is from ◊ *Ordovician* to Recent; they have little geological importance, although where they do occur, it is in abundance.

Echinoidea. ◊ *Echinodermata.*

Eclogite (Eklogite). A metamorphic rock having a chemical composition similar to that of a ◊ *basic* igneous rock, but which has crystallised or recrystallised under conditions of high temperature and high pressure. The normal constituents of an eclogite are a ◊ *garnet* (usually pyrope), a ◊ *pyroxene*, characteristically omphacite, and sometimes a corresponding ◊ *amphibole*. ◊ *Sphene,* ◊ *zoisite,* and ◊ *magnetite* are common accessories. The rock is generally granulose, and fairly coarse-grained. (Cf. ◊ *Charnockite.* ◊ *Facies, metamorphic.*)

Ecological niche. The environment in which an organism lives to which it is better suited than any other organism, and which suits it better than any other environment. The more specialised an animal is to its particular environment, the less likely it is to be displaced by another organism, and the more likely it is to be extinguished by evolutionary or environmental change.

Ectinite. A metamorphic rock which has developed isochemically, i.e. without introduction or loss of material. (Cf. ◊ *Migmatite;* ◊ *Metasomatism.*)

Edenian. A stratigraphic stage name for the North American basal Upper ◊ *Ordovician.*

Edge coal. A highly inclined seam of ◊ *coal.*

Edgewise conglomerate. A ◊ *conglomerate* consisting of flat fragments packed together parallel to one another, standing more or less 'on edge' in the bed as a result of current action.

Edrioasteroidea. ◊ *Echinodermata.*

Eemian. A stratigraphic stage name for the European Upper Pleistocene (Saale/Weichsel Interglacial) (◊ *Tertiary System*).

Effusive. A synonym of ◊ *extrusive.*

Eifelian. A stratigraphic stage name for the base of the European Middle ◊ *Devonian.*

Ejectamenta. Solid material flung from a volcanic vent. (◊ *Pyroclastic rocks.*)

Eklogite. ◊ *Eclogite.*

Elaeolite. A synonym for the mineral ◊ *nepheline.*

Elaterite. A 'rubbery' ◊ *bitumen.* (◊ *Hydrocarbon minerals.*)

Elbaite. A green variety of the mineral tourmaline (◊ *Cyclosilicates*).

Eleutherozoa. ◊ *Echinodermata.*

Ellipsoid. A solid figure of which all plane sections are either

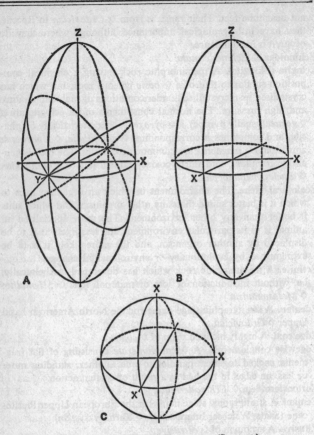

48. Ellipsoids. X, Y, and Z are axes. (See text.)

ellipses or circles, used for illustrating the variation of a property. The major axis of such a solid represents the maximum value of a variable, and with the minor axis, which represents the minimum value of the variable, defines a plane. The intermediate axis is chosen so as to lie perpendicular to this plane. There are three cases ($\diamond$ Fig. 48): (A) Three axes of different values, giving rise to axial plane sections which are ellipses. This kind has two circular cross-sections having the intermediate axis as a common diameter and symmetrically placed with respect to the other axial planes. (B) An ellipsoid with a single circular cross-section, which lies perpendicular to an axis which is greater or less than the radius of the circular cross-section. This may be regarded as being derived from case 1 by the intermediate value becoming equal to either the maximum or minimum value, so that the two circular cross-sections of case 1 become coincident. (c) All axial plane sections are circular, i.e. the ellipsoid has 'degenerated' into a sphere.

This method of representation is particularly suited to two sets of geological phenomena:

(1) The optical properties of crystals. The directions in the ellipsoid can represent refractive index, wave velocity, colour absorption, etc. The ellipsoid bears a convenient relationship to the crystallographic ◊ *axes*; cubic crystals require case 3; tetragonal, hexagonal and trigonal crystals require case 2; and all other systems, case 1. (◊ *Indicatrix*.)

(2) In structural geology the three directions can be used to represent values of ◊ *stress* or ◊ *strain*, and if this is done (assuming case 1), the circular cross-sections represent the planes of maximum stress or strain, and in general coincide with directions of maximum deformation or rupture. It should be noted that for most structural purposes the strain ellipsoid is the one used; the stress ellipsoid is much less commonly required.

Eltonian. A stratigraphic stage name for the base of the ◊ *Ludlovian* in Britain (Silurian System).

Elutriation. The process by which a granular material can be sorted into its constituent particle sizes by means of a moving stream of fluid (usually air or water). Elutriators are extensively used in studies of sediments for determining ◊ *particle size* distribution. Under certain circumstances the wind, rivers, and streams may act as natural elutriating agents.

Eluvial deposits. Weathered material ('float') which is still at, or near, its point of formation. The term is especially applied to deposits of economic substances. (Cf. ◊ *Placer deposits*; *Alluvial deposits*; *Colluvial deposits*.)

Eluviation. The process of leaching in a ◊ *soil*, which mainly removes iron and calcium.

Elvan. A miner's term from south-west England, applied to any ◊ *dyke* of ◊ *granite*, ◊ *microgranite*, or quartz ◊ *porphyry*. The term was also used (in the form 'blue elvan') for ◊ *dolerites* or other basic (or dark intermediate) dyke rocks.

Emerald. ◊ *Gem*-quality beryl (◊ *Cyclosilicates*), bright green in colour owing to the presence of chromium.

Emery. A rock consisting of finely granular ◊ *corundum* and ◊ *magnetite*, which results from the thermal metamorphism of a ferruginous bauxite (◊ *Laterite and bauxite*).

Emplacement. A term used to describe the development of a body of igneous rock within an ◊ *envelope* of another rock without having any genetic implication. For example, one can refer to a granite as being emplaced without implying specifically that it was forcibly intruded, ◊ *stoped* its way into position or arrived by ◊ *replacement* (◊ *Granitisation*).

Emsian. A stratigraphic stage name for the top of the European Upper ◊ *Devonian*.

Enantiomorphic. A crystallographic term applied to forms which are similar but not congruent, i.e. cannot be superimposed even by rotation and/or inversion. Enantiomorphic forms are mirror-images of one another, and are generally referred to as being left-handed or right-handed. Quartz crystals are an example.

Enantiotropic change. A change of one ◊ *polymorph* into another which can be reversed by changing the conditions (e.g. quartz ⇌ tridymite ⇌ cristobalite; ◊ *Silica group of minerals*).

Enclave. In effect, a synonym of ◊ *xenolith*.

Encrinital. Crinoidal; used mainly as a descriptive term for ◊ *limestones* made up of crinoid (◊ *Echinodermata*) debris.

Endo- (prefix). Within.

Endobionts. Those forms living within the bottom sediments of the oceans.

Endogenetic. A term used to describe the processes and materials which originate within the Earth. (Cf. ◊ *Exogenetic*.)

Endometamorphic, Endomorphic. Terms used to describe the changes undergone by an igneous rock as a result of its assimilating the ◊ *country rock* being invaded.

Engineering geology. The application of the principles and methods of geology for the purposes of civil engineering operations. Broadly speaking there are two main divisions: (1) The study of raw materials, e.g. aggregates, masonry, etc. (2) The study of the geological characteristics of the immediate area where engineering operations are to be carried out, e.g. the ◊ *ground-water* characteristics, the load-bearing capacity of rocks, the stability of slopes, the ease or otherwise of excavation, etc. (◊ *Rock mechanics*.)

Englacial. Literally, within a glacier (e.g. englacial streams, en-glacial moraine).

Enrichment. The processes by which the relative amount of one constituent mineral or element contained in a rock is increased. This may be due either to the removal of other constituents selectively, or to the introduction of increased amounts from an external source. The process may be: (1) Mechanical, e.g. the transport of light debris, such as quartz, away from heavier material, such as gold or cassiterite. (2) Chemical, e.g. downward percolating solutions containing copper may convert chalcopyrite, $CuFeS_2$ (34·5% copper), to covellite, CuS (66·4% copper), the copper ions in solution replacing the iron originally present. This process is often referred to as supergene enrichment ($\Diamond$ *Secondary enrichment*).

Enstatite. $\Diamond$ *Pyroxenes.*

Entrenched meander. $\Diamond$ *Rejuvenation.*

Entropy. The normal concept of entropy, as used by physicists and chemists, has been applied to sediments as a measure of their degree of uniformity. Uniform sediments are said to have a high entropy, while mixed sediments have a low entropy. It should be noted that to some extent the entropy of a sediment depends upon the choice of end-members of the system, e.g. if one defines a set of end-members as clay-grade, silt-grade, and sand-grade ($\Diamond$ *Particle size*), an arkose ($\Diamond$ *Arenaceous rocks*) might show a high entropy, whereas if the end-members are selected as quartz grains, feldspar grains and lithic grains, an arkose might well show a much lower entropy.

The term is also used to describe columnar sections of strata; a sequence of shales, quartzites, and limestones will have a high entropy, whereas a series of argillaceous sandstones, impure lime-stones, etc. will have a low entropy.

Entropy is capable of being expressed in precise mathematical terms, and entropy maps (with isopleths of entropy) may be constructed.

Envelope. (1) The rock bodies surrounding an intrusive mass of igneous rock; more or less synonymous with $\Diamond$ *country rock* plus $\Diamond$ *metamorphic aureole* (the thermally metamorphosed part of the envelope).

(2) Structurally, the outer layers of a $\Diamond$ *fold*, especially a recumbent fold. (Where a recumbent fold has a crystalline core the outer sedimentary portion is the envelope.)

Environment. (1) The place in which an animal or plant lives, e.g. in burrows, in sand, in shallow, marine or tropical waters. The environment may be continental (◊ *Environments, continental*) or marine (◊ Fig. 49; *Littoral*; *Continental shelf and continental slope*; *Bathyal*; *Abyssal*). (◊ *Adaptive radiation*; *Habitat*.)

(2) The term has come to be commonly applied also to the conditions of deposition of a sediment (cf. ◊ *Facies*).

Environments, continental. Non-marine ◊ *environments* which may

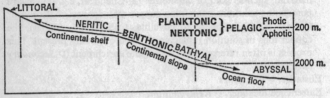

49. Marine environments.

be either terrestrial (dry land) or aquatic. Terrestrial environments are largely defined by climatic zones. Aquatic environments are sub-divided as follows:

Fluvial	{ Estuarine Deltaic	rivers, fresh water or brackish water
Paludal		swamps
Paralic		a lacustrine environment affected by marine conditions
Lacustrine Limnic Lagoonal	}	lakes lagoons, brackish water

Eocene. The epoch of the ◊ *Tertiary* period between the Palaeocene and Oligocene epochs.

Eozoic. A synonym of ◊ *Precambrian*.

Eozoon. A pseudo-fossil, consisting of rather finely banded calcite and serpentine. Formerly thought to be a kind of primitive colonial animal, it is now known to be inorganic.

Epeiric sea. A shallow inland sea.

Epeirogeny. The uplift or depression of continental or sub-continental land masses as a result of widespread adjustments of level. Epeirogenic movements are generally even in character, producing little more than tilting, slight warping, and minor faulting of the rocks. (Cf. ◊ *Orogeny*.)

Epi- (prefix). On, upon.

Epibole (Peak zone). The stratum in which a species reaches its acme. (◊ *Stratigraphic nomenclature.*)

Epicentre. The point on the surface of the Earth above the focus of an ◊ *earthquake.*

Epicontinental. Situated within the limits of a continental mass; e.g. an epicontinental sea is one covering part of a continent, one covering the ◊ *Continental shelf.*

Epidiorite. A granular metamorphic rock derived from a basic igneous rock but displaying the minerals of a ◊ *diorite*, i.e. hornblende and intermediate plagioclase. The following diagram shows the sequence of changes involved:

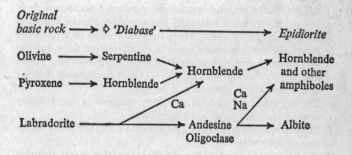

Epidiorites are normally thermally metamorphosed rocks. For regionally metamorphosed equivalents, ◊ *Hornblende schist.*

Epidosite (Epidotite). A rock consisting of ◊ *epidote* and quartz which may be metamorphic or hydrothermal in origin.

Epidotes. A group of rock-forming ◊ *silicate* minerals containing both SiO_4 and Si_2O_7 silicon-oxygen units. The SiO_4 and Si_2O_7 units link chains of AlO_6 (sometimes AlO_4OH_2) octahedra. Both ortho-rhombic and monoclinic varieties occur and twins are not uncommon.

A general formula may be written $R_2''R_3'''O.Si_2O_7.SiO_4(OH)$ where R'' is generally Ca but substitution by Mn'' and Ce is possible, and R''' can be Al, Fe''' and Mn'''.

The five most important minerals of the group are:

Zoisite (orthorhombic)
Clinozoisite ⎱ $Ca_2Al_3O.Si_2O_7.SiO_4(OH)$

Epidote (s.s.) $Ca_2Fe'''Al_2O.Si_2O_7.SiO_4(OH)$

Piemontite $Ca_2(Mn'''Fe'''Al)_3O.Si_2O_7.SiO_4(OH)$

Allanite (Orthite) $(CaMn''Ce)_2(Fe''Fe'''Al)_3O.Si_2O_7.SiO_4(OH)$

Except where indicated the minerals are monoclinic.

A number of minor varieties have been named, especially of piemontite, where the Mn content is rather variable.

In general minerals of the epidote group occur as crystals elongated parallel to the b-axis, and with a basal cleavage. Zoisite and clinozoisite tend to be grey, green or brownish in colour (colourless in thin section). Epidote (s.s.) is generally various shades of green, from light yellow-green to almost black, and is green in thin section. Piemontite is generally reddish-black, and pink or violet in thin section. Allanite is brown to black (yellow-brown in thin section). Their hardness is 6.

All the epidote minerals are low-temperature types, occurring in low to medium-grade regionally metamorphosed rocks, and occasionally in igneous rocks, where they are formed as a late-stage product – they sometimes occur in quartz-epidote veins, pegmatites and as ◊ *amygdale* minerals in lavas. Sometimes joint planes may be found coated with epidote (s.l.) with metasomatic epidotes (s.l.) developed on either side of the joint.

Detrital grains of epidotes are fairly frequent in the ◊ *heavy mineral* assemblages of some sandstones.

Rocks consisting entirely of epidotes (s.l.) (rare) are called epidotites or epidosites.

Epifauna. Organisms which live attached to another, larger organism constitute an epifauna. A typical example is the growth of corals and bryozoa on brachiopod shells, a growth which may develop during the life of the brachiopod or after its death. The attachment is quite haphazard, and no special relationship such as parasitism or symbiosis is necessarily implied.

Epigenetic. A term used to describe a process occurring at or near the surface of the Earth. (Cf. ◊ *Hypogene*.)

Epitaxy, Epitaxial. Certain minerals are capable of occurring in the form of an overgrowth of one crystal on another. If the crystal structures of the two minerals are arranged in some specially related orientation, the overgrowth is said to be epitaxial. It is not necessary (although it often occurs) that the minerals should be closely similar in structure, e.g. ◊ *rutile* (TiO_2) grows on ◊ *haematite* (Fe_2O_3), which have markedly different structures. In the case of grossularite (◊ *garnet*) and ◊ *idocrase* overgrowths, part of the garnet and part of the idocrase structures are identical.

Epitheca (Coelenterata). ◊ Fig. 21.

Epithermal. A term applied to low-temperature (100–200°C.) ◊ *hydrothermal* processes. (Cf. ◊ *Mesothermal*; ◊ *Hypothermal*.)

Epoch. ◊ *Stratigraphic nomenclature.*

Equigranular. A ◊ *textural* term applied to a rock in which the constituent grains are all of about the same size.

Era. ◊ *Stratigraphic nomenclature.*

Erathem. ◊ *Stratigraphic nomenclature.*

Erian. A stratigraphic stage name for the North American Middle ◊ *Devonian.*

Erosion. That part of the process of ◊ *denudation* which involves the wearing away of the land surface by the mechanical action of transported debris. The agents of ◊ *transportation* are by themselves only capable of minute wearing action upon the rocks, but when they contain particles of weathered material they become powerful agents of destruction. (◊ *Cycle of erosion; Glaciers; Wind erosion; Marine erosion; Valley profile.* Cf. ◊ *Weathering.*)

Erratic. A large pebble, cobble, or boulder which has been transported some distance from its source. The term is commonly applied to glacially transported blocks. The tracing of such erratics back to their sources may yield important information concerning the direction of movement of the ice. It has been suggested that certain erratic blocks found in sediments in circumstances which preclude direct transport by a glacier may have been transported by icebergs, by being attached to giant seaweeds, or as the stomach stones of certain reptiles.

Escarpment. The steeper slope of a ◊ *cuesta.*

Esker (Osar). A term applied to long, winding ridges of sand and gravel, unrelated to the surrounding topography, found in previously glaciated regions. They originated within or beneath the ice, either from continuous deposition at the mouth of a subglacial stream as the ice retreated, or from infilling of the tunnels of these streams before recession.

Essential minerals. Minerals whose presence or absence decides the name (i.e. the position in a classification) of a rock. An essential mineral need not be present in large quantities, e.g. a trace of nepheline or a trace of quartz is sufficient to place a ◊ *syenite* in the ◊ *unsaturated* or ◊ *oversaturated* class respectively. In the same way, the presence or absence of plagioclase feldspar will determine whether the rock is to be called a ◊ *picrite* or a ◊ *peridotite.* (Cf. ◊ *Accessory minerals.*)

Essexite. The name widely used for a variety of ◊ *alkali gabbros.* The original rock, from Massachusetts, has been shown to be metamorphosed, and certain Norwegian essexites ('Oslo-essexites') contain no feldspathoid, which is generally regarded as essential. Current usage applies the name to a labradorite, titan-augite, olivine, nepheline and/or analcite rock. Some alkali feldspar may be present.

Esterian. A stratigraphic stage name for the mid Middle Pleistocene (◊ *Tertiary System*) in Europe.

Etching, Etch marks. When crystal faces are attacked by suitable reagents, small regular-shaped pits develop, owing to the reactivity of the crystal to the solvent being different in different directions. These etch marks (which may occur as a result of natural processes) conform to the symmetry of the crystal and hence are of use in assigning crystals to their appropriate ◊ *crystal system*. In particular, when a ◊ *form* is common to two or more classes within a system, the etch marks will serve to distinguish between them. Fig. 50 shows etch marks developed on

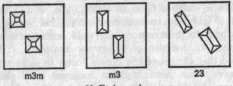

m3m   m3   23

50. Etch marks.

the cube faces of galena (spinel) type, pyrite-type and tetrahedrite-type crystals (classes m3m, m3, 23, of the ◊ *Cubic system*). The term is also applied to the rough, pitted surfaces of some sand grains.

Eu- (prefix). Well, well-developed.

Eucrite. ◊ *Gabbro.*

Eucrystalline. A term applied to those igneous rocks which are ◊ *holocrystalline* and well-crystallised. (Cf. ◊ *Dyscrystalline.* ◊ *Texture.*)

Euhedral. A term applied to grains displaying fully developed crystal form. (◊ *Texture.*)

Eulysite. A metamorphic rock rich in iron and manganese. Typical minerals include hedenbergite and Fe-hypersthene (◊ *Pyroxenes*), fayalite and Mn-fayalite (◊ *Olivine group of minerals*), FeMn ◊ *garnet*, etc.

Eury- (prefix). Tolerant.

Eurypterida. ◊ *Arthropoda.*

Eustatic. Pertaining to absolute changes in sea level, i.e. to world-wide changes and not local changes produced by local movements of the land or the sea floor.

Eutaxitic. A term applied to the structure of certain welded tuffs and other ◊ *pyroclastic rocks*. It refers to the streaky nature of the rocks, the streaks and bands being notably discontinuous, as opposed to the more or less continuous layers of flow-banded lavas (◊ *Flow structure*). 'Parataxitic' has been suggested as a term for extreme elongation of the streaks, giving a close approximation to true flow-banding.

Euxinic (lit. 'pertaining to the Black Sea'). A term used to describe an environment characterised by the presence of large volumes of stagnant water which is de-oxygenated, giving rise to reducing conditions. The characteristic sediments are black carbonaceous pyritic muds. Such conditions exist at the present in the deeper parts of the Black Sea; perhaps parts of the British ◊ *Rhaetic* deposits represent a fossil example.

Evaporate. A synonym of ◊ *evaporite*.

Evaporite. A sediment resulting from the evaporation of saline water. Most evaporites are derived from bodies of sea water, but under special conditions lakes may give rise to bodies of borate

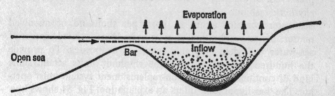

51. The bar theory of evaporite deposition. Increasing density of shading implies increasing density and concentration of salt solution.

minerals, sodium carbonate minerals, and other similar types. There are two views as to the origin of thick beds of evaporites. One suggests formation in a ◊ *sabkha* environment; the other suggests formation by the evaporation of an enclosed body of saline water. Evaporites of the second kind begin to form when sea water is concentrated to about 50% of its original volume. The minerals are formed in the reverse order of their solubilities, i.e. the least soluble first. It is probable that the original character of most evaporite deposits has been destroyed by replacement through the agency of circulating fluids; e.g. polyhalite (K_2SO_4.

$MgSO_4.2CaSO_4.2H_2O$) is not a primary evaporite as it is formed by the reaction of appropriate solutions with $CaSO_4$ minerals. Experimental work shows that the temperature at which crystallisation takes place and the concentrations of the various ions present control the nature of the actual substances deposited, e.g. pure solutions containing calcium and sulphate ions will deposit gypsum ($CaSO_4 . 2H_2O$) below 42°C., and anhydrite ($CaSO_4$) above this temperature. The presence of quite a small proportion of sodium chloride (NaCl) depresses this temperature considerably.

A typical evaporite sequence is as follows:

4. Potash and magnesium salts
   (kainite, carnallite, sylvite, kieserite)
3. Rock salt (halite)
2. Gypsum or anhydrite
1. Calcite and dolomite

Because of the high solubility of beds 4 and 3, it commonly happens that these beds are re-dissolved during the next invasion by the sea. If, however, a layer of impervious sediment is deposited above the evaporite sequence, the highly soluble layers may be preserved. In this way several sequencs may be deposited in a ⬦ *rhythmic* fashion.

Sea-water contains about 31 parts per thousand of dissolved matter; hence evaporation of even quite large isolated masses of sea-water will yield only thin layers of evaporites. To provide the hundreds of feet which some rhythmic units attain, some kind of continuous evaporation-replenishment system must operate. The 'bar hypothesis' offers an explanation. Fig. 51 shows how fresh sea-water enters the basin where evaporation is occurring, resulting in concentration of the salts in solution until crystallisation and deposition takes place. Owing to local variations in the concentration, different salts may be deposited in different parts of the basin. It has been shown that the Gulf of Kara Boghaz, on the Caspian Sea, conforms to the theoretical requirements of the bar hypothesis to a marked extent.

Major evaporite fields of considerable economic importance occur at Stassfurt, Germany; Cheshire, North Yorkshire, and South Durham, England; Arizona and New Mexico, U.S.A.; California, U.S.A. (borates); and Chile (nitrates); etc.

(⬦ *Red beds*; *Salt domes*.)

Ex- (prefix). Without.

Exfoliation (onion-skin weathering). A process of ◊ *weathering* in which thin sheets of rock split off owing to differential expansion and contraction during heating and cooling over the diurnal temperature range.

Exhalation. The evolution of volcanic gases and vapours at the surface of the Earth.

Exinite. ◊ *Coal.*

Exo- (prefix). Out of, outside.

Exogenetic. The term used to describe processes or events which take place at or very near the surface of the Earth, e.g. ◊ *weathering,* ◊ *erosion,* etc. (Cf. ◊ *Endogenetic.*)

Exotic. ◊ *Autochthonous.*

Exposure. A place where rocks can be seen *in situ,* i.e. not as detached boulders or fragments.

Extinction. When the vibration directions of the ordinary and extraordinary rays of an ◊ *anisotropic* mineral are parallel to the vibration directions of the polars in a petrological microscope, no light reaches the eye and the mineral is said to be in extinction. This is because light which can pass the ◊ *polariser* (vibrating E–W) also passes the mineral (following one of its vibration directions), but is stopped by the ◊ *analyser,* which has a N–S vibration direction. This phenomenon occurs every 90° of rotation of the mineral. (◊ *Polarised light.*)

STRAIGHT EXTINCTION occurs when a mineral is in extinction with its ◊ *cleavage* or elongation parallel to the crosswires of the eyepiece.

INCLINED EXTINCTION occurs when a mineral is in extinction with its cleavage or elongation at an angle to the crosswires. The angle between the cleavage and the crosswires is known as the extinction angle.

SYMMETRICAL EXTINCTION: a mineral showing a square or rhombic cross-section or two cleavage directions which extinguishes so that the crosswires form the diagonals of the square or rhombus, or bisects the angle between the two cleavage directions, is said to have symmetrical extinction.

The determination of the three preceding types of extinction depends on the occurrence of cleavage or traces of crystal form.

UNDULOSE EXTINCTION occurs when a crystal has been bent. It appears as a band or series of bands of darkness crossing a single crystal unit during rotation. It should be carefully distinguished from variation in extinction angle due to compositional zoning. (◊ *Zoned crystal.*)

EXTINCTION CROSSES, which remain stationary when the microscope stage is rotated, are shown by ◊ *spherulites* consisting of radially arranged fibres of a mineral. The arms of the cross correspond to the vibration directions of the polarising system. The fibres constantly pass into and out of alignment with these directions during rotation, so that there are always some in extinction to form the cross.

Extraordinary ray. ◊ *Vibration directions.*

Extrusive. A term applied to igneous rock which has flowed out at the surface of the Earth, as opposed to intrusive (◊ *Intrusion*). Hence the term is synonymous with ◊ *volcanic* and effusive.

Fabric. The term is commonly used, at least in part, as a translation of the untranslatable German word 'Gefüge'. The fabric of a rock is the total of all the features of ◊ *texture* and ◊ *structure*, ranging from the orientation of individual grains to a regional joint or cleavage system. It is most commonly used of metamorphic rocks which have developed a ◊ *preferred orientation* of mineral grains – a ◊ *tectonite*. The measurement of the degree of orientation and the kind of orientation developed is the object of petrofabric analysis. By comparing fabrics developed under known stress conditions (either experimentally produced or deduced from field evidence), it is possible to use petrofabric analysis as evidence to support a particular structural interpretation in areas of otherwise unknown structure. (◊ *Lineation*; *Cleavage*.)

Face, crystallographic. A single flat surface of a naturally developed crystal, which exists in fixed angular relation to adjoining faces (◊ *Interfacial angle*). It may be regarded as one element of a ◊ *form*. The occurrence of a face is controlled by the ◊ *atomic structure* of the compound, faces occurring most frequently in directions parallel to planes in the lattice which contain a maximum number of atoms. The shape and size of faces is of no significance crystallographically, except in so far as they control ◊ *habit*.

Face, to. ◊ *Young. to.*

Facet. (1) A face on a crystal which is small in proportion to the crystal's size, often truncating an edge or corner.

(2) Any plane surface which truncates an object, e.g. a faceted pebble is one that has flat surfaces cutting across the rounded form (cf. ◊ *Dreikanter*).

(3) A polished flat surface on a gemstone.

Faceted spur (Truncation spur). A ridge with a flat truncation at its nose.

Facies. The sum total of features such as sedimentary rock type, mineral content, ◊ *sedimentary structures*, ◊ *bedding* characteristics, fossil content, etc. which characterise a sediment as having been deposited in a given environment. Facies which are particularly characterised by rock type are referred to as lithofacies, whereas those especially characterised by their fauna are called biofacies. It should be noted that a particular lithofacies may be ◊ *diachronous*, and this may be only detectable if the fossil evidence is adequate. Study of the distribution of facies leads to the reconstruction of past geographies (◊ *Palaeogeography*; *Faunal province*).

Facies, metamorphic. An assemblage of metamorphic rocks which are considered to have been formed under similar conditions of temperature and pressure. It is assumed that mineralogical variations within a facies are due to variations in the composition of the parent rocks. The assumption is also made that the metamorphism has taken place isochemically. The original definitions by Eskola of the seven major facies now recognised were established by work on metamorphosed basic igneous rocks, and hence the names given are derived from such rocks. Later workers have established many sub-facies, often of only local significance.

Facies fossil/fauna. A fossil species or fauna restricted to a certain sedimentary environment. (◊ *Trace fossils*.)

Faecal pellets. Small masses of the excreta of invertebrates, usually no larger than 2–3 mm. in diameter (◊ *Pellets*). Generally they are ovoid, but occasionally they may be rod-shaped. It is probable that faecal pellets are present in most sediments but are rarely recognised as such because of their dispersion amongst other material. Occasionally, however, thin beds or lenses are found consisting almost entirely of these pellets. (Cf. ◊ *Coprolite*.)

Fahlerz. A mining term for the grey copper ores, including ◊ *tetrahedrite*.

False-bedding. ◊ *Cross-bedding*.

Famennian. A stratigraphic stage name for the top of the European Upper ◊ *Devonian*.

Fan, alluvial. A low heap of material, having a roughly triangular plan, in the form of a section of a cone. Alluvial fans are typically formed where rivers or streams with a high velocity have their speeds suddenly checked by reaching a flatter region, e.g. where a mountain stream enters the plains. Outwash fans develop where the melt waters of glaciers flow through a breached ◊ *moraine*.

Fanglomerate. A ◊ *conglomerate* formed by the ◊ *lithifaction* of an alluvial ◊ *fan*.

Farlovian. A stratigraphic stage name for the European Upper ◊ *Devonian* (Old Red Sandstone facies).

Fasciculate. ◊ *Coelenterata* and Fig. 23. (Phacelloid is a synonym.)

Fasciole (Echinodermata). ◊ Fig. 45.

Fault. A fracture in rock along which there has been an observable amount of displacement. Faults are rarely single planar units; normally they occur as parallel to sub-parallel sets of planes along which movement has taken place to a greater or lesser extent. Such sets are called fault or fracture-zones.

Various terms are used to describe the attitude of the fault plane and the nature of the movement upon it ($\Diamond$ Fig. 52):

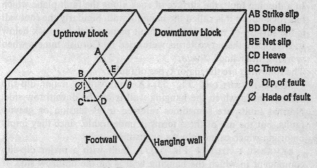

AB	Strike slip
BD	Dip slip
BE	Net slip
CD	Heave
BC	Throw
θ	Dip of fault
$\varnothing$	Hade of fault

52. Fault nomenclature.

A fault plane may be vertical, or may have a $\Diamond$ *dip*. The hade is the angle which the fault plane makes with the vertical plane – the complement of the angle of dip.

The term $\Diamond$ *strike* is applied to a fault plane in precisely the same way as to a $\Diamond$ *bedding plane*. However, it is customary to describe a fault according to the relationship of the fault-strike and bedding-strike: (a) Fault-strike parallel or sub-parallel to the bedding-strike = strike fault. (b) Fault-strike approximately at right-angles to the bedding-strike (i.e. nearly parallel to the dip) = dip fault. (c) Fault-strike making a well-defined angle with the bedding-strike = oblique fault.

If the direction of movement on the fault plane is parallel to the dip of the fault (i.e. upwards or downwards) the fault is said to have a dip-slip movement. If the direction of movement on the fault plane is parallel to the strike of the fault (i.e. sideways), the fault is said to have a strike-slip movement. If the direction of movement along the fault plane is in any other direction, the fault is said to have an oblique-slip movement. An oblique slip can always be resolved into dip-slip and strike-slip components.

The blocks on either side of a fault display relative movement with respect to each other. It is convenient to talk about up-thrown or downthrown sides of a fault – the throw being the measure of the vertical component of the displacement of the upthrown and downthrown blocks. It is emphasised that, in general, it is very difficult to determine which side of a fault has actually moved, or whether, in fact, both sides have moved. The horizontal component of the displacement is known as the heave.

In a dipping fault, the surface of rock along the fault plane which has rock above it is called the hanging wall. Similarly the footwall is the surface of rock along the fault plane which has rock below it. Originally these two terms were used by Cornish miners when working an inclined ◊ *vein*.

The following are the major types of fault:

  NORMAL FAULT (◊ Fig. 53). A fault with a major dip-slip component in which the hanging wall is on the downthrow side. Normal faults are sometimes referred to as tension or gravity faults, but the use of these terms is undesirable, since they imply an origin which cannot always be justified.

  REVERSE FAULT (◊ Fig. 53). A fault with a major dip-slip component in which the hanging wall is on the upthrow side. If

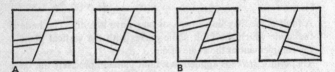

53. (A) A normal fault ; (B) A reverse fault.

the dip of such a fault is low, then the term thrust or thrust fault is used (see below). Reverse faults are sometimes called compressional faults, but the use of this term is also undesirable (see 'normal fault', above).

  TEAR FAULT (WRENCH FAULT, TRANSCURRENT FAULT) (◊ Fig. 54). A fault in which the movement is dominantly strike-slip (i.e. horizontal). The terms dextral and sinistral are applied to tear faults to describe the apparent direction of movement. If one stands on the outcrop of a particular rock, facing the fault plane, then for dextral movement the corresponding outcrop on the other side of the fault will be displaced to the right hand, and for sinistral movement will be displaced to the left hand.

Rotational movement between the two fault blocks will result

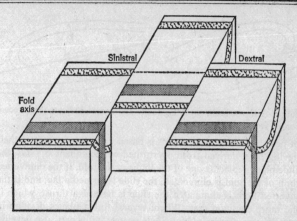

54. Tear faults.

in varying throws being recorded along the strike. Two main types may be recognised:

HINGE FAULT (◊ Fig. 55), where the displacement increases from zero to a maximum along the strike (this type often passes into a flexure of some kind).

PIVOT FAULT (SCISSOR FAULT) (◊ Fig. 55), where one block appears to have rotated about a point on the fault plane

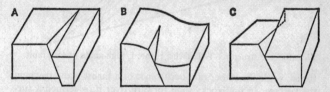

55. (A) A hinge fault; (B) A monocline passing into a fault; (C) A pivot fault.

such that for part of its length the fault is normal with a decreasing throw, and for the remainder of its length is a reverse fault with an increasing throw, the position of no displacement being the point around which rotation appears to have taken place.

THRUST FAULTS (◊ Fig. 56) should perhaps be called slides – a more general term for both thrusts and lags. This classification

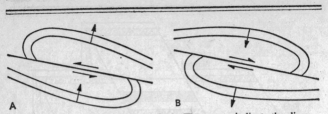

56. (A) A thrust fault; (B) A lag fault. The arrows indicate the direction of younging.

of near-horizontal dislocations is based upon the concept of their having been derived from overturned or recumbent ◊ *folds*, the slide plane replacing one of the limbs of the fold. If the uninverted limb of the fold is eliminated, the slide is termed a lag, and if the inverted limb is eliminated the slide is termed a thrust, which is the more common case. Where several thrust planes develop in parallel sets, a series of high-angle reverse faults may also develop between pairs of thrust planes, giving rise to what is known as imbricate or schuppen structure (◊ Fig. 57). A thrust which develops otherwise than from a recumbent fold is termed a shear

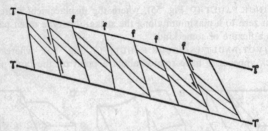

57. Imbricate structure. T–T, thrust plane; f, high-angle reverse fault.

thrust, while the rather rare occurrence of a thrust plane reaching the surface, so that rocks slide forward over it, is termed a surface thrust.

Patterns of faults develop according to the ◊ *stress* system involved, and may be parallel, en échelon, radial, concentric, or in two directions (intersecting) to give a trellis pattern.

A downthrown area between two parallel faults is produced by trough faulting, and if this occurs on a large scale a ◊ *rift valley* or graben results. An upthrown area between two parallel faults is called a horst. A parallel series of faults with a repeated downthrow in the same direction is termed step faulting. (◊ Fig. 58.)

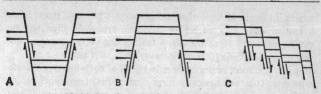

58. (A) A graben; (B) A horst; (C) Step faults.

Riedel faults (fractures) are branching faults developed in the cover by movement in the basement. Faults which are intimately associated with the development of folds are termed longitudinal crestal faults, cross faults and marginal faults.

The terms antithetic and synthetic are sometimes applied to fault planes which dip in the opposite direction to, or in the same direction as, the bedding. These terms have, however, now been replaced by a simple statement concerning the direction and amount of dip or hade of the fault plane.

Fault breccia. During the process of faulting, the rocks are commonly broken up to a greater or lesser extent. If the rocks on either side of the fault plane are hard, the fragments produced will be large and angular and are termed fault breccia. These fragments are often subsequently cemented by secondary calcite, silica, etc. and are sometimes mineralised. Softer rocks, and more intense movement involving the more resistant rocks, may produce a quantity of fine ◊ *rock flour* or fault gouge. For intense dislocation under metamorphic conditions ◊ *Mylonites.* (Cf. ◊ *Crush breccia.*)

Fault gouge. ◊ *Rock flour.*

Faunal province. A region characterised by an assemblage of organisms which is distinct from those of neighbouring regions. The term is generally applied to fairly large areas, e.g. the Australasian Faunal Province, characterised by marsupials. Rarely, single or small groups of oceanic islands may be regarded as a faunal province, e.g. the Galapagos Islands. The establishment of faunal provinces in the fossil record depends upon a synthesis of world-wide stratigraphical and palaeontological information. (Cf. ◊ *Facies.*)

Faunizone (Assemblage zone, Coenozone). A zone in which the strata are characterised by a faunal assemblage. (◊ *Stratigraphic nomenclature.*)

Fayalite. ◊ *Olivines.*

Feather edge. The thin edge of a wedge-shaped body of sediment,

171

where it adjoins an area of non-deposition, e.g. a land-mass. Such a feather edge may be highly indented, crenulate, digitated, or even 'perforated' (fenestrated). A feather edge is commonly overlain unconformably (◊ *Unconformity*) by a later bed.

Feldspars. The most important single group of rock-forming silicate minerals. They are basically three-dimensional framework structures (tektosilicates) in which all four of the oxygen atoms of any SiO_4 tetrahedra are shared with adjoining tetrahedra. This structure produces a fully 'balanced' lattice, and, in order to permit insertion of cations, replacement of silicon by aluminium is essential – at least one in four silicons being replaced, rising to a maximum of one in two.

The feldspars are either monoclinic or triclinic, and four chemically distinct groups exist: potassium feldspars ($KAlSi_3O_8$); sodium feldspars ($NaAlSi_3O_8$); calcium feldspars ($CaAl_2Si_2O_8$); barium feldspars ($BaAl_2Si_2O_8$). The barium ones are exceedingly rare and are not further considered.

PLAGIOCLASE GROUP. The sodium and calcium feldspars form a continuous series of solid solutions which are termed the plagioclase feldspars. They are triclinic. The range from pure sodium-plagioclase (albite, abbreviated to Ab) to pure calcium-plagioclase (anorthite, abbreviated to An) is divided into sections, traditionally named as follows:

	%Ab	%An
	100	0
Albite		
	90	10
Oligoclase		
	70	30
Andesine		
	50	50
Labradorite		
	30	70
Bytownite		
	10	90
Anorthite		
	0% Ab	100% An

Plagioclases almost always exhibit multiple lamellar ◊ *twinning*. Their optical and physical properties vary progressively with their composition and thus can be used in identification.

The sodium-plagioclases are found mainly in ◊ *alkali*-rich rocks (granites, syenites, etc.), andesine and oligoclase are found in ◊ *intermediate* rocks (diorites), while labradorite, bytownite and anorthite are characteristic of ◊ *basic* rocks (gabbros) and ◊ *anorthosites*. (◊ Appendix.)

ALKALI-FELDSPAR GROUP. At high temperatures there is a continuous series of solid solutions between potash-feldspar and soda-feldspar, but at low temperatures only limited solid solution is possible (for exsolution phenomena, ◊ *Perthite*).

decreasing temperature →

(1) Sanidine is the name given to high-temperature potassium feldspar or potassium-sodium feldspar. Anorthoclase is a high-temperature sodium-potassium feldspar.

(2) Orthoclase is the general name for monoclinic potassium feldspar. (◊ Appendix.)

(3) Microcline is the general name for triclinic potassium feldspar. (◊ Appendix.)

(4) Adularia is a very low-temperature monoclinic potassium feldspar.

Sanidine and orthoclase show simple twinning, while microcline shows complex twinning in two directions at right angles to one another. Adularia is rarely twinned.

Alkali-feldspars are mainly found in alkali igneous rocks (granites, syenites, etc.).

All feldspars show very good cleavage parallel to the basal plane, and a second cleavage parallel to the side pinacoid. Apart from occurring in igneous rocks as noted above, feldspars are also found abundantly in metamorphic rocks. In sediments, they are commonly found in those formed in arid environments (◊ *Arenaceous rocks*). A few cases of ◊ *authigenic* growth of feldspars in sediments have been recorded.

Potassium feldspars obtained from ◊ *pegmatites* are used in the glass-making and ceramic industries. Plagioclases obtained from ◊ *anorthosites* are widely used as household abrasives.

The spelling 'feldspar' is preferred to the common 'felspar', as the original usage of the word makes it quite clear that 'field-stone' was intended and not 'rock-stone'.

Feldspathoids. A group of rock-forming ◊ *silicates* characterised by three-dimensional silicate lattices (tektosilicates). Chemically they all contain Na and/or K and their special characteristic is that they are ◊ *unsaturated*, i.e. they can never occur in association with free quartz in a rock. They are most usually cubic or hexagonal. Two main groups can be established: the simple feldspathoids, with silicate as the only anionic group; and the complex feldspathoids, in which one or more different anionic groups are present in addition to the silicate.

SIMPLE FELDSPATHOIDS. Two groups exist with differing degrees of undersaturation, the nepheline-kalsilite series, based $(AlSiO_4)$, and the leucite-analcite series, based on $(AlSi_2O_6)$. Nepheline is ideally $NaAlSiO_4$, but the natural material always contains some potassium. It is hexagonal (carnegieite is an artificial, high-temperature, cubic form) and is widespread in undersaturated rocks and also in certain metasomatic environments (◊ *Fenites*). Kalsilite $(KAlSiO_4)$, a potassium-analogue, is exceedingly rare, as is kaliophilite, which is a ◊ *polymorph*. Leucite is $KAlSi_2O_6$ and usually contains some Na. It is cubic (pseudocubic) and occurs characteristically in undersaturated lavas but never in ◊ *plutonic* rocks. Leucite is cubic above 625°C. but tetragonal below this temperature. Analcite $(NaAlSi_2O_6, H_2O)$ is cubic and is commonly regarded as a ◊ *zeolite* but its obvious relationship to leucite suggests that it ought properly to be classified as a feldspathoid. Conversion of analcite to leucite by K-◊ *metasomatism* has been recorded. Analcite occurs mainly with other zeolites but also quite frequently as a primary constituent of certain igneous rocks. (For nepheline, leucite, and analcite, ◊ Appendix.)

COMPLEX FELDSPATHOIDS

(1) Sodalite Group

> Sodalite $Na_8(Al_6Si_6O_{24})$ Cl_2 (◊ Appendix)
> Nosean $Na_8(Al_6Si_6O_{24})SO_4$
> Haüyne $(Na,Ca)_{4-8}(Al_6Si_6O_{24})(SO_4,S)_{1-2}$

All these are cubic. Sodalite and haüyne are commonly blue; lazurite, a main constituent of lapis-lazuli, is a sulphur-rich haüyne. Ultramarine is a synthetic lazurite. These minerals are commonly associated with nepheline and/or leucite, but lapislazuli is a metamorphosed limestone.

(2) Cancrinite Group

Cancrinite is hexagonal, with a composition of: $(Na,Ca)_{7-8}$ $(Al_6Si_6O_{24})(CO_3,SO_4)_{1.5-2}$. $1-5H_2O$. Vishnevite is similar to cancrinite, but the sulphate radical exceeds the carbonate radical. Cancrinite is often yellow and occurs in association with nepheline in plutonic rocks, sometimes as an alteration product of nepheline. Almost pure nepheline rocks (which must not be called nephelinites – ◊ *Alkali basalt*) sometimes occur and are used as a raw material in the glass industry. Leucite has been worked as a source of potash salts.

Felsic. An acronymic word derived from *fel*dspar and *si*lica, and used to describe light-coloured silicate minerals such as quartz, feldspar, and feldspathoids. (Cf. ◊ *Mafic*.)

Felsite. A fine, evenly grained acid or intermediate igneous rock, forming dykes and veins both in the ◊ *country rocks* and in the parent plutonic mass (cf. ◊ *Microgranite*; ◊◊ *Aplite*).

Felsitic. A texture term for granular ◊ *cryptocrystalline* aggregates formed by the ◊ *devitrification* of glassy material; also often used for the groundmass of porphyritic types.

Felspar. ◊ *Feldspars*.

Femic. An acronymic word made up from *fe*rro and *m*agnesian, for a particular group of arbitrarily calculated minerals in the ◊ *CIPW Classification*. It is not synonymous with ◊ *mafic*, but corresponds approximately to ◊ *ferromagnesian*. (Cf. ◊ *Salic*.)

Fence diagram. If three or more horizontal sections are constructed across a geological map, their interrelationship can very often be best appreciated by drawing them in perspective in their correct relative orientation to one another. It is common practice when doing this to arrange that the various sections have a point or points in common, and it may therefore come about that the final diagram has the appearance of an enclosure bounded by fences. The technique is especially well adapted to displaying data from a series of boreholes. A similar result is obtained if a series of sections are joined end to end and presented in perspective in the form known as a ribbon diagram. (◊ Fig. 59.)

Fenite. The rocks surrounding a ◊ *carbonatite* plug commonly show ◊ *metasomatic* alteration, especially the development of nepheline. This was first described from the Fen district in Norway and the term fenite (the process is fenitisation) was used. Fenites commonly contain nepheline (◊ *Feldspathoids*), ◊ *perthite*, and ◊ *aegirine*, and often show a gneissose texture.

Fergusite. ◊ *Alkali syenite*.

Ferroan. Iron-bearing.

Ferrohortonolite. ◊ *Olivines*.

Ferromagnesian minerals. Rock-forming silicate minerals which contain essential iron and/or magnesium. It is a convenient group-name for the common minerals ◊ *olivine*, ◊ *hypersthene*, ◊ *augite*, ◊ *hornblende*, and ◊ *biotite*. (Cf. ◊ *Femic*; ◊ *Mafic*.)

Ferrosilite. ◊ *Pyroxenes*.

Festoon. An ◊ *island arc*.

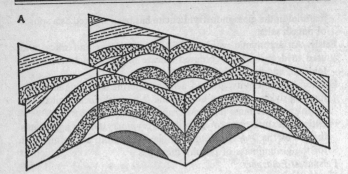

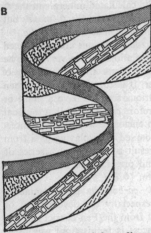

59. (A) A fence diagram; (B) A ribbon diagram.

Festoon-bedding. A type of ◊ *cross-bedding* in which deposition takes place in small troughs and gulleys, the bedding surfaces of the sediments being approximately parallel to the walls of the trough. In a thick series of sediments which display festoon-bedding, successive troughs develop by the erosion of earlier ones, with the partial removal of the contained sediment; thus, in a section at right angles to the line of the channels, a complex pattern is observable. In a section parallel to the length of the channels, a type of small-scale current-bedding can be observed. (◊ *Ripple marks*.)

Fibrolite. Sillimanite (◊ *Aluminium silicates*).

-fic (suffix). Becoming.

Filter-pressing. ◊ *Magmatic differentiation.*

Fingerlakian. A stratigraphic stage name for the base of the North American Upper ◊ *Devonian.*

Fiords (Fjords). Long, steep-sided coastal inlets which have developed as a result of intense glaciation of a previously existing river system in a mountainous area near the sea. Valleys were first guided, and then over-deepened, by the ice, until in many cases the floors were well below sea level. Fiords typically have a threshold (or rock ◊ *bar*) at their seaward end, which is probably the site of the terminal ◊ *moraine,* or the result of a decrease in the ice's erosive power as it reached the sea.

Fire clay. A fossil ◊ *soil* of ◊ *argillaceous* type, found associated with coal seams. It is especially useful as a refractory material.

Fire fountain. A phenomenon which sometimes occurs during the course of a volcanic eruption. It consists of a jet of incandescent gas carrying liquid lava, violently erupting under pressure from a narrow orifice. The jets have been known to reach a height of 400 m. Occasionally several fire fountains coalesce to form a fire curtain. Fire fountains have been observed only in association with volcanoes producing basaltic rocks and they are clearly a function of the high gas content and low viscosity of this type of lava.

Firn. Snow so compacted that no air remains in it. It represents a stage in the formation of glacier ice (◊ *Glaciers and glaciation*).

Fissure eruption. An eruption of ◊ *volcanic* material especially ◊ *lava* through a linear vent or crack.

Fjords. ◊ *Fiords.*

Flagstone. A fissile micaceous sandstone (◊ *Arenaceous rocks*).

Flame structure. ◊ *Injection structure.*

Flandrian. A stratigraphic stage name for the British and European postglacial Pleistocene (Holocene) (◊ *Tertiary System*).

Flap. ◊ *Fold,* Fig. 75.

Flaser structure. The development of a streaky patchy structure in a granular igneous rock, produced by dynamic metamorphism. Flaser gabbros are not uncommon, and usually show a crude pseudo-flow structure around patches of granular material. (Cf. ◊ *Augen.*)

Flat. A more or less horizontal ore body characteristically replacing a bedded sediment, generally limestone. A cross-cutting ◊

vein which acted as a feeder may sometimes be observed. Flats commonly occur at the junction between layers of differing lithology. (◊ *Replacement* (4).)

Flint. A variety of ◊ *chert* occurring primarily in the Upper Cretaceous and as detrital pebbles in the Tertiary. It has a conchoidal fracture.

Flinty-crush rocks. Partially fused rocks resulting from intense ◊ *cataclastic* activity. (◊◊ *Mylonite*.)

Float. ◊ *Eluvial* minerals occurring in ◊ *placer deposits*.

Float stone. (1) A variety of pumice. (◊ *Pyroclastic rocks*.)
(2) (Chordata). ◊ Fig. 18.

Flocculation. The change which takes place when the dispersed phase of a ◊ *colloid* forms a series of discrete particles which are capable of settling out from the dispersion medium. In geological processes, flocculation is almost inevitably a result of a colloidal solution mixing with a solution containing electrolytes, e.g. sea water. Flocculated particles are generally sub-crystalline and only acquire a crystalline character upon ◊ *diagenesis*.

Flood plain. A flat tract of land bordering a river, mainly in its lower reaches, and consisting of ◊ *alluvium* deposited by the river. It is formed by the sweeping of the ◊ *meander* belts downstream, thus widening the valley, the sides of which may become some miles apart. In time of flood, when the river overflows its banks, sediment is deposited along the banks and in the channel itself. The process elevates the river channel above the level of the flood plain. These raised banks are known as levées, and the lower land behind them as back swamps. Levées may constitute a danger when towns are situated in such areas, owing to the increased likelihood of serious flooding. (◊◊ *Cycle of erosion*; *Delta*; *Ox-bow lake*; *Rejuvenation*; *River terrace*.)

Flos ferri. Stalactitic ◊ *aragonite*, $CaCO_3$ (orthorhombic).

Flowage. The ◊ *deformation* which results from the stressing of material beyond its elastic limit, i.e. irreversible deformation. A typical example is the thinning of ◊ *incompetent beds* over crests of anticlines.

Flow banding. ◊ *Flow structure*.

Flow cleavage. ◊ *Cleavage, rock*.

Flow structure, Flow. The structure arising when directional movement of a liquid (e.g. a ◊ *lava*) containing crystals causes these crystals to take up a parallel orientation. If this results in mineralogical or textural banding, the structure is referred to as flow banding, as is seen in many ◊ *rhyolites*. Flow structure is not confined to volcanic rocks. A parallel arrangement of tabular crystals (igneous lamination) is sometimes seen in intrusive rocks, and is caused by movement of the magma into a fissure or cavity. The term flow structure is also loosely used of metamorphic rocks, especially where micaceous layers are folded and contorted round ◊ *porphyroblasts*. (◊ *Eutaxitic*; *Trachyte*; *Layered igneous structures*.)

Fluidisation. A mechanism invoked to explain certain peculiar igneous ◊ *intrusions*, including tuff dykes and pipes, and net-veining of one igneous rock by another, generally an acid one net-veining a basic one. The process involves the entrainment of finely powdered rock material in an uprising mass of very hot gas. From industrial observations it is known that mixtures of this kind behave essentially as a liquid. The stream of gas and particles actively erodes, and may explosively brecciate, pre-existing rocks which it intrudes. When the gas escapes the particles 'freeze' and ◊ *frit* together to form a solid mass. In many cases the particulate nature of the rock is destroyed by this welding process.

D. L. Reynolds, *American Journal of Science*, Vol. 252, 1954.

Fluorescence of minerals. Certain minerals (more particularly, certain varieties of certain minerals) display a remarkable luminescence when irradiated by ultra-violet light of an appropriate wave-length. This phenomenon can sometimes be observed to a very limited extent in some varieties of the mineral ◊ *fluorite* in daylight (owing of course to the small amount of ultra-violet which penetrates to ground level). The name fluorescence is derived from this phenomenon. Commercial use is made of this property in prospecting for ◊ *scheelite*, $CaWO_4$.

Fluorine dating. A method of determining whether vertebrate remains from the same horizon of ◊ *Quaternary* deposits are contemporaneous. It is based on the assumption that fluorine will be absorbed from ◊ *ground water* by bones and teeth at a constant rate (the absorption is due to the conversion of hydroxy-apatite to fluor-apatite); thus bones of the same age from the same deposits should have the same percentage of fluorine within them. It does not follow that either absolute dating or correlation between different deposits is possible, since the amount of fluorine present in ground water varies considerably over short distances. The method serves only to establish the contemporaneity of vertebrate material from one horizon.

Fluorite (Fluorspar). A vein mineral, CaF_2, found also in ◊ *pneu-*

matolytic deposits, and ◊ *pegmatites*. Blue John is a blue orna-mental variety. (◊ Appendix.)

Fluorspar. ◊ *Fluorite*.

Flute cast. A groove eroded by ◊ *turbulent flow* and subsequently filled with coarse sediment. (◊ *Sedimentary structures*.)

Flysch. A term which originated in Switzerland and is strictly ap-plicable only to sediments associated with the Alpine ◊ *orogeny*. It is used to describe sediments produced by the erosion of up-rising and developing ◊ *fold* structures, which are subsequently deformed by later stages in the development of these same fold structures. In Switzerland, such sediments are marine, and con-sist of ◊ *argillaceous rocks*, impure sandstones, ◊ *breccias*, and ◊ *conglomerates*. The whole assemblage is ◊ *syntectonic* (cf. ◊ *Molasse*). Swiss geologists are strongly of the opinion that 'flysch is a term which cannot be exported'.

Focus. The point of origin of an ◊ *earthquake* in the Earth's crust.

F'oids. A contraction of ◊ *feldspathoids*.

Fold, Folding. A flexure in rocks; that is, a change in the amount of dip of a bed, and also often a change in the direction of dip. (◊ Fig. 60.) If the flexure takes the form of an arch, the fold is

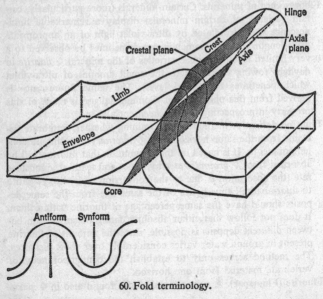

60. Fold terminology.

termed an anticline (more generally, an antiform: see below), while a flexure in the form of a trough is a syncline (more generally, synform).

The line along which a change in the amount and/or direction of dip takes place is known as the hinge line, and on many folds this coincides with the position of maximum curvature. The area adjacent to the hinge line is known as the hinge area or nose of the fold. The limb of a fold is the part which lies between one hinge and the next, and the angle between the limbs is called the interlimb angle.

A fold is said to close in the direction in which the limbs converge. As seen on a map, one says that a fold 'closes to the N.W. etc.'. As seen in a vertical section, a fold may be said to close upwards (antiform) or downwards (synform). For an antiformal fold the direction of closure is the direction of plunge (see below). A plane which joins the hinge lines of the successive beds in a fold is the axial plane. Fold axis is a term which has been used by different authors with different meanings, but from a practical point of view the best definition appears to be that it is a line parallel to the hinge. This may be expressed alternatively as the trace of the intersection of the axial plane on any bed constituting the fold. It should be noted that the trace of the axial plane on the ground surface coincides with the axis only for folds of certain specific attitude. A crest line marks the highest points on the same bed in an antiformal fold. Similarly, a trough line marks the lowest points on the same bed in a synformal fold. The crest plane and trough plane are the planes connecting the crest lines and trough lines respectively of the successive beds of a fold. Axial planes, crest planes, and trough planes are not always flat and may be curved in various ways, and also the axial plane of a fold is not necessarily coincident with the crestal plane.

A fold is said to plunge if the axis is not horizontal (◊ Fig. 61) – the amount of plunge being the angle between the axis and a horizontal line lying in a common vertical plane. Pitch is sometimes wrongly used as synonymous with plunge; strictly pitch is the angle between a horizontal line and the axis measured on the axial plane; unless the axial plane is upright, pitch and plunge do not coincide.

FOLD TYPES. If the axial plane bisects the fold, the fold is said to be symmetrical or upright. If the axial plane has a dip, the fold is described as inclined or asymmetrical, while if the axial

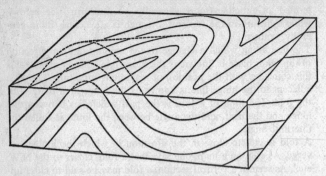

61. Plunging folds.

plane is horizontal or sub-horizontal the fold is said to be a recumbent fold or nappe. (◊ Fig. 62.) If in an inclined fold the two limbs tend to dip in the same direction at different amounts (i.e. one limb is inverted), the fold is said to be overturned. Many folds which are described as recumbent are large-scale structures,

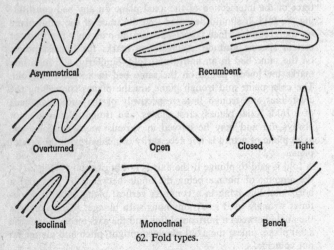

62. Fold types.

and the axial planes are often flexed to varying extents. Folds in which the interlimb angle is greater than 70° are said to be open or gentle, if between 30° and 70°, closed, and below 30°, tight.

If the two limbs are parallel, the fold is said to be isoclinal. The fold profile is the trace of the fold seen on a plane normal to the axis: unless the axis is horizontal this will differ from the trace seen on a vertical plane.

The flexure resulting from a sudden increase in the dip of a bed to a near vertical position, followed by a flattening to the original dip without a change in the direction of dip, is termed a monocline or anticlinal bend. Similarly, a flattening of dip without change in direction is known as a structural bench or terrace, or a synclinal bend.

A antiform
S synform
Older
Younger

63. The position of older and younger rocks in downward facing antiforms and synforms.

In the normal type of anticline the older rocks will occupy the core of the fold and the younger rocks the outer portion, while in a syncline the younger rocks occupy the core and the older rocks the outer portion. In areas of intense deformation, folds have been found which are apparently anticlines except that they have younger rocks in the core. This type of structure is termed an antiform or a downward-facing syncline. Similarly the term synform or downward-facing anticline is applied to a synclinal type of structure with the oldest rocks forming the core. (◊ Fig. 63.)

An anticlinal structure which plunges in all directions (quaquaversal dip) is a dome, while a syncline which dips inwards in all directions is known as a structural basin. (◊ Fig. 64.) Domes and

A   B

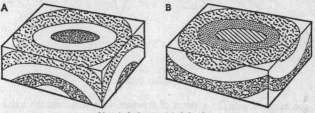

64. (A) A dome; (B) A basin.

basins are collectively termed periclinal structures, although some authors would confine this term to domes. Very large-scale regional folds may have their limbs folded on a smaller scale into a series of minor flexures: a large-scale fold-complex of this kind is called an anticlinorium or synclinorium, depending upon its form (cf. drag fold, below).

Cylindrical folds are those whose profile is essentially semi-circular and remains constant when traced along their axes. A parallel or concentric fold is one in which the successive semi-

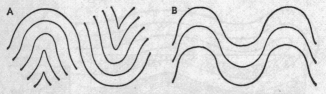

65. (A) Parallel folds; (B) Similar folds.

circles have a constant centre and a regularly increasing radius, while a similar fold maintains the same radius but shifts the centre along the axial plane. ($\Diamond$ Fig. 65.) In general, folds are rarely completely concentric or similar, but show characteristics of both types. Angular (chevron, zig-zag, concertina, or accordion) folds have straight limbs and sharp hinges; in fact, in many cases

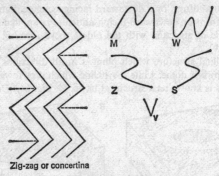

Zig-zag or concertina

66. Angular folds.

the hinge is actually a point of fracture, in which case the axial plane becomes a plane of weakness. The letters M, V, W, Z have been used to describe these folds, as opposed to the S-shaped cylindrical folds. (◊ Fig. 66.) The concepts of parallel and similar folds may be applied to angular folds also, and all gradations exist between angular folds and those having a semicircular profile.

Box folds (◊ Fig. 67) are those having an approximately rectangular cross-section. They appear to change their profile as they are traced up or down their axial plane. Conjugate folds, which are superficially similar to box folds, maintain their profile

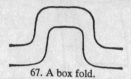

67. A box fold.

along their axial planes. They consist of a pair of folds in which the axial planes dip towards one another so that they intersect in a line parallel to the fold axis.

Non-cylindrical folds do not maintain their profile along their axial direction. The simplest type is a conical fold, while the terms inconstant and 'wild' fold have been used to describe more complex types.

Most folds are caused by orogenic movements. In areas which

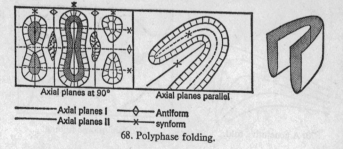

Axial planes at 90°   Axial planes parallel

———— Axial planes I   —◊— Antiform
———— Axial planes II   —×— synform

68. Polyphase folding.

have had a complex tectonic history, more than one sets of folds may develop, giving rise to what has been called cross-folding (polyphase folding), in which the axial planes of the folds and/or the axes of the folds intersect one another at an angle (◊ Fig. 68). Where cross-folding has taken place, the pattern of folds as seen on the ground may be very complex. If two anticlines belonging to different sets of folds coincide, the result is a culmination, but where an anticline and syncline coincide the result is a depression; these two terms are especially applied to refolded recumbent folds (◊ Fig. 69).

A generative fold (◊ Fig. 70) is one which increases in size as it

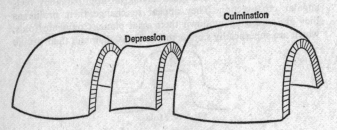

69. Culmination and depression.

is traced upwards along the axial plane. Some generative folds are non-orogenic in origin.

During folding, ◊ *competent* and ◊ *incompetent* beds may behave differently, giving rise to different styles of folding. This is known as disharmonic folding (◊ Fig. 71). If the competent bed is broken

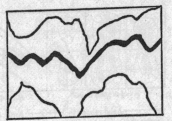

70. A generative fold.   71. Disharmonic folding.

into a series of discontinuous units with compensating flowage of the incompetent beds, the folding is described as disjunctive. During the process of folding, there is almost inevitably ◊ *bedding-plane slip*, and where this occurs between competent and incompetent layers it is commonly found that the incompetent strata are deformed into small subsidiary folds termed drag or parasitic folds ($\diamond$ Fig. 72). In a very large number of cases the direction and amount of pitch of the drag folds is the same as that of the major fold (Pumpelly's Rule). Drag folds which conform to this rule (and indeed any minor fold which conforms to the direction of a major fold structure) are said to be congruous folds; those that do not are termed incongruous.

TYPES OF FOLDING. Buckle folding results from lateral compression of a sheet. During buckling, points on the Earth's surface come closer together (crustal shortening). Its effects may be

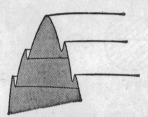

72. Drag folds.

most readily seen in incompetent beds, as these tend to deform by shear folding (see below) or flowage (see below).

Bending is folding which takes place by vertical forces of different intensities acting in different places; it does not involve gravity or shear movement, but characteristically produces rather broad structures.

Shear folding is produced by differential movement along close-spaced cleavage planes. This type of folding is characteristic of slates and phyllites, and may occur in incompetent beds which are involved in other types of folding. The effect may readily be seen by drawing a line across the edge of a pack of cards and then pushing on an edge at right angles to the marked one with a pencil or finger. ($\diamond$ Fig. 73.) This type of folding is also sometimes referred to as Gleitbretter folding.

Flow folding ($\diamond$ Fig. 74) develops in highly incompetent beds which behave more as a viscous fluid than a solid rock. Typical rocks in which this occurs are rock salt, gypsum, and soft clays. The contortions which arise during flow folding are often highly disharmonic. Some types of $\diamond$ *ptygmatic* folding are probably of this type.

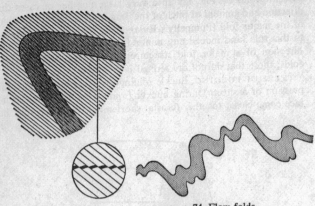

73. Shear folds   74. Flow folds.

Certain types of folding are not produced by orogenic forces, and to these the name non-diastrophic may be given. The following are a few major types:

(a) Cambering occurs where competent beds form the capping of hills, overlying incompetent beds. The incompetent beds flow

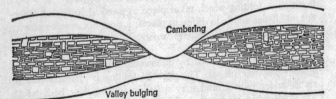

Cambering

Valley bulging

75. Cambering and valley bulging.

outwards into the valley, causing the competent beds to down-warp. (◊ Fig. 75.)

(b) Valley bulging is a rather similar effect, which occurs where incompetent material is forced up into a valley by the weight of the hill masses on either side, which become 'turned up' at the edges in the process. (◊ Fig. 75.)

(c) Folds produced by hill creep arise from movement of debris down a hillside. They take the form of deformation of strata adjacent to the plane of soil-movement. The ◊ *dips* taken in such regions are often highly inaccurate. (◊ *Gravity transport.*)

(d) Gravity collapse structures develop occasionally where erosion of incompetent beds has allowed slightly more competent beds to move under gravity and become contorted into various types of fold (knee, flap, cascade, etc.). (◊ Fig. 76.)

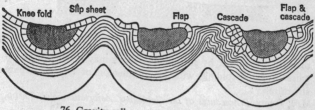

76. Gravity collapse structures due to folding.

(e) Supratenuous folds are produced when deposition takes place over a ridge. They may also develop through differential compaction of sediment around such a ridge, and may be accentuated if movement takes place during sedimentation in such circumstances. (◊ Fig. 77.)

(f) During sedimentation on an inclined surface, deformation of the sediments may occur by sliding of the layers down slope

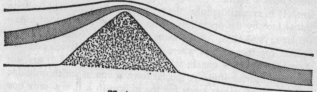

77. A supratenuous fold.

under the influence of gravity. This may give rise to the curious phenomenon of folded beds interbedded with unfolded beds.

(g) Diapiric or piercement folds develop during the upward movement of the mass of rock forming a ◊ *diapir*, the strata surrounding it being strongly deformed. (◊ Fig. 78.) It should be noted that although diapiric folding associated with ◊ *salt domes* is non-diastrophic (and to a large extent this is also true of igneous

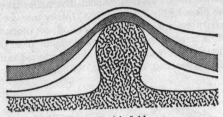

78. A diapiric fold.

diapirs) piercement by incompetent beds may occur during intense orogenic folding, with the production of typical diapiric structure.
M. J. Fleuty, *Proceedings of the Geologists' Association*, Vol. 75, pp. 461–92, 1964.

Foliation. Parallel orientation of platy minerals or mineral banding in rocks. (Cf. ◊ *Schist*; ◊ *Flow structure*.)

Fool's gold. Iron pyrites (◊ *Pyrite*); the name has probably also been applied to ◊ *chalcopyrite*.

Footwall. ◊ *Fault*.

Foramen. ◊ *Brachiopoda* and Figs. 11 and 12; (Chordata) ◊ Fig. 18.

Foraminifera. ◊ *Protozoa*.

Fore deep. A deep oceanic trench on the convex side of an ◊ *island arc*.

Foreset beds. ◊ *Delta*.

Form. (1) In crystallography, the group of faces generated from an initial face by the operation of the symmetry elements of a particular crystal class. All the faces comprising a form are always present together, except where the crystal is mal-developed.

A general form is one in which the ◊ *indices* (h k l) are unrestricted in magnitude. A special form is one in which only one possible set of values exists for the indices (h k l), e.g. only one octahedron (1 1 1) is possible in the cubic system. Certain forms are neither special nor general – part of their index is variable and part fixed, e.g. in a prism (h k l), l must always be zero; and in trisoctohedra of the cubic system, h must always equal k. This type of form may be referred to as a restricted form.

An open form is an assemblage of faces which cannot enclose a volume of space, and hence cannot exist in the absence of a second form. A closed form is an assemblage of faces which can enclose a volume of space.

(2) A term used in a rather loose way in palaeontology with the sense of 'a minor variety'.

Formation. ◊ *Stratigraphic nomenclature.*

Forsterite. A mineral of the ◊ *olivine* group.

Fossil. (1) (noun). An organic trace buried by natural processes, and subsequently permanently preserved. The term 'organic trace' is here used to include skeletal material, impressions of organisms, excremental material, tracks, trails, and borings. Human artifacts are not regarded as fossils.

Fossils may be preserved in the following forms:

(a) With unaltered hard parts: animals having skeletal material which is stable over a considerable period of time are commonly found unaltered, e.g. calcitic shells, teeth, and aragonitic shells in the Tertiary and Mesozoic only.

(b) With unaltered soft parts: these are very rarely preserved. Mammoths have been found in the Siberian ◊ *permafrost*, but these are considered to be in a state of temporary preservation. Other examples are insects in amber (dehydrated only), and the sepia sacs of belemnites (◊ *Mollusca, Cephalopoda*).

(c) With hard parts altered by petrification (s.s.), the infilling of pore spaces in shells, cellular plant material, or bones, by one of a number of different substances, e.g. silica, calcite, limonite, pyrite, etc., or by replacement, either coarse or molecular, of the original skeletal material by either similar or different substances, e.g. calcareous shells may be replaced by crystalline calcite or by silica. Coarse replacement destroys fine detail while molecular replacement preserves it.

(d) With altered soft parts: by complete removal, leaving an impression of the form, e.g. Precambrian jellyfish; or by carbonisation – organic tissues transformed to a thin film of carbon, e.g. graptolites (◊ *Chordata*), and fin-membranes of fish are commonly preserved thus. Carbonisation of plant remains is considered to be an alteration of the hard parts.

(◊ *Casts and Moulds.*)

(2) (adjective). The term 'fossil' is widely used to indicate great age, extinction, or 'having been dug up'. 'Fossil horse' implies an ancient, extinct, variety of horse; 'fossil rainprint' – an ancient rainprint; and 'fossil peneplain' – an ancient plain which has been covered and revealed by renewed erosion.

(◊ *Derived fossil; Facies fossil; Index fossil; Remanié fossils; Trace fossils; Zone fossil.*)

Fossil man. Many discoveries have been made concerning man's ancestry, especially in the last ten years, and on the basis of these discoveries a detailed picture of the evolution of man has been compiled. Up to 1945 only isolated discoveries had been made and no composite picture could be established.

The earliest hominids probably lived in Africa and were derived from the bipedal savannah-dwelling anthropoid apes, like Ramapithecus. Two main hominid groups were present during the late Pliocene/early Pleistocene, the primitive Australopithecinae, and the more advanced forms representing the genus Homo. The Australopithecinae may well have been in direct competition with the early true man such as Homo habilis, but although present in S. Africa for a period of possibly 500,000 years they apparently gave rise to no descendants.

The genus Homo diverged rapidly, with Homo habilis evolving into a sub-species of Homo erectus as Java and Pekin man together with African and European forms of Chellean man. It was the species Homo erectus which finally gave rise to modern man (Homo sapiens sapiens) via Swanscombe and Neanderthal man (Homo sapiens neanderthalensis). There is considerable evidence that Homo sapiens sapiens and Homo sapiens neanderthalensis coexisted and that many of the physiognomic features exhibited by the latter are still to be found in some races of Homo sapiens sapiens.

The most recent discoveries suggest that some modifications of the above scheme may become necessary, particularly in the relationships of the earliest hominids (Ramapithecus, Kenyapithecus and Australopithecus).

Fossil plants. ◊ *Palaeobotany.*

Foyaite. ◊ *Alkali syenite.*

Fracture. In the broadest possible sense the word fracture may be applied to any break in a material. However, in geology the term usually implies breakage along a direction or directions which are not ◊ *cleavage* or fissility directions; this applies to both rocks and minerals.

Fracture cleavage. ◊ *Cleavage, rock.*

Fracture zone. A zone in which ◊ *faulting* has taken place.

Franconian. A stratigraphical stage name for the North American mid Upper ◊ *Cambrian.*

Franklinite. ◊ *Spinel group of minerals.*

Frasnian. A stratigraphical stage name for the base of the European Upper ◊ *Devonian.*

Free face. An outcrop of bare rock on a hillside. (◊ *Slope.*)

Fritting. The process by which a granular material becomes co-herent by virtue of partial melting at grain contacts, or diffusion of material across grain contacts, causing adhesion. The process can occur either as a result of metamorphism or in the hotter parts of volcanic ash deposits. (◊ *Pyroclastic rocks*; *Fluidisation*).

Front. A name for a hypothetical upsurge of chemical activity sup-posed to take place during igneous activity. The concept supposes that fluids carrying a concentration of one or more elements rise up through the rocks forming the crust of the Earth, react-ing with them and driving out other elements in advance of the uprising 'front'. Thus in ◊ *granitisation* a concentration of sodium and potassium ions is postulated as responsible for the ejection of calcium, magnesium, and iron ions in the form of a 'basic front'. (◊ *Metasomatism*.)

Füchsel, G. Christian (1722–73). German doctor who introduced the term 'formation' to stratigraphy, and whose detailed study of the Triassic had a great influence on the stratigraphic methods of his countrymen.

Fuchsite. A mineral of the ◊ *mica* group containing chromium. A variety of muscovite.

Fulgurite. A ramifying or branching tube consisting of fused silica or silicates, formed by the action of lightning striking sandy soil, e.g. dunes, beaches, etc. Fulgurites should not be confused with the siliceous or other concretions which occasionally form around the roots of plants and are somewhat similar.

Fuller's earth. (1) An aluminium-poor montmorillonite clay. It has the property of absorbing oil and was originally used for de-greasing (fulling) fleeces. (◊ *Argillaceous rocks*; *Clay minerals*.) (2) A local stratographic name for a division of the upper part of the Middle Jurassic in Britain which contains substantial beds of Fullers' Earth.

Fumarole. A hot spring emitting ◊ *volatiles*.

Fusain. ◊ *Coal*.

Fusibility scale. A list of minerals arranged in order of their fusi-bility. Von Kobell's scale of fusibility is:

1. Stibnite (525°C.)
2. Natrolite (965°C.) or chalcopyrite.
3. Almandine garnet (1,200°C.)
4. Actinolite (1,296°C.)
5. Orthoclase (1,200°C.)
6. Bronzite (1,380°C.)

Fusibility is not the same as melting point. A good conductor of heat, e.g. native copper, fuses less readily than a poor conductor with the same melting point. Fusibility is usually determined with the aid of a simple blow-pipe flame.

Fusiform. Spindle-shaped.

Gabbro. A coarse-grained (plutonic) ◊ *basic* igneous rock consisting of basic plagioclase (labradorite to anorthite; ◊ *Feldspar*), a ◊ *pyroxene* (augite and/or hypersthene) and, very commonly, ◊ *olivine* in substantial amounts. ◊ *Hornblende,* ◊ *biotite,* and quartz may occur in accessory amounts, and ◊ *magnetite* and ◊ *ilmenite* are common accessories. With increasing ◊ *ferromagnesian minerals,* gabbros grade into picrites and other ◊ *ultrabasic* types; with increasing feldspar they grade into ◊ *anorthosites*; with the introduction of alkali-feldspars and/or feldspathoids, gabbros become olivine ◊ *monzonites* and ◊ *alkali gabbros.* (For the diorite–gabbro boundary, ◊ *Diorite*.)

Gabbros are the coarse-grained equivalents of the volcanic ◊ *basalts* and hypabyssal ◊ *dolerites*.

Chemically gabbros are low in silica and high in Mg and Ca. Na and K are very low, while Fe ranges from low values in the feldspathic types to exceedingly high ones in the Fe-rich ferrogabbros.

The table below sets out the composition of major rock types included under the general heading gabbro:

	Plagioclase			
Pyroxene ↓	Labradorite		Bytownite–Anorthite	
	No olivine	With olivine	No olivine	With olivine
Augite	Ortho-gabbro	Olivine gabbro	Eucrite ↑	Olivine eucrite ↑
Augite and orthopyroxene	Hypersthene gabbro	Olivine hypersthene gabbro		
Orthopyroxene	Norite	Olivine norite	Hypersthene eucrite ↓	Olivine hypersthene eucrite ↓
No pyroxene	(Anorthosite)	Troctolite	(Anorthosite)	Allivalite

195

Gabbros are commonly equigranular and porphyritic types are rare. Varieties in which a coarse type of ◊ *ophitic texture* occurs are known. (For microgabbro, ◊ *Dolerite*.)

Apart from the names set out in the chart above the following terms are sometimes used: bojite, a gabbro containing primary amphibole – either barkevicite or hornblende – as opposed to many so-called hornblende gabbros in which the amphibole is secondary. Hyperite is an obsolete name for gabbro containing hypersthene and augite; leucogabbro and leuconorite are feldspar-rich varieties; melagabbro and melanorite are pyroxene and/or olivine-rich varieties. Beerbachite is a gabbro ◊ *aplite*.

Gabbros commonly occur in the form of either large ◊ *lopoliths*, e.g. Bushveld, South Africa; Stillwater, Montana; and Duluth, Minnesota; or ◊ *ring* complexes, e.g. Ardnamurchan, Scotland; or ◊ *layered* complexes, e.g. Skaergaard, Greenland.

Gahnite. ◊ *Spinel group of minerals.*

Galena. PbS, the most important ore of lead, found in ◊ *hydrothermal* veins and as a ◊ *replacement mineral*. (◊ Appendix.)

Gamachian. A stratigraphic stage name for the top of the Upper ◊ *Ordovician* in North America.

Gangue. That part of an ◊ *ore* deposit from which a metal or metals is not extracted, i.e. it is not the material which is the objective in working the ore deposit. The term is also loosely used for the waste material from the process of separation and concentration of ores. Common gangue minerals include quartz, ◊ *calcite*, ◊ *fluorite*, ◊ *siderite*, and ◊ *pyrite*. It should be noted that the gangue minerals of one ore body may be the ore mineral of another, e.g. pyrite may be separated when it occurs with lead and zinc, and referred to as gangue, while in other places pyrite is worked profitably. Some non-metallic gangue minerals are valuable in their own right, e.g. ◊ *barytes* and ◊ *fluorite*.

Gannister. An arenaceous seat earth found below coal seams. It is a pure silica sand, low in alkalies, iron, and aluminium, and containing frequent carbonised traces of plant roots (Stigmaria). ◊ *Fireclay* is the argillaceous equivalent. As a source of pure silica gannisters have a slight economic significance.

Gargasian. A stratigraphic stage name for the European Upper ◊ *Aptian* (Cretaceous System).

Garnets. Minerals which have a structure consisting of independent SiO_4 tetrahedra and AlO_6 octahedra, which are linked by cations in 8-coordination. Garnets are externally cubic in form, although the Ca-garnets are probably cubic only at high temperatures and invert to a twinned tetragonal form at ordinary temperatures. The general formula for the garnets is:

$$R''_3R'''_2 Si_3O_{12}$$

where R'' can be Fe'', Mg, Mn'', or Ca, and R''' can be Fe''', Al or Cr.

Garnets are commonly divided into two groups – pyralspite and ugrandite. Pyralspite is an acronym of the names of the three major varieties included: PYRope (Mg Al garnet, usually with Fe), ALmandine (Fe'' Al) and SPessartite (Mn'' Al). Similarly the name ugrandite is derived from Uvarovite (Ca Cr), GRossularite (Ca Al), and ANDradite (Ca Fe'''). In each of these two groups there is a series of solid solutions between the end-members but there appears to be a gap between the two groups unrepresented in nature. Melanite is black andradite with Ti. Schorlomite is similar but with more Ti. Demantoid is a clear, green andradite. Hessonite (cinnamon stone) is a yellow or brownish-red variety of grossularite.

Garnets characteristically form crystals of trapezohedral or rhombdodecahedral habit. They have a high refractive index, no cleavage, and a wide range of colours. Occasionally very large crystals of almandine are found (up to 30 cm. diameter).

The commonest environment for garnet formation is in metamorphic rocks. Almandine is characteristic of medium and high grade regionally (and more rarely thermally) metamorphosed pelitic rocks, while grossularite and andradite are characteristic of metamorphosed impure limestones (◊ *Skarn*). Pyrope is an essential constituent of ◊ *eclogites* and also occurs in certain ◊ *ultrabasic* rocks which are assumed to have crystallised under high pressures. Garnets sometimes occur in granites and pegmatites, possibly as a result of ◊ *assimilation* of ◊ *country rock*. They also occur in a wide variety of lavas and ◊ *pyroclastics*. Primary melanite occurs in undersaturated igneous rocks. Chrome-garnets are found in ultrabasic rocks, skarns and in chromite-rich rocks. Garnet is a resistant mineral, and detrital grains are commonly found in sediments.

Grossularite containing hydroxyl (in partial replacement of SiO_4 units) is known as hydrogrossularite. It is found in skarns and hydrothermal replacement bodies and is probably commoner than is generally realised.

Clear, flawless crystals of garnet are cut and polished as semiprecious gemstones; garnet is also valuable as an abrasive.

Garnierite. Nickel-bearing ◊ *serpentine*.

Gas cap. An accumulation of gas above an ◊ *oil* pool.

Gas streaming. A possible cause of ◊ *differentiation*, in which the upward streaming of gases transfers ions or early crystals and leads to the separation of an igneous magma into two or more fractions.

Gastrolith. A ◊ *stomach stone*.

Gastropoda. ◊ *Mollusca*.

Geanticline. (1) The broad uplifted area bordering a ◊ *geosyncline* which supplies sediments for its infilling.

(2) A broad anticlinal uplift which develops within a ◊ *geosyncline* at an early stage of the orogenic phase. It is commonly associated with vulcanism and there is some suggestion that ◊ *island arcs* may be geanticlinal in character.

Gedinnian. A stratigraphic stage name for the base of the European Lower ◊ *Devonian*.

Gedrite. An orthorhombic ◊ *amphibole*.

Geikie, Archibald (1835–1924). Scottish geologist, later Director of the ◊ *Geological Survey*, who wrote some of the earliest geological textbooks. He was famed during his life for his account of the volcanic areas of Great Britain, and for his advocacy of the predominance of fluvial and glacial erosion in moulding the landscape of Scotland. His brother James was also a geologist.

Gel. A coagulated ◊ *colloid* of a jelly-like nature.

Gem, Gemstone. Artificially polished fragments (usually faceted or sometimes with a smooth, curved, finish) of certain minerals, used for decorative purposes, mainly in jewellery. The best-known gems are the so-called 'precious stones' – ◊ *diamonds* (crystalline carbon), ◊ *rubies* and ◊ *sapphires* (crystalline aluminium oxide = ◊ *corundum*), and ◊ *emeralds* (beryllium aluminium silicate with chromium = ◊ *beryl*). ◊ *Topaz*, ◊ *aquamarine*, peridot (◊ *Olivine*), rubellite (◊ *Tourmaline*), ◊ *zircon*, ◊ *garnet*, ◊ *spinel*, etc. are referred to as semi-precious stones. Gem minerals are hard, relatively free from cleavage, and occur as transparent crystals. Many other minerals have been used as gems, even though they do not conform to this definition, e.g. ◊ *opal*, ◊ *fluorite*, ◊ *jade*, ◊ *agate*.

Genotype. The ◊ *type* species of a genus.

Geo- (prefix). Earth.

Geochemical prospecting. A method of searching for concealed bodies of metallic ores by means of chemical techniques. ◊ *Ground water* passing through an ore body will take into solution ions of the various elements present. This water may: (1) Pass into the local drainage system. In certain circumstances sediments (especially clay minerals) may absorb ions from stream water. (2) Pass into the soil, where the water may accumulate in an ◊ *aquifer*; evaporate, leaving the dissolved ions behind in the soil; or become absorbed by ◊ *clay minerals*. (3) Pass into vegetation, which extracts the ions and transpires the water. Plants are, however, commonly selective in the ions that they will extract from ground waters. Seasonal effects and the stage of development of the plants may control their intake of ions from solution. Soil and stream sediments may also contain fine particles of the ore minerals weathered from the outcrop.

Geochemical prospecting depends for its success upon: (1) Sufficiently sensitive and reliable techniques for detecting low concentrations of the elements in which the prospector is interested. The development of modern reagents which give characteristic coloured reaction-products with metals enables not only the detection but the quantitative estimation of an element to be done in the field. (2) Careful sampling techniques and pre-analysis handling of the samples. Samples usually consist of: (a) water from streams, rivers, swamps, springs, and occasionally wells or boreholes; (b) sediments from streams, rivers, lakes, etc., including alluvium, terrace gravels, etc.; (c) soil, from the surface, shallow pits, or auger holes; (d) samples of vegetation, which are usually 'ashed' before analysis.

The geochemical prospector searches for a geochemical anomaly, i.e. a sudden increase in quantity of a particular element over a certain limited area. As an example, soil from above a vein carrying zinc blende (ZnS) may carry about 100 ppm of Zn, compared with the normal soil content of less than 10 ppm.

Geochemistry. 'The primary purpose of geochemistry is on the one hand to determine the composition of the Earth and its parts, and on the other to discover the laws which control the distribution of individual elements' (V. M. Goldschmidt, 1933). The term was introduced by the Swiss chemist Schonbein in 1838.

With the more refined techniques of modern analytical chemistry, detailed studies can be made of the distribution of many elements, often present in quantities of only a few parts per million. This has resulted in information becoming available concerning not only the distribution of elements in rock types but also their distribution amongst the major crustal units on the one hand, and in various lattice types (◊ *Silicates*) on the other. Refinements of technique have also made it possible to establish the relative abundance of the individual isotopes of the elements.

The geochemist initially dealt only with the crustal rocks, but nowadays extends his work to the Earth as a whole, to the planets and moons of the Solar System, to the stars and interstellar space, under the all-embracing title cosmochemistry.

From the available data three generalizations can be made:

(1) Heavy elements, i.e. those having atomic weights greater than nickel (58·71), have a lower abundance than the lighter elements. It can be shown that the eleven commonest elements (on a cosmic basis) all have atomic weights less than nickel.

(2) Elements of even atomic number are, in general, more abundant than those of odd atomic number. In particular, elements of even atomic number are more abundant than those with the odd atomic numbers on either side (Oddo and Harkins Rule), e.g. Fe^{26} is much more abundant than Mn^{25} or Co^{27}.

(3) Hydrogen is between one thousand times and one hundred times as abundant as any other element in the universe.

TRACE ELEMENTS. The discoveries of the X-ray crystallographers made it clear to geochemists that the ionic radius of an element was of the utmost importance when considering its geochemical characteristics. The problem of the way in which trace elements (very small quantities of non-essential elements occurring in addition to, or in place of, essential elements in a crystal) occur in crystals is largely explicable in terms of ionic radius. If a space in a lattice occurs, any available atom, providing that it has the right ionic radius, can occupy the space, regardless of its chemical affinities and providing its charge does not differ by more than 1 from that of the element normally occupying this space. This kind of replacement of one ion by another is called diadochic substitution. (◊ *Ionic radius ratio.*)

Three types of occurrence of trace elements can be postulated:

(a) Camouflage, where the trace element and the common element have the same charge, e.g. $Ga^{3+} \rightleftharpoons Al^{3+}$

(b) Capture, where the trace element substitutes for a common element with a lower charge, e.g. $\begin{matrix} Ba^{2+} \rightleftharpoons K^{1+} \\ Pb^{2+} \rightleftharpoons K^{1+} \end{matrix}$

(c) Admission, where the trace element replaces a common element with a higher charge, e.g. $Li^{1+} \rightleftharpoons Mg^{2+}$.

GEOCHEMICAL CLASSIFICATION. From a study of the composition of minerals and rocks, the sea and other waters, and the atmosphere, elements can be divided into various groups, based upon their association, i.e. elements become concentrated in a preferred environment. The following divisions are recognised:

Chalcophile Elements having a strong affinity for sulphur, characterised by the sulphuric ore minerals.

Lithophile Elements having a strong affinity for oxygen, concentrated in the silicate minerals.

Siderophile Elements with weak affinities for oxygen and sulphur, but soluble in molten iron; they are presumed to be concentrated in the core of the Earth and are found on metal phases of meteorites.

Atmophile The elements of the atmosphere (nitrogen and the 'inert' gases).

Biophile The elements concentrated in living matter (carbon, phosphorus).

An element may occur in more than one group. The terms lithosphere, atmosphere, biosphere and hydrosphere are used

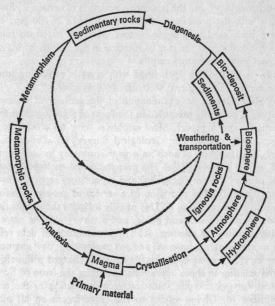

79. The geochemical cycle.

to define geochemical environments. (For examples of geochemical differentiation, ◊ *Sedimentary rocks*.) Certain elements (e.g. chlorine, bromine, iodine) become concentrated in sea water, while others (e.g. fluorine, iron) are rapidly removed.

GEOCHEMICAL CYCLE. One of the most fruitful concepts of geochemistry is the geochemical cycle. This results from a consideration of the way in which an element, commencing in a primary ◊ *magma*, may circulate through a sequence of geochemical environments. (◊ Fig. 79.) As shown, the cycle is a generalised one; it will be realised that the cycle for an individual element may be much more complicated. Fully detailed geochemical 'balance sheets' have not yet been drawn up for the majority of the elements.

Geochronology. The measurement of time intervals on a geological scale, either the duration in years (or 10^6 years) of some interval, or the dating of an event in years (or 10^6 years) 'before the present'. Besides ◊ *radioactive dating*, the counting of ◊ *varve* units in varved sediments, counting of annular rings in trees (dendrochronology), and the possible correlation of certain ◊ *rhythmic* units with the precession of the equinoxes (a 26,000-year cycle) have been used with some success. Methods based on rates of sedimentation and the rate of increase in the salinity of sea water do not yield satisfactory results.

Geode. A cavity in a rock lined with crystals projecting towards the centre. (Cf. ◊ *Vugh*; ◊ *Druse*; ◊ *Amygdale*.)

Geological Survey. An examination of the geology of an area. A survey is generally presented in the form of a geological map and sections, and a report. Most countries have an official government body called 'The Geological Survey' (cf. Ordnance Survey). The function of such an organisation may include some, or all, of the following: (a) The preparation of geological maps of the country (on varying scales) and explanatory reports to accompany them. (b) The provision of a service of geological information to those requiring it. This usually includes the publication of more general works and guides and a survey may be associated with a geological museum. (c) The coordination of data relating to mining, water supply, oil and gas exploration, civil engineering works, etc. In some countries the survey is charged with enforcing laws relating to these operations, including the issue of licences. Most surveys compile statistical records relating to the mining industry. (d) Giving advice to the Government on all matters relating to geology and allied sciences and their application.

Some Geological Surveys organise geological work in other countries, often as part of a 'Technical Assistance Programme'. In Britain, the Home Geological Survey is combined with Overseas Geological Survey in the Institute of Geological Sciences.

Geological thermometry. Geological thermometry is a term applied to a group of techniques designed to determine the temperatures at which various geological processes took place. It is obvious that the direct method of measuring temperatures at which modern processes go on, e.g. the temperature of a modern lava, the temperature of sea water precipitating aragonite, will yield results of reasonable reliability. Similarly, for fairly simple processes in reproducible environments, experimental determinations of temperatures will give good results. However, in order to determine temperatures of subsurface events, e.g. thermal metamorphism, crystallisation of granite, deposition of ore veins, indirect techniques must be employed. Some of these techniques are as follows:

(1) The temperature at which one form of a compound inverts to another is measured, e.g. low quartz → high quartz (◊ *Silica group of minerals*).

(2) Decomposition temperatures are measured. In many cases decomposition temperatures vary with pressure, water-vapour concentration, etc.

(3) Temperature of formation of eutectic intergrowths is measured. This method suffers from the same disadvantages as (2).

(4) Many minerals contain microscopic cavities which are filled or partly filled with liquid. It must be assumed that, in the partly filled cavities, at the moment of formation the liquid completely filled the bubble. Thus, by heating the material and observing the cavities under the microscope until the liquid just fills them, a temperature of formation can be obtained. This method has yielded remarkably good results, although certain precautions are necessary.

(5) Isotope ratios. Observation has shown that the ratio of the oxygen isotopes O^{16} / O^{18} found in modern shells is a function of the temperature of the sea water in which they lived. The use of this method is limited because of post-burial changes in the shell material. However, belemnites (◊ *Mollusca*, Cephalopoda) have yielded consistent results in a few cases.

(6) Decrepitation. An extension of method (4) above involves heating the material until the bubbles burst, which can be detected by using a very sensitive microphone, and high-gain ampli-

fier, 'set up' in the furnace. The noise of the bursting bubbles can be detected and the temperature corresponding to a 'sound maximum' recorded. The method is particularly valuable for opaque ◊ ore minerals.

(7) Concentration of trace elements (◊ Geochemistry). The amount of a particular trace element which occurs in equilibrium in a mineral is often a function of temperature, e.g. Ti in magnetite, Sc in biotite.

Earl Ingerson, Crust of the Earth (ed. A. Poldervaart), Geological Society of America, Special Paper 12, 1955.

Geology. The study of the Earth as a whole, its origin, structure, composition, and history (including the development of life), and the nature of the processes which have given rise to its present state. The word was first used in 1778 in the work of Jean André de Luc (a Swiss-born scientist who lived at Windsor for much of his life as adviser to Queen Charlotte) and at much the same time in the work of the Swiss chemist, S. B. de Saussure.

Geology comprises the following branches: (1) Crystallography, mineralogy, ◊ petrology and ◊ geochemistry, which are concerned with the materials and composition of the Earth. (2) Structural geology and ◊ geophysics, which are concerned with the form and disposition of the various units of which the Earth is built. (3) ◊ Stratigraphy and historical geology. (4) ◊ Palaeontology. (5) Physical geology (including ◊ geomorphology), which involves the study of the processes affecting the Earth.

Geomorphology. The description and interpretation of land forms.

Geophysical prospecting. The study of variations in the values of the physical parameters of the crust of the Earth with the object of gaining information, for economic purposes, about sub-surface structure or the presence of mineral bodies. Methods employed involve the study of (a) ambient fields and (b) the effects of applied fields. Examples of (a) are: gravitational; magnetic; electrical (both large-scale telluric currents and localised currents due to the spontaneous potential associated with certain ore bodies); natural radioactivity; temperature gradients and heat flow. Examples of (b) are: seismic (by the application of shock waves); electrical (both by the direct application of DC or AC currents – resistivity methods – and by electro-magnetic induction); radioactivity (by excitation by a suitable source).

Techniques include aerial surveys, both land and marine surface surveys, and surveys in boreholes.

The objectives of geophysical prospecting include the search for ores of economic importance, the delineation of potential oil- and gas-bearing structures, the location and nature of underground water supplies, the estimation of reserves of natural constructional materials, the investigation of shallow structural conditions affecting civil engineering works, and the determination of the physical properties of foundation materials.

Geophysics. The section of Earth science in which is studied all the relevant physical phenomena which have a bearing on the structure, physical conditions, and evolutionary history of the Earth as a whole. Branches include the studies of: seismology – the behaviour of earthquakes, and the shock waves which they produce; the gravity field and rotation of the Earth; the magnetic field of the Earth; temperature gradients and rate of heat flow; vulcanology – the behaviour of volcanoes and their products; natural radioactivity; oceanography; meteorology and climatology.

Many of the techniques of study have common ground with those used in ◊ *geophysical prospecting*, and improvements in instrumentation and interpretation have followed from research in both fields, with mutual benefits. Present theories concerning the structure, composition, physical properties, and evolution of the Earth have arisen from a synthesis of the results of research in geophysics, geochemistry, petrology, astronomy, and cosmogeny.

Georgian. Obsolete stratigraphic stage name for the North American Lower ◊ *Cambrian* (◊ *Waucoban*).

Geosynclines. Major structural and sedimentational units of the ◊ *crust of the Earth*. They consist of elongated basins which become filled with very great thicknesses of sediment, and, because of this fact, progressive subsidence of the basin floor must have occurred. Volcanic rocks are found intercalated with sediments. The accumulated pile of sediments is subsequently strongly deformed by ◊ *orogenic* forces into a fold-mountain chain, and during this process the lower portions of the sedimentary pile may become highly metamorphosed and granite emplacement may take place. It has been suggested that the Gulf of Mexico is a modern example.

In a classification of geosynclines the type referred to above is an orthogeosyncline; the following are further types which have been recognised and named:

EUGEOSYNCLINE. One which has thick sediments with an abundance of volcanic rocks, forming some distance from the ◊ *kraton*.

MIOGEOSYNCLINE. A thinner development of sediments and no volcanic rocks, forming adjacent to the kraton.

TAPHROGEOSYNCLINE. Synonym of ◊ *rift valley*.

PARAGEOSYNCLINE. A geosyncline that lies within the kraton.

ZEUGOGEOSYNCLINE. A parageosyncline with marginal uplifts.

AUTOGEOSYNCLINE. A parageosyncline *without* marginal uplift.

The term geosyncline is also used for other types of basin which do not necessarily conform to the above definition.

Geotechnical processes. A term used in engineering geology for processes used to modify the properties of soils and incoherent rocks to make them suitable for engineering operations.

Geotumor. A regional uplift produced by the swelling of part of the ◊ *crust of the Earth* caused by local heating. Geotumors were originally suggested as an explanation for the origin of fold mountains. A magma blister is a small geotumor. It is suggested that folding takes place by ◊ *gravity gliding* of the cover down the sides of the geotumor.

Geyser (Icelandic *geysir*, 'spouter' or 'gusher'). The ejection, sometimes violent, of superheated water and steam from underground sources in active or recently active volcanic regions. A geyser consists of a central tube with ramifying branches leading from it. ◊ *Juvenile water* accumulates in this system, and, owing to the pressure-head of the column of water, superheating of the liquid takes place. The extent to which this can occur is a function of the height of the column of water. When the critical temperature is exceeded, the liquid passes into the gaseous (steam) phase, whereupon the whole mass boils vigorously and fountains into the air. Small geysers may throw water to a height of a few metres, large ones may erupt 60–70 m. Many geysers show remarkable regularity of eruption, the time between eruptions remaining constant over long periods. This is presumably a function of the time taken for water to fill the system, and the rate at which heat is being evolved from the earth.

Three examples of geyser regions are Iceland, New Zealand, and the Yellowstone National Park, Wyoming, U.S.A. Calcareous or siliceous material is often deposited by the ejected water and forms masses of ◊ *sinter*, e.g. the Pink and White terraces of Rotorua, New Zealand.

(◊ *Volatiles*; *Volcanoes and vulcanicity*; *Volcanic products*.)

Ghost stratigraphy. In certain granite masses large numbers of

◊ *xenoliths* occur which, when mapped, show a continuation of the stratigraphy and structure of the ◊ *country rocks* to a remarkable degree. In some cases the xenoliths even maintain their correct orientation. This phenomenon is known as ghost stratigraphy, and is often regarded as evidence in favour of the formation of granite by the replacement of country rock. The classic example is the Donegal Granite (described by Read and Pitcher, *Quarterly Journal of the Geological Society*, 1958).

Gilbert, Grove Karl (1843–1918). American geologist whose work in the Henry Mountains was the basis of his theory of river development and erosion of valleys, which formed the starting-point for the development of modern theories. He was the first to give an account of volcanic intrusions, which he termed 'laccolites', and their deformation of the country rock. He also introduced the term 'hanging valley'.

Gilbertite. Fluorine-rich muscovite (◊ *Micas*, Dioctahedral: *Pneumatolysis*, Greisening).

Gipping. A stratigraphic stage name for the base of the British Upper Pleistocene (◊ *Tertiary System*.)

Givetian. A stratigraphic stage name for the top of the European Middle ◊ *Devonian*.

Glabella (Arthropoda). ◊ Fig. 3.

Glacial breaching. Glacial breaching occurs when the volume of ice in a ◊ *glacier* increases to a point where the original valley system can no longer contain it and the pre-glacial watersheds are breached by the escaping ice. In many cases after the ice melts, these diffluent channels remain only as eroded cols on the valley sides; but where the rock is less resistant and ice action vigorous, new valleys and troughs may be formed which permanently modify the former drainage pattern by development of lakes or diversion of the headwaters of the pre-glacial rivers.

Glaciers and glaciation. The term 'glaciation' embraces both the processes and the results of erosion and deposition arising from the presence of an ice mass on a landscape, and is applied to these phenomena whether on a continental scale or within the confines of a valley. Ice masses originate as compacted snow in small hollows above the level at which snow melts in summer – the snow-line – so that with every snowfall each year the ice mass grows. The snow is compacted into névé and finally, when all the air is removed, into firn. Once the ice has sufficient depth and can exert enough pressure it will begin to move from its point of origin.

Three main types of glacier may be distinguished:

(1) MOUNTAIN OR VALLEY GLACIERS, whose source is the snow lying in that part of the mountain range above the snow-line. Generally they start with accumulated snow in ◊ *cirques* at valley heads which gradually coalesce to form a glacier.

(2) PIEDMONT GLACIERS, which develop when valley glaciers emerge from their confining channels and spread out over a lowland area.

(3) ICE SHEETS AND ICE CAPS, which spread over their supply areas. Greenland and Antarctica possess the only existing continental ice sheets, but smaller ice caps (jökulls) cover considerable areas of Iceland and Spitzbergen.

Glaciers move at rates varying between a few centimetres and some tens of metres a day, the speed being affected both by the nature of the ice and the retarding effects of the ice-rock interface at the sides and base of the glacier. This also determines the erosional power of the glacier; thus a continental ice sheet will have relatively little effect on the landscape by comparison with a glacier moving in a steeply descending valley. Constriction by more resistant rock valley sides can cause pressure ridges, while where the valley opens out or steepens crevasses result. (A particular type of crevasse occurs in the cirque – the ◊ *bergschrund*.) Glaciers terminate where the rate of loss of ice by melting and ablation is equal to the forward advance of the glacier.

Ice erodes mainly by abrasion of rock surfaces and plucking away of larger blocks where the bed rock is well-jointed – both processes being reinforced by the load of debris carried. The results of glacial erosion can be seen in striated rock surfaces; the characteristic U-shaped valleys of glaciated highlands with their cirques, ◊ *hanging valleys*, truncated spurs, ◊ *fiords*, and rock bars separating deep basins later filled with water or alluvium; and the ◊ *crag and tail* and ◊ *roche moutonnée* features of glaciated lowland areas.

Deposition by a glacier or ice sheet is the shedding of the load (derived from erosion) of rock material which the glacier carries. There are two types: (a) direct deposition of debris beneath the ice or at an ice front at the lower limit of ice advance, which may result in forms such as ◊ *drift*, ◊ *drumlins*, ◊ *erratics*; (b) fluvioglacial deposition by streams flowing within the glacier, and by melt-waters during times of advance and retreat. Surface (supraglacial) streams and those flowing within the glacier (englacial) leave few traces after the ice has melted, but subglacial streams are more active and leave behind layers of sediment sometimes in channels cut into the drift or bedrock. Outwash plains of drift spread out beyond the terminal or end-◊*moraines*, washed there by streams flowing through the moraine. As a result there is a gradation of particle size decreasing with distance from the moraine. Where streams emerge from a glacier at a high level, a steep-sided alluvial cone (kame) is formed against the ice. A kame terrace may be formed in this way in a valley on either side of the glacier. (◊ *Esker*.) Kettleholes are left where masses of stagnant ice remain, having become detached as the ice retreated. Eventually this ice melts, leaving depressions in the drift which covered them. If these detached blocks had crevasses within them, these would be filled by drift, thus leaving behind short ridges, or 'crevasse infillings', when the ice melted.

With the recession of continental ice sheets, lakes frequently form as temporary features. Into these lakes is deposited material derived from seasonal meltwaters. This produces layered deposits known as ◊ *varves* – one varve being equivalent to one season's melting.

The term periglacial is used of the areas on the margins of glaciers, past and present, where certain processes and features are likely to be found, for example, ◊ *nivation*, ◊ *solifluxion*, ◊ *loess*, ◊ *ventifacts*.

Glance. A miner's term for minerals possessing a high reflectivity, often specular, e.g. lead glance (◊ *galena*) and iron glance (◊ *haematite*, 'specularite').

Glass, Glassy. The word is used in geology specifically for rocks or parts of rocks which do not consist of discrete crystalline units and are wholly without crystalline structure. Igneous glasses such as obsidian (◊ *Rhyolite*) and ◊ *tachylite* result from rapid cooling of molten material. Glassy ◊ *meteorites* are referred to as tektites. A few examples of rocks fused to a glass during intense faulting or shearing are known, the material being called pseudo-tachylite or buchite. Complete melting of sediments by thermal metamorphism is rare, but the natural silica glass lechatelierite has been recorded.

Glauconite. ◊ *Micas* (Dioctahedral) and Appendix.

Glauconitic sandstone. ◊ *Arenaceous rocks*.

Glaucophane. A mineral of the ◊ *amphibole* group, $Na_2(Mg,Fe'')_3$ $(Al,Fe''')_2Si_8O_{22}(OH,F)_2$, found in metamorphic rocks (e.g. glaucophane schists) and in igneous rocks, particularly soda-rich granites. (◊ Appendix.)

209

Glei. A waterlogged ◊ *soil* or soil-horizon in which reduction may take place.

Glenmuirite. ◊ *Teschenite.*

Glimmerite (Micaite). An ◊ *ultrabasic rock* consisting almost entirely of the ◊ *micas* phlogopite and/or biotite.

Globularites. ◊ *Devitrification.*

Glomeroporphyritic. A ◊ *porphyritic* texture in which groups of crystals (either all of the same mineral or of different minerals) are scattered through a finer-grained ground mass. Cumulophyric is a synonym.

Gneiss. A term applied to banded rocks formed during high-grade regional metamorphism. Included under this heading are a number of rock types having different origins. Gneissose banding consists of the more-or-less regular alternation of schistose (◊ *Schist*) and granulose (◊ *Granulite*) bands. The schistose layers consist of ◊ *micas* and/or ◊ *amphiboles*; ◊ *pyroxene* is a rather rarer constituent. The granulose bands are essentially quartzofeldspathic and may or may not show a preferred orientation. Gneisses are generally fairly coarse-grained rocks. The thickness of the bands may vary from a millimetre or so up to several centimetres, and may or may not be consistent. In some gneisses the quartzo-feldspathic bands locally develop large clots or 'eyes' of very coarse crystals; these structures are called 'augen' (German, 'eyes') and the rocks, 'augen-gneisses'. Various types of gneiss are recognised according to the presumed original rock or the mode of formation, e.g. paragneiss (from a sedimentary parent), orthogneiss (from an igneous parent), ◊ *lit-par-lit* gneiss, injection gneiss and segregation gneiss, and ◊ *migmatitic* gneiss. The term 'gneissose' is occasionally applied to granites in which the biotite crystals, although dispersed, are aligned. All gradations between this and a normal banded gneiss can be observed (◊ *Granitisation*). However, some gneissose granites would be more accurately described as flow-banded granites (◊ *Flow structure*). (For rocks consisting mainly of hornblende which have a gneissose character, ◊ *Hornblende schist*.)

The term gneiss is occasionally used in a stratigraphic sense e.g. specifically as in 'Lewisian gneiss', or, more generally, as in the term 'Basement gneiss'.

Goethite. ◊ *Limonite.*

Gold. A rare native metal, Au, found in ◊ *hydrothermal* veins, ◊ *placer deposits*, and ◊ *banket*. (◊ Appendix.)

Gondwanaland. The name given to the hypothetical southern hemisphere 'super-continent' consisting of South America, Africa, Madagascar, India, Arabia, Malaya and the East Indies, New Guinea, Australia andAntarctica, prior to its break-up under the forces causing ◊ *continental drift*. It corresponds to ◊ *Laurasia* in the northern hemisphere. (◊ *Pangaea*.)

Gondwana Series. Stratigraphic name for the Indian equivalent of the ◊ *Karroo* Series.

Goniatites. ◊ *Mollusca* (Cephalopoda).

Goniometer. A device for measuring ◊ *interfacial angles* of crystals. Two main types are in use: (1) The contact goniometer, for use with large crystals. It consists of a protractor with a pivoted arm which is placed in contact with crystal faces. Results are, in general, of a low order of accuracy. (2) The optical goniometer, suitable for small crystals with brilliant reflecting faces. Various versions of this type exist, all depending upon the ability of a crystal to reflect a beam of light which is directed on to it from a collimator. The reflection is located by means of an observing telescope. The crystal is rotated from one reflecting position to the next and the angle of rotation measured. Optical goniometers capable of a very high level of accuracy and precision are available.

Gorge. Any deep and narrow vertically sided valley. A ◊ *canyon*.

Gossan. The leached and oxidised near-surface part of a vein containing sulphides, especially iron-bearing ones, e.g. ◊ *pyrite*, ◊ *chalcopyrite*, ◊ *bornite* etc. The gossan (or 'iron hat') consists essentially of a mass of hydrated iron oxides (◊ *limonite*) from which copper (and any other metals) and sulphur have been removed by downward percolating waters. This process occasionally concentrates highly insoluble elements, such as gold, in the gossan. The characters of the cellular mass of 'limonite' (◊ *boxwork*) formed may give a clue to the nature of the underlying sulphide minerals. (◊ *Secondary enrichment*.)

Gothlandian. A stratigraphic name (from the Island of Gothland, Sweden), a synonym of ◊ *Silurian*, formerly used by Continental authors. It is now obsolete.

Gouge. ◊ *Rock flour*.

Graben. A downthrown block between two parallel faults. (◊ *Rift valley*.)

Grade, Grading, Graded bedding. (1) Of a river. A river or stream may be said to be 'graded' or 'at grade' when it is transporting all its load. Grade implies the idea of a balance, on average, between

211

erosion and deposition, or between transporting power and amount of load. Hence a graded stream may be regarded as downcutting at an infinitesimally slow rate and, taken over a short term (tens of years), is neither downcutting nor infilling its channel. The term 'grade' is also applied to the ◊ *long-profile* of a river when it is smooth and without major irregularities, e.g. waterfalls, rapids, and lake basins. Indeed a river cannot be graded if these irregularities are present, since they themselves lead to rapid downcutting, or deposition. However, it is possible for a river to be graded without its profile being a continuous smooth curve.

(2) Of ore. The quality of an ore; in effect, the metal content (cf. ◊ *Tenor*).

(3) Graded sediment. A sedimentary unit which displays a sorting effect with the 'coarsest' material at the base and the 'finest' material at the top. 'Coarse' and 'fine' in this sense may be fine sand grading up into clay, or large pebbles grading up into coarse sand. (Geologists use 'graded sediment' to mean a well-sorted sediment, i.e. particles all more or less of one size. Civil engineers use the term 'graded aggregate' to mean a mass of fragments exhibiting a wide range of particle size, which the geologist would describe as ungraded or poorly graded.) The phenomenon can be used as a method of determining the orientation (◊ *Way-up criteria*). (◊ *Turbidite*.)

(4) Metamorphic grade. The level which a metamorphic process has reached. (◊ *Metamorphism*; *Facies, metamorphic*; *Zone, metamorphic*.)

(5) ◊ *Particle size*.

Grain size. ◊ *Particle size*; *Igneous rocks*.

Granite. A coarse-grained igneous rock consisting essentially of quartz (20–40%), alkali ◊ *feldspar*, and very commonly a ◊ *mica*, biotite and/or muscovite. A number of accessory minerals may be present, including ◊ *apatite*, ◊ *zircon*, and ◊ *magnetite*. Granites which have been altered by hydrothermal or pneumatolytic action may contain a variety of minerals, e.g. ◊ *tourmaline*, ◊ *topaz*, ◊ *kaolin*, etc. (◊ *Hydrothermal processes*; *Pneumatolysis*). Granites may also contain minerals which are regarded as 'contaminants', derived by assimilation of thermally metamorphosed country rock, e.g. ◊ *garnet*, ◊ *cordierite*, ◊ *andalusite*. Small amounts of calcium-bearing plagioclase (◊ *Feldspars*) may occur in forms transitional to ◊ *granodiorite* and ◊ *adamellite*. Soda-rich granites may contain ◊ *aegirine* (rockallite) and/or ◊ *riebeckite*.

Granites characteristically contain a high proportion of silica, often more than 70%, and relatively high soda and potash. ($Na_2O + K_2O$ ranges from 5 to 12%.) The MgO is usually less than 1%, and calcium, except in varieties grading toward adamellite and granodiorite, is also low. The ratio of potassium to sodium is used in classification (◊ *Alkaline*). Granites containing one feldspar are said to be hypersolvus granites, and those with two feldspars subsolvus. The ◊ *colour index* of granites is invariably low. Granites consisting only of quartz and alkali feldspar are termed alaskites.

By decreasing content of free quartz, granites pass into quartz syenites and thence to ◊ *syenites*.

Adamellite is a variety of granite containing a calcium-bearing plagioclase, usually oligoclase, and a potassium feldspar, in roughly equal amounts. Other minerals usually present are biotite and hornblende, with sphene a common accessory. Adamellites pass by decreasing quartz content into quartz monzonites and thence to ◊ *monzonites*.

Granites occur exclusively as intrusive bodies, and may occur in almost any form, e.g. ◊ *dykes*, ◊ *sills*, ◊ *plugs*, ◊ *bosses*, ◊ *ring complexes*, ◊ *cauldron subsidences*, etc. The largest masses of granite are however ◊ *batholiths*. It is fairly well established that granites occurring in different ways have different origins (◊ *Granite series*).

Granite is the plutonic equivalent of the volcanic ◊ *rhyolite*, while adamellite is the plutonic equivalent of the volcanic rhyodacite (toscanite).

(◊ *Microgranite*; *Granitisation*; *Granodiorite*.)

Granite porphyry. ◊ *Microgranite*.

Granite series. A scheme suggested by Read in 1949 to present the relationships in time and space of the various types of ◊ *granite*. Four main groups were recognised, as follows:

1. Early deep-seated granites (◊ *autochthonous* granites), formed by *anatexis*, associated with migmatites and fairly high-grade regionally metamorphosed rocks. They are commonly syntectonic (◊ *Synkinematic*).

2. Slightly later and formed at a slightly higher crustal level the para-autochthonous granites, which have moved upwards only a short distance from their place of origin. They are associated with regionally metamorphosed rocks and often have a surrounding injection complex (contact ◊ *migmatites*).

3. At a higher level still and later in time are found the forcefully

intruded magmatic granites. These commonly deform the ◊ *country rock* and develop a well-marked ◊ *metamorphic aureole*. They are generally late tectonic in age.

4. Finally, at the highest crustal levels, the mechanically emplaced granites (◊ *Pluton*) are found. These are described as having a 'permitted ◊ *emplacement*'. They include some ◊ *cauldron subsidences* and small ◊ *bosses* with negligible metamorphic aureoles, and are commonly associated with ◊ *acid* or ◊ *intermediate* ◊ *volcanic rocks*. This type is generally post-tectonic.

It is essential to realise that the concept of the granite series relates to the occurrence of granites in orogenic belts; the granite masses associated with volcanism (e.g. Arran and Skye, Scotland, in the Tertiary Volcanic District) are not included in the scheme. Also inherent in the idea is a belief in the concept of ◊ *granitisation*.

H. H. Read, *The Granite Controversy*, 1957.

Granitic texture. An irregular, granular texture as found in non-porphyritic granites.

Granitisation. The name given to a group of theories which have as a common basis the concept that pre-existing rocks can be transformed to ◊ *granite* or granite ◊ *magma* by the action of energetic fluids of appropriate composition arising from the depths (◊ *Metasomatism*). The precise source of these fluids and their exact mode of operation are matters of dispute. Some geologists believe that they arise from the ◊ *sial*, others that they represent a volatile, alkali-rich effusion from the ◊ *sima*. Various explanations of the way in which the fluids (ichors, emanations) operate range from the 'dry' diffusion of appropriate ions through the atomic lattice of existing minerals to a 'wet' reaction of some kind, in which Na and K are introduced into the system, forcing out Ca and Mg – the so-called 'basic front'. (Cf. ◊ *Anatexis*, ◊ *Syntexis*. ◊ *Granite series*.)

Granoblastic. A term originally applied to metamorphic rocks which are equigranular. The term has also been applied to ◊ *diagenetic* processes which give rise to an equigranular texture by recrystallisation. This extension of the usage seems undesirable.

Granodiorite. A coarse-grained acid igneous rock consisting of quartz (20–40%), calc-alkali feldspar, and various ◊ *ferromagnesian minerals*, dominantly ◊ *hornblende* and ◊ *biotite*. Small amounts of alkali feldspar may also occur, while ◊ *sphene*, ◊ *apatite*, and ◊ *magnetite* are the most important accessory minerals. Granodiorites differ from ◊ *granites* in having a lower percentage of silica, and a higher calcium and magnesium content. The textures are essentially the same as those of granites, except that ◊ *graphic texture* does not seem to occur.

By decreasing the amount of quartz, granodiorites pass into quartz diorite (tonalite), and thence to ◊ *diorite*. It should be noted that many of the rock names in the acid igneous rock group are loosely used, and many so-called tonalites are granodiorites, and almost all 'hornblende-biotite-granites' are granodiorites. Trondhjemite may be defined as a potash feldspar-free granodiorite, usually very low in ferromagnesian minerals.

Many of the great ◊ *batholiths* described as granitic are really granodioritic, and a large proportion of granite bodies contain some granodiorite. Granodiorites are the plutonic equivalents of the volcanic dacites (◊ *Rhyolites*). (For microgranodiorite, ◊ *Microgranite*.)

Granophyre. A ◊ *microgranite* which displays granophyric texture (◊ *Graphic texture*).

Granophyric texture. Small-scale ◊ *graphic texture*.

Granular texture. The texture of a rock consisting of mineral grains of approximately equal size.

Granule. A rock fragment with a diameter of between 2 mm. and 4 mm., i.e. larger than a sand grain and smaller than a pebble. (◊ *Particle size*.)

Granulite. A metamorphic rock of regional metamorphic origin, having a ◊ *granular texture*. It usually consists of ◊ *feldspars*, ◊ *pyroxenes*, and ◊ *garnets*; streaked or banded varieties are called leptites (cf. ◊ *Hornfels*). Granulites do not always show an obvious ◊ *fabric* as does a ◊ *schist* or ◊ *gneiss*, but detailed examination may show the existence of a preferred orientation in the constituent minerals.

Granulitic texture. A texture found in certain ◊ *dolerites* or ◊ *basalts*; an alternative to an ◊ *ophitic* texture. It consists of granular crystals of augite and/or olivine between laths of plagioclase. 'Intergranular texture' has much the same meaning.

Graphic texture. In certain ◊ *pegmatites*, ◊ *granites*, and ◊ *microgranites* – and more rarely in other rocks – there occurs an intergrowth of quartz and alkali feldspar which presents on flat surfaces the appearance of Hebrew, cuneiform, runic, or hieroglyphic writing. This results from the simultaneous crystallisation of the two minerals, in many cases (but not always) from a eutectic mixture. Small-scale graphic texture is called micrographic or granophyric texture. (Cf. ◊ *Perthite*; ◊ *Symplektite*.)

Graphite. A soft black form of native carbon, C (hexagonal), found in metamorphic rocks, crystalline limestones, igneous rocks, veins and pegmatites. (◊ Appendix.)

Graptolite, Graptoloidea. ◊ *Chordata.*

Grauwacke (Greywacke). ◊ *Arenaceous rocks.*

Gravel. This term is used in geology with a precise meaning of a grain size, and not implying loosely compacted coarse sediment. In general, the term is applied to grains larger than coarse sand and finer than pebbles, i.e. 2 mm. to 4 mm. The term, however, is also used stratigraphically and by civil engineers in a much looser sense, e.g. river gravels and glacial gravels which may range from medium sand to boulders. (◊ *Particle size.*)

Gravity collapse structure. ◊ *Fold* and Fig. 76.

Gravity (gliding) tectonics. The operation of gravitational forces to produce faulting, thrusting, and folding by the movement of the cover down the slope of an uplifted region. (◊ *Décollement.*)

Gravity transport. Rock material may move under the influence of gravity either as a movement of weathering products down a slope, or as a mass movement of rock along joint planes, bedding planes, etc.

(1) MOVEMENT OF WEATHERING PRODUCTS. Two separate categories are recognised, according to the speed of movement, material involved, and the role of water in the process: (a) Slow movements (creep), e.g. soil creep, talus creep. Slow downhill movement of the soil or talus cover resulting in the displacement of objects supported by the soil. The process is accelerated by an access of water, and in sub-arctic regions quite rapid soil movement may take place when the ground water, frozen during the winter, melts in the spring. The term solifluction (solifluxion) is commonly used for this process. (b) Rapid movements (◊ *flows*), e.g. mud flows, earth flows, soil flows, rock avalanche. These movements generally take place suddenly, and involve a large quantity of water. The high ratio of water to soil or clay particles causes the material to behave as a liquid, and earth flows, mud flows, and soil flows result. Such flows are highly efficient eroding agents and may transport large boulders considerable distances. Lahars are mud flows resulting from torrential downpours of rain, mass melting of snow and ice, or the breaching of a crater lake acting upon loose ◊ *pyroclastic* material on the flanks of a volcanic cone. The term is also applied to the deposits resulting from such flows. A rock avalanche is the result of the de-stabilising

of a mass of rock fragments, e.g. a scree slope, generally as a result of a sudden excess of water.

(2) MASS MOVEMENTS. In this type of movement sliding takes place on a definite plane which may be a structural plane, e.g. bedding, joints, schistosity, etc., or a curved shear plane, giving rise to slumps such as are found in clays. The first type is referred to as a landslide or landslip and the most common form involves movement along a lubricated bedding plane, often at the interface of permeable and impermeable rock types. Slumping in clays involves a rotary movement on a curved shear surface. (◊ Fig. 80.)

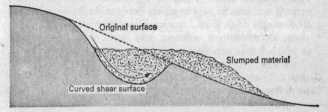

80. Gravity transport (slumping).

Greenough, George Bellas (1778–1855). English geologist, pupil of ◊ *Werner*, later an M.P., who founded the Geological Society of London. He also produced a geological map of England and Wales shortly after William ◊ *Smith*.

Greensand. (1) ◊ *Arenaceous rocks*.

(2) A lithostratigraphic unit of the ◊ *Cretaceous* in Britain and north-west Europe.

Greenstone. A field term for any slightly altered basic igneous rock, e.g. ◊ *diabase* (English sense), ◊ *epidiorite*, etc.

Greisen, Greisening. ◊ *Pneumatolysis*.

Grenville. A stratigraphic division of the Canadian ◊ *Precambrian* Lower ◊ *Ontarian*.

Greywacke (Grauwacke). ◊ *Arenaceous rocks*.

Grike (Gryke). A cleft in the bare limestone pavements of ◊ *karst scenery*.

Grit. A term applied to ◊ *arenaceous rocks* in which the particle shape is angular to sub-angular. The term is also used by quarry-men, in England, for a ◊ *limestone* containing abundant shell fragments. Stratigraphic usage of the term 'grit' does not necessarily conform to either of these definitions (e.g. Millstone Grit).

Groove (Mollusca), Fig. 105.

Grooves. ◊ *Sedimentary structures.*

Grorudite. A soda-rich variety of aegirine ◊ *microgranite.*

Grossularite. A mineral of the ◊ *garnet* group.

Groundmass. The finer-grained material constituting the main body of a rock, in which is set larger units (crystals, pebbles, etc.). In ◊ *porphyritic* texture, for example, the larger ◊ *phenocrysts* are set in a fine-grained groundmass.

Ground water. Water occupying openings, cavities, and spaces in rocks. There are two main sources of such water: ◊ *juvenile* water, which rises from a deep, magmatic source, and ◊ *meteoric water*, which is due to rainfall having soaked into the underlying rock. Water may be held in spaces between the grains of a rock (◊ *Porosity*) or in joints, cleavage, bedding planes, etc. (◊ *Pervious*). Occasionally a sediment may retain some of the water in which it was deposited, and this is called ◊ *connate water.* (◊◊ *Springs*; *Artesian structure.*)

Group. ◊ *Stratigraphic nomenclature.*

Growth line. (Brachiopoda) ◊ Fig. 11; (Mollusca) ◊ Fig. 101.

Grunerite. A monoclinic mineral of the ◊ *amphibole* group.

Gryke. ◊ *Grike.*

Guadalupian. A stratigraphic stage name for the North American Middle ◊ *Permian.*

Guano. ◊ *Phosphatic deposits.*

Guard (Mollusca). ◊ Fig. 106.

Guettard, Jean Étienne (1715–86). French geologist who was one of the first to study the total distribution of rocks, minerals, and fossils, and to attempt to represent them on maps and determine laws to account for their distribution. His propositions that the same strata were to be found in Brittany and south-west England, that fossils were past marine creatures, and that the Auvergne had once been a volcanic region were received with considerable amazement and disbelief by his contemporaries. But when he suggested that the earth must be much older than the Bible stated for landscape changes to have been caused by normal forces, and that some fossil forms were of extinct creatures, he was forced to recant his views publicly by his theological compatriots.

Gulfian. A stratigraphic stage name for the North American Upper ◊ *Cretaceous.*

Gull. A tension gash associated with cambering (◊ *Fold*, Fig. 75). The fissure tapers downwards and is filled with material from above.

Gum copal. ◊ *Hydrocarbon minerals.*

Gumbo. A special type of ◊ *soil* which yields a characteristic sticky mud when wet. The name is in use in the western and south-western United States, where it is also used for any 'sticky' clay, whether a soil or not.

Gumbotil. A clay-rich glacial deposit or leached glacial deposit which has ◊ *gumbo* properties.

Gutenburg Discontinuity. ◊ *Discontinuity.*

Guyot. A submarine table (i.e. flat-topped) mountain. They occur in the deep oceans, especially the Pacific, and may perhaps be the foundations of coral atolls. (◊ *Reef.*)

Gypsum. An ◊ *evaporite* mineral, $CaSO_4.2H_2O$, found in clays and limestones; sometimes associated with sulphur. (◊ Appendix.)

Gzhelian. A stratigraphic stage name for the Russian lower and mid ◊ *Stephanian* (Carboniferous System).

Habit. A term referring to the relative development of individual ◊ *faces* and ◊ *forms* in a crystal, e.g. in Fig. 81 crystal A has a maximum development of prisms, and displays prismatic habit; crystal B has a maximum development of pyramids and displays pyramidal habit; C has the basal plane developed giving a tabu-

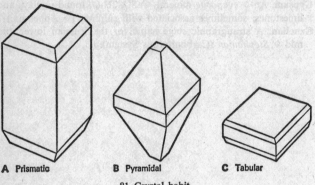

A Prismatic   **B** Pyramidal   **C** Tabular

81. Crystal habit.

lar habit. All three crystals develop the same faces, but to different extents.

If a large number of crystals of a particular substance are examined, certain habits will be found to be very common, e.g. in quartz the prism plus rhombohedra is predominant. Some habits, however, are exceedingly rare, e.g. in quartz the basal pinacoid is very uncommon, while the trigonal trapezohedra are generally rare. It can be shown that the relative abundance of a face is a function of the number of atoms in the lattice (◊ *Unit cell*) which lie in planes parallel to the face in question; thus, in quartz, the prism and the rhombohedron directions contain numerous atoms, while the basal pinacoid direction passes through relatively few atoms.

Habitat. The ◊ *environment* in which a plant or animal lives. The term is generally applied to the environment to which the organism is most closely adapted and hence in which it is most successful.

Hade. The angle which a ◊ *fault* plane makes with the vertical plane.

Hadrynian. A Canadian ◊ *Precambrian* Era.

Haematite. An important ore mineral of iron, Fe_2O_3, found as an accessory in igneous rocks, in ◊ *hydrothermal* veins and replacements, and in sediments. (◊ Appendix.)

Halite. An ◊ *evaporite* mineral, NaCl, common salt. (◊ Appendix.)

Hall, Sir James (1761–1832). Scottish baronet often thought of as the founder of experimental geology. A friend of ◊ *Hutton*'s, he provided proofs of Hutton's theory of the igneous origin of rocks by his experiments with differential heating and cooling of glass, and later of natural rocks and minerals.

Hall, James (1811–98). American geologist, for some time head of the New York State Geological Survey, who spent 62 years establishing a record of all the fossils below the Coal Measures in the whole of the U.S.A. This work also gave him the evidence for his theory of the sedimentary origin of the Appalachians, which evolved into a generally applicable theory of mountain-building based upon the elevation of geosynclines which is fundamental to modern orogenic ideas.

Hälleflinta. A metamorphosed acid volcanic or pyroclastic rock, with characteristic 'flinty' or 'horny' appearance, sometimes banded or ◊ *porphyroblastic*. A kind of ◊ *hornfels*.

Hanging valley. A valley formerly occupied by a tributary valley ◊ *glacier*, which, because of its small size, failed to deepen the originally accordant tributary valley at the same rate as the main glacier deepened the main valley. After deglaciation these tributary valleys were left high above the main valley floor, and any drainage they now possess reaches the main valley stream by means of a waterfall. It is possible, but rare, for differential erosion by a river and its tributaries to give rise to a hanging valley without glacial intervention.

Hanging wall. ◊ *Fault* and Fig. 52.

Hardness. A property of minerals which is determined by reference to an empirical scale of standard minerals. Mineral hardness is a 'scratch' (scelerometric) hardness, as opposed to the engineers' indentation hardness. The scale in general use by mineralogists is due to Mohs, and is as follows:

Mineral	Common equivalents
10. Diamond	
9. Corundum	
8. Topaz	Hard file
7. Quartz	Penknife
6. Orthoclase	Window glass
5. Apatite	Teeth, copper coin, brass pin
4. Fluorite	
3. Calcite	
2. Gypsum	Fingernail
1. Talc	

The steps between various minerals are by no means equal, e.g. diamond is about ten times as hard as corundum, whereas corundum is only about 10% harder than topaz. It should also be appreciated that hardness of non-cubic minerals varies with direction. This is rarely of sufficient magnitude to be detected by the rather crude tests in general use, but one or two minerals, e.g. kyanite and topaz, have widely differing hardnesses (kyanite has a hardness of 5 parallel to its length, and 7 across its length). It is doubtful if measurement of hardness defines a single property; it probably involves an integration of several properties, which it is almost impossible to measure separately. Despite this, it is one of the most important tests available to the mineralogist.

Hard-pan. A layer of strongly cemented material occurring in unconsolidated sediments, often found a short distance below the surface. Such layers are usually formed as a result of ◊ *groundwater* action, and one consequence of their development is to limit the downward percolation of rainwater. The cementing material may be calcareous, siliceous, or (as in iron pan) ferruginous. Duricrust is a more or less synonymous term.

Harker diagram. A type of ◊ *variation diagram*.

Harzburgite. An ◊ *ultrabasic rock* composed of ◊ *olivine* and ortho◊*pyroxene*.

Hatchettine. A ◊ *hydrocarbon mineral*.

Hausmannite. ◊ *Spinel group of minerals*.

Hauterivian. A stratigraphic stage name for the European mid Lower ◊ *Cretaceous*.

Haüy, René-Just (1635–1703). French physicist and mineralogist who was the first to make a detailed study of the crystallographic forms of minerals on the basis of geometric laws. (This work originated from the happy accident of dropping some calcite crystals which he was examining.) The mineral haüyne (◊ *Feldspathoids*) was so named in his honour.

Haüyne. ◊ *Feldspathoids.*

Haüynophyre. ◊ *Trachyte.*

Hawaiite. ◊ *Andesite.*

Head. A synonym of ◊ *combe rock.*

Heat flow. The rate at which heat is lost from the inner zones of the Earth to the atmosphere. The average value is $1·25 \times 10^{-6}$ calories per cm.^{2} per second, but higher values have been recorded in certain localities (especially in some areas of the ocean basins), which it is suggested may be due to uprising convection currents in the ◊ *mantle.*

Heat gradient. The increase in temperature which is observed as one descends to deeper levels in the Earth's ◊ *crust.* The average gradient is 1°C. per 30–35 m. of depth but in volcanic areas gradients as high as 1°C. per 10 m. have been recorded, and in certain other areas, notably the Witwatersrand goldfields, South Africa, gradients as low as 1°C. per 100 m. are encountered.

Heave. The measure of the horizontal displacement between the upthrown and downthrown sides of a ◊ *fault.* (◊ Fig. 52.)

Heavy liquid. A dense liquid used in separating minerals of differing densities. The two commonest are bromoform, $CHBr_3$ (density 2·9) and methylene iodide, CH_2I_2 (density 3·32). Acetylene tetrabromide ($C_2H_2Br_4$) has almost exactly the same density as bromoform and some advantages in use. Clerici's solution (density 4), a saturated solution of thallium malonate plus formate, and Thoulet's liquid (density 3·18), a saturated solution of mercury potassium iodide, are also used. For very dense minerals, melts of mercurous nitrate (density 4·3) and thallium silver nitrate (density 4·6) are employed, but are unpleasant to use (they are poisonous) and attack some minerals, notably sulphides.

Heavy mineral, Heavy mineral separation. By use of a ◊ *heavy liquid* a mixture of mineral grains of different densities may be separated into two fractions: those which sink, the 'heavy minerals', and those which float, the 'light minerals'. By manipulating the density of the liquid, further separation of the fractions may be achieved. The term 'heavy mineral' is commonly applied to minerals which sink in bromoform (density 2·9). The process of heavy mineral separation is of major importance in the study of arenaceous rocks, since it enables a small proportion of dense minerals, e.g. tourmaline (◊ *Cyclosilicates*), ◊ *zircon*, ◊ *hornblende*, ◊ *sphene*, etc., to be separated from a great bulk of low-density material, mainly quartz. The process is also used for separating pure samples of minerals from crushed igneous and metamorphic rocks for further investigation.

Heavy spar. Common name for the mineral ◊ *barytes*, $BaSO_4$.

Hedenbergite. ◊ *Pyroxenes*.

Helderbergian. A stratigraphic stage name for the base of the North American Lower ◊ *Devonian*.

Helicitic. A textural term originally app ed to S-shaped or reversed-S(?)-shaped trails of inclusions in ◊ *poikiloblastic* crystals, especially ◊ garnets and ◊ *staurolite*, found in regionally metamorphosed rocks. The use of the term has been extended to

82. Helicitic structure.

include any arrangement of inclusions representing a ◊ *relict* texture preserved from an earlier metamorphism. Similar S-shaped trails produced by rotation during metamorphism should not, strictly speaking, be referred to as helicitic. (◊ Fig. 82.)

Helikian. A Canadian ◊ *Precambrian* Era.

Helvetian. A stratigraphic stage name for the European Middle Miocene (◊ *Tertiary System*).

Hemera. A time interval determined by the period of maximum development of a particular plant or animal species. (◊ *Stratigraphic nomenclature.*)

Hemi- (prefix). Half.

Hemichordata. ◊ *Chordata*.

Hemicrystalline. A ◊ *texture* term applied to rocks which contain both crystals and glass.

Hemihedral. The term applied to the class of a crystal system which has reduced symmetry, so that the general form has one half the number of faces of the corresponding form in the ◊ *holohedral* class, e.g. in the ◊ *cubic system* the hemihedral general form has 24 faces, compared with 48 in the holohedral class.

Hemimorphic. A term used to describe classes of crystals which lack both a centre of symmetry and a plane of symmetry perpendicular to the c-◊*axis*. This results in sets of faces terminating the c-axis prism forms being differently developed, e.g. tourmaline.

Hemimorphite. A zinc mineral, $Zn_4Si_2O_7(OH)_2.H_2O$, of the soro◊ *silicate* group found in the oxidised zones of zinc deposits associated with ◊ *smithsonite*. (◊ Appendix.)

Hemingfordian. A stratigraphic stage name for the North American equivalents of the Lower ◊ *Vindobonian* + Upper ◊ *Burdigalian* (Tertiary System).

Hemphillian. A stratigraphic stage name for the North American equivalents of the Upper ◊ *Pannonian* + ◊ *Astian* (Tertiary System).

Hercynite. ◊ *Spinel group of minerals.*

Hermatypic. A term applied to a ◊ *reef*-building organism.

Hessonite. ◊ *Garnet.*

Hetero- (prefix). Different.

Heteropygous. A term applied to trilobites (◊ *Arthropoda*) which have a cephalon larger than the pygidium.

Hexacoralla. ◊ *Coelenterata* (Anthozoa).

Hexagonal system. Some authorities include under the term 'hexagonal' both the hexagonal (s.s.) and the ◊ *trigonal* systems. The more usual practice of separating them is here followed. The system is divided into seven classes, as in the table on the next page.

Hexagonal crystals are referred to a system of four axes, three arranged mutually at 120° in a horizontal plane with the fourth axis vertical and perpendicular to the plane. (For the distribution of positive/negative signs, ◊ *Axes, crystallographic.*) The parameters employed for the horizontal axes are all equal, but that for the vertical axis is either greater or smaller than the horizontal ones.

(For the international symbols, ◊ *Symmetry.* ◊◊ *Crystal system.*)

Hill creep. The movement of loose material down a hillside slope. (◊ *Fold.*)

Hilt's Law. The broad hypothesis that, other things being equal, deeper ◊ *coals* are of higher ◊ *rank* than those above them.

	International symbol	Centre	Plane of symmetry	Axes of symmetry rotation	Axes of rotary inversion	Example
1	6/mmm	C	7	6 ii, 1 vi	—	Beryl
2	$\bar{6}$m2	—	4	3 ii —	1 $\bar{\text{vi}}$	Benitoite (BaTiSi$_3$O$_9$)
3	6mm	—	6	— 1 vi	—	Zincite (ZnO)
4	622 (62)	—	—	6 ii, 1 vi	—	β-Quartz, Kalsilite
5	6/m	C	1	— 1 vi	—	Apatite
6	$\bar{6}$	—	1	— —	1 $\bar{\text{vi}}$	No known example
7	6	—	—	— 1 vi	—	Nepheline

Hinge. ◊ *Fold* and Fig. 60.

Hinge fault. ◊ *Fault* and Fig. 55.

Hinge line (Brachiopoda). ◊ Fig. 11.

Hinge teeth. ◊ *Mollusca* (Lamellibranchiata) and Fig. 103.

Historical geology. A term approximately synonymous with ◊ *stratigraphy*, although its implications are wider, in that it deals with the whole sequence of events which make up the history of the Earth.

Hog's back. A ◊ *cuesta* in which the dip slope and scarp slope are both approximately 45°.

Holo- (prefix). Complete, entirely.

Holocene. ◊ *Quaternary*.

Holocrystalline. Of rocks, entirely crystalline. (◊ *Texture*.)

Holohedral. The class of a crystal system in which the maximum symmetry is displayed, and the general ◊ *form* has the maximum number of faces possible, e.g. 48 in the ◊ *cubic system*.

Holohyaline. Of rocks, entirely of glass. (◊ *Texture*.)

Holotheca (Coelenterata). ◊ Fig. 23.

Holothuroidea. ◊ *Echinodermata*.

Holotype. The single ◊ *type* specimen of a species, selected in order to show its main characters.

Holsteinian. A stratigraphic stage name for the top of the British Middle Pleistocene (◊ *Tertiary System*).

Homeochilidium (Brachiopoda). ◊ Fig. 12.

Homeodeltidium (Brachiopoda). ◊ Fig. 12.

Homeomorphy (Homoeomorphy). Morphological similarities shown by genetically distinct members of the same phylum. It is most likely to occur in phyla which have a limited number of structural elements (e.g. corals). (Cf. ◊ *Homomorphy*.)

Homo- (prefix). Same.

Homomorphy. Superficial similarities in the morphology of members of different phyla, e.g. the Rudistid lamellibranchs (◊ *Mollusca*) and corals. (Cf. ◊ *Homeomorphy*.)

Homonym. If two distinct organisms or groups of organisms are given the same name then the name becomes a homonym. The more recent of the two names is referred to as the junior homonym and is usually invalid. (Cf. ◊ *Synonym*.)

Hooke, Robert (1635–1703). A very inventive and versatile British scientist who was one of the first to work with a microscope. His most important geological contribution was concerned with the marine origin of 'figured stones' – fossils – and their significance in establishing the past history of the earth. He was also one of the first to suggest that denudation together with transport and cementation of the eroded material could be an important element combined with uplift (in the form of earthquakes and volcanoes) in governing the nature of the Earth's surface.

Hopkins, William (1793–1866). English physicist and mathematician who tried to prove by mathematical calculation that the Earth was a solid rigid mass and that all structures including valleys were the result of crustal uplift. He suggested the term 'physical geology' for the study of this relationship.

Horizon. A time-plane recognisable in rocks by some characteristic feature such as flora, fauna, or lithology.

Horn. A feature formed by the coalescence of two or more ◊ *cirques*.

Hornblende. A mineral of the ◊ *amphibole* group, $NaCa_2(Mg,Fe'')_4(Al,Fe''')(Si,Al)_kO_{22}(OH,F)_2$, widespread in metamorphic and igneous rocks. (◊ Appendix.)

Hornblende schist. A rock consisting essentially of orientated crystals of ◊ *hornblende*, giving rise to a linear schistosity (◊ *Schist*). Other minerals which may occur include ◊ *plagioclase*, ◊ *quartz*, ◊ *epidote*, calcium ◊ *garnet*, ◊ *biotite*, ◊ *calcite*, and ◊ *sphene*. The term amphibolite is synonymous with hornblende schist, but has been used loosely in the past to include rocks which are more accurately described as ◊ *epidiorites*. Hornblende schists appear to be derived mainly from ◊ *basic* (and possibly ◊ *ultrabasic*) igneous rocks, but in a few cases highly impure argillaceous ◊ *dolomites* may have been the parent rocks.

From the petrofabric point of view there are two types: (1) Those in which the c-axes of hornblende are more or less parallel to one another, but the prismatic cleavage is randomly orientated. (2) Those in which both the c-axes of hornblende and the prismatic cleavage are more or less regularly orientated.

It should be especially noted that hornblende schists consist mainly of hornblende, but hornblende gneisses are in general ordinary quartzo-feldspathic gneisses containing hornblende-rich layers.

Hornblendite. A ◊ *perknite*. (◊ *Ultrabasic rocks*.)

Hornfels. A medium or fine-grained granulose rock produced by thermal metamorphism. It shows no cleavage, schistosity, or parallel alignment of minerals due to metamorphism, although relict ◊ *fabrics* (either depositional or compressional) may persist. Material which is totally enclosed in an igneous rock as a ◊ *xenolith* may be partially or wholly fused, and is then called a buchite (◊ *Mylonite*). Hornfelses are described by prefixing the names of significant minerals or mineral groups, e.g. garnet-hornfels, pyroxene-hornfels, ◊ *calc-silicate hornfels*. The term is occasionally used adjectivally, e.g. hornfelsed slate. (◊ *Metamorphic aureole*.)

Hornitos. Very small lava flows (◊ *Volcano*). (Cf. ◊ *Driblet cone*.)

Hornstone. A very fine-grained volcanic ash (◊ *Pyroclastic rocks*). Perhaps the name is derived from honestone, referring to the use of these rocks as sharpening hones; it seems unlikely that the name is derived from their appearing horny.

Horse. A large mass of ◊ *country rock* enclosed within a mineral vein. Horses are especially common in veins occupying fault fissures. (A miners' term.)

Horst. An upthrown area between two parallel ◊ *faults*. (◊ Fig. 58.)

Hortonolite. ◊ *Olivines*.

Hoxnian. A stratigraphic stage name for the top of the British Middle Pleistocene (the interglacial between Lowestoft and Gipping Tills) (◊ *Tertiary System*).

Hudsonian. A Canadian ◊ *Precambrian* orogeny.

Humic coals. ◊ *Coal*.

Humite minerals. Minerals which somewhat resemble ✧ *olivine* but which contain hydroxyl and fluorine ions. Humite is $3Mg_2SiO_4$. $Mg(OH,F)_2$; chondrodite, perhaps the commonest member of the group, is $2Mg_2SiO_4.Mg(OH,F)_2$. They occur in thermally metamorphosed limestones.

Hums. Small detached hills of limestone formed at a late stage of the development of ✧ *karst scenery*.

Humus. The organic part of ✧ *soil*.

Huronian. A stratigraphic division of the Canadian ✧ *Precambrian* Pre-✧-*Algonkian*.

Hutton, James (1726–97). Scottish geologist who originated some of the fundamental principles of modern geology. He was famed in his lifetime for his plutonic theory of the origin of the Earth, which emphasised heat as the principal agent in elevating land masses. This was in complete opposition to ✧ *Werner's* popular neptunean theory, and indeed seemed absurd to many of his contemporaries – especially such suggestions as granite having once been molten. In his theory he also described all the processes of denudation and deposition and suggested that these were sufficient explanation of the configuration of the surface if they had operated over millions of years. Disturbed by the charge of atheism levelled against him for this timeless concept of natural change – 'We find no vestige of a beginning – no prospect of an end' – he attempted to rewrite his theory, producing more evidence, but his unreadable style did nothing to increase understanding, which was only achieved after his death by ✧ *Playfair*.

Hyaline. Glassy, i.e. non-crystalline. (✧ *Texture*.)

Hyalo- (prefix). Glassy.

Hyalocrystalline. A ✧ *texture* term applied to rocks which contain both crystals and glass.

Hyalopilitic. A variety of ✧ *pilotaxitic* texture in which the felt of crystals or crystallites is embedded in a glassy base.

Hyalosiderite. ✧ *Olivines*.

Hybrid. A term applied to rocks formed by the mixing of two magmas or the ✧ *assimilation* of one igneous rock by another.

Hydraulic limestone. A limestone containing sufficient siliceous and aluminous impurities to yield a true cement (i.e. one which will set under water) when calcined.

Hydro- (prefix). Water.

Hydrocarbon minerals. A term usually taken to mean the naturally occurring solid varieties of carbon–hydrogen (sometimes also with oxygen) compounds, i.e. the natural 'organic' minerals. Two groups may be recognised – ◊ *coal*, and the ◊ *bitumens*, waxes, and resins. (For liquid petroleum, ◊ *Oil* and *Natural gas*.)

◊ *Asphalt* and its relatives, and the natural waxes, appear to be residues from the volatilisation of the lower boiling-point liquid hydrocarbons from a natural oil deposit. Seepages of asphalts may form asphalt lakes or ◊ *tar pits*, but the material also occurs occupying joints and fissures, in porous sandstones, and even in lenses in sediments. Elaterite is a peculiar 'rubber-like' bitumen. The natural waxes include ozokerite, albertite, and hatchettine, all of which appear to be oil residues which have seeped or been squeezed into joints, fissures, or cavities in the rocks. They have been worked commercially as a source of candlewax. Many names have been proposed for minor varieties of these rather indeterminate minerals.

The resins include amber and gum copal. They are fossilised resin masses produced by coniferous trees and preserved in sediments. Amber is found in ◊ *Oligocene* estuarine deposits around the Baltic coast, while gum copal is found in ◊ *Quaternary* and Recent sediments in New Zealand.

Hydrogeology. The study of the geological factors relating to the Earth's water.

Hydrology. The study of all waters in and upon the Earth. It includes underground water, surface water, and rainfall, and embraces the concept of the hydrological cycle.

Hydrolysates. A geochemical division of the ◊ *sedimentary rocks* corresponding approximately to the ◊ *argillaceous rocks*.

Hydropore (Echinodermata). ◊ Fig. 43.

Hydrosphere. The total body of water of the Earth, i.e. the oceans, rivers, lakes, underground and atmospheric water. (Cf. ◊ *Lithosphere*.)

Hydrostatic pressure. ◊ *Pressure*.

Hydrothermal processes. The name given to any processes associated with igneous activity which involve heated or superheated water (cf. ◊ *Pneumatolysis*). Water at very high temperatures is an exceedingly active substance, capable of breaking down silicates and dissolving many substances normally thought of as insoluble. Two main types of hydrothermal activity may be considered:

(1) ALTERATION PROCESSES. These include ◊ *serpentinisation* of olivine and rhombic-pyroxenes, chloritisation of ferromagnesian minerals (◊ *Chlorite group*), ◊ *saussuritisation*, ◊ *uralitisation*, and ◊ *propylitisation*. Kaolinisation is the most

important hydrothermal process, resulting in the production of the ◊ *clay mineral* kaolin from the feldspars in a granite. A typical reaction may be expressed:

$$K_2O \cdot Al_2O_3 \cdot 6SiO_2 + CO_2 + 2H_2O \rightarrow$$

Feldspar $\quad\quad\quad K_2CO_3 + Al_2O_3 \cdot 2SiO_2 \cdot 2H_2O + 4SiO_2$

$\quad\quad\quad\quad\quad\quad$ (soluble) $\quad$ Kaolin $\quad\quad\quad\quad$ Silica

This is precisely the same as the example given for the chemical weathering of feldspars by sub-aerial processes, but there is no doubt that hydrothermal action is responsible for some kaolin deposits, e.g. those of Devon and Cornwall.

(2) DEPOSITION. It is generally held that many ore deposits are deposits from hydrothermal solutions, e.g. Cu, Pb, and Zn. These deposits may fill fissures (◊ *Vein*; ◊ *Ore body*) or may replace existing rocks. It has been objected that the sulphides of metals such as Cu, Pb, and Zn are so extremely insoluble in water that their transport in solution is improbable. However, no satisfactory alternative has been proposed and it seems possible that the presence of other ◊ *volatile* substances associated with water (F, CO_2), together with the known high chemical

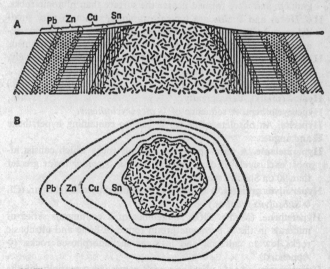

83. The zoning of mineral deposits around an igneous intrusion. (A) Section; (B) Plan.

activity of water at high temperatures, may account for most of the difficulties which have been raised. The temperature and pressure at which a hydrothermal ore body is deposited controls the form and mineralogy of the deposit. (◊ *Hypothermal*; *Mesothermal*; *Epithermal*.)

Hydrothermal ore deposits are commonly found arranged concentrically around an igneous mass, the high-temperature (hypothermal) minerals being closest to or actually within the igneous rock, and the low-temperature (epithermal) ones being formed farthest away (◊ Fig. 83). In the case of an inclined contact of the igneous with the ◊ *country rock*, a zonal arrangement may appear in depth. There is a tendency for hydrothermal deposits to be concentrated at the extreme top of an igneous mass and many vein systems are associated with minor upward projections (◊ *Cupolas*) on the upper surface of the igneous bodies.

Hydrozoa. ◊ *Coelenterata.*

Hyp- (prefix). Under, nearly; sometimes 'hypo-'.

Hypabyssal rocks. A term applied to intrusive igneous rocks which have crystallised under conditions intermediate between ◊ *plutonic* and ◊ *volcanic*. In general, hypabyssal rocks are medium-grained, and were formed nearer the surface than plutonic rocks. ◊ *Dykes* and ◊ *sills* are typical examples, although very small dykes and sills may be fine-grained and very large ones coarse-grained, in which cases it is not possible to distinguish specimens of the rock from the volcanic or plutonic equivalents respectively, in the absence of information about the field occurrence.

Hypautomorphic. Having some trace of crystal form. (◊ *Texture*.)

Hyper- (prefix). Very, greatly.

Hypercyclothem. A sequence of ◊ *megacyclothems*.

Hyperite. An obsolete name for ◊ *gabbro* containing hypersthene and augite.

Hypermelanic. A term applied to igneous rocks which consist almost exclusively of dark minerals with a ◊ *colour index* greater than 90 on Shand's scale.

Hypersolvus granites. ◊ *Granites* containing only one feldspar. (Cf. ◊ *Subsolvus granites*.)

Hypersthene. $(Mg,Fe'')SiO_3$. A member of a continuous series of minerals in the ◊ *pyroxene* group found in basic and ultrabasic rocks low in calcium, and in some metamorphosed rocks. (◊ Appendix.)

Hypersthenite. An ◊ *ultrabasic* rock consisting essentially of the orthorhombic ◊ *pyroxene* mineral hypersthene.

Hypidiomorphic. A term used to describe the ◊ *texture* of a rock in which the mineral grains show some trace of crystal form.

Hypo-, hyp- (prefix). Under, nearly.

Hypocrystalline. A term used to describe rocks which contain both crystals and glass. (◊ *Texture.*)

Hypogene. A term used to describe processes originating within the Earth, especially the formation of mineral deposits by ascending fluids. (Cf. ◊ *Supergene.*)

Hypohyaline. A synonym of ◊ *hypocrystalline.*

Hypostome (Arthropoda). ◊ Fig. 3.

Hypothermal. A term used to describe ore deposits of ◊ *hydrothermal* origin, formed at high temperatures (300° to 500°C.).

Ice age. A period of time during which glacial ice spreads over regions which are not normally ice-covered. The emphasis is on a large-scale regional spread of the ice, not merely a local advance of valley glaciers due to minor climatic fluctuations. (For examples, ◊ *Tertiary System* (Pleistocene): ◊ *Interglacial period*; *Glaciers and glaciation*.)

Ice cap ◊ *Glaciers and glaciation*.

Iceland spar. A variety of the mineral ◊ *calcite*, $CaCO_3$; optically clear cleavage rhombs.

Icenian. A stratigraphic stage name for the British Lower Pleistocene (◊ *Tertiary System*).

Ice sheet. ◊ *Glaciers and glaciation*.

Ichors. One of the names given to the fluids by which, it has been suggested, rocks may be transformed to granite or granite magma during ◊ *granitisation*.

Idio- (prefix). Of itself.

Idioblastic. A term used to describe the ◊ *texture* of a metamorphic rock in which the grains display fully developed crystal form.

Idiomorphic. A term used to describe the ◊ *texture* of an igneous rock in which the grains display fully developed crystal form.

Idocrase (Vesuvianite). A mineral, $Ca_{10}Al_4(Mg,Fe'')_2(Si_2O_7)_2(SiO_4)_5(OH)_4$, formed by contact metamorphism of impure limestone. (◊ Appendix.)

Igneous lamination. ◊ *Layered igneous structures*.

Igneous rocks. One of the three main groups of rocks, often regarded as the primary source of the material comprising the Earth's surface (◊ *Metamorphism*: *Sedimentary rocks*). They characteristically appear crystalline, i.e. a mass of interlocking crystalline units, although non-crystalline super-cooled ◊ *glasses* do occur. Most igneous rocks appear to have crystallised from a silicate melt, the ◊ *magma* (but ◊ *Granitisation*; ignimbrites under *Pyroclastic rocks*; *Fluidisation*).

Igneous rocks may be extrusive on the Earth's surface (◊ *Volcanic*) or intrusive into the rocks forming the ◊ *crust of the Earth* (◊ *hypabyssal* and ◊ *plutonic*). It is possible to find intrusive rocks at the Earth's surface only after the cover of other rocks has been eroded.

The classification of igneous rocks is based upon the following criteria :

(1) Degree of ◊ *saturation* (◊ *Acid rock*; *Intermediate rocks*; *Basic rocks*; *Ultrabasic rocks*).

(2) ◊ *Colour index*.

(3) Feldspar character (◊ *Alkali*; *Calc-alkali*; *Monzonites*).

(4) Grain size. Qualitatively the terms ◊ *aphanitic*, ◊ *phaneritic*, and ◊ *hyaline* are suitable for general descriptions. It is however convenient to attach arbitrary grain-size limits as follows :

Very coarse	Greater than 3 cm.
Coarse	5 mm.–3 cm.
Medium	1–5 mm.
Fine	Less than 1 mm.
Glassy	Effectively, a grain size of zero

Grain sizes of rocks are always determined on the general ◊ *groundmass*, and ◊ *phenocrysts* are ignored. The terms coarse, medium, and fine are not necesarily synonymous with plutonic, hypabyssal, and volcanic.

(5) ◊ *Texture*.

A large number of igneous rock classifications have been suggested, varying from quantitative through semi-quantitative and pseudo-quantitative to purely qualitative ones, and based upon chemical composition, mineralogical composition, or a combination of both.

The naming of igneous rocks is not very systematic; a large number of names based on places or persons have been suggested often for relatively minor variants. The ideal system would appear to be a limited number of fundamental names, modified as required by a name or names of significant mineral(s).

(◊ *Alkali basalt*; *Alkali gabbro and alkali dolerite*; *Alkali syenite*; *Andesite*; *Anorthosite*; *Aplite*; *Appinite*; *Basalt*; *Carbonatite*; *Diorite*; *Dolerite*; *Gabbro*; *Granite*; *Granodiorite*; *Lamprophyre*; *Microgranite*; *Monzonites*; *Pegmatite*; *Pyroclastic rocks*; *Rhyolite*; *Serpentine*; *Syenite*; *Trachyandesite*; *Trachybasalt*; *Trachyte*; *Ultrabasic rocks*. For altered igneous rocks, ◊ *Pneumatolysis*; *Hydrothermal processes*.)

Ignimbrite. ◊ *Pyroclastic rocks*.

Ijolite. ◊ *Alkali syenite*.

Illite. A ◊ *clay mineral*.

Illuviation. A zone of deposition of leached material in a ◊ *soil*.

Ilmenite. A mineral, $FeTiO_3$, found as an accessory in basic igneous rocks, as detrital deposits, and in veins. ($\Diamond$ Appendix.)

Imbricate structure. (1) $\Diamond$ *Fault* (Thrust fault).

(2) A $\Diamond$ *fabric* found in $\Diamond$ *conglomerates* and pebble beds. When these consist of fragments having a noticeable elongation, deposited under the influence of a powerful current, the long axes of the pebbles tend to lie more or less parallel with one another 'leaning' in the direction of current flow. ($\Diamond$ Fig. 84.) It has been

84. Imbricate structure.

suggested that under certain circumstances sand grains might show a micro-imbricate structure.

Impermeability. $\Diamond$ *Permeability*.

Impersistent. Geologists invariably use 'impersistent' as the converse of 'persistent', rather than 'non-persistent'. It carries the connotation of discontinuity (especially in an areal sense) and spasmodicity.

Impervious. $\Diamond$ *Pervious*.

Implication. A synonym of granophyric or $\Diamond$ *graphic texture*. (Cf. $\Diamond$ *Symplectic*.)

Impregnation. The infilling of pores in a rock by mineral matter, or the $\Diamond$ *replacement* of existing pore-filling material. Not equivalent to cementation ($\Diamond$ *Diagenesis*); the term is usually used with the sense of a later event, especially the introduction of an ore mineral or a liquid, e.g. oil.

Inarticulata. $\Diamond$ *Brachiopoda*.

Incised meanders. $\Diamond$ *Rejuvenation*.

Inclusion. A portion of one substance enclosed within another. For example, a fragment of an early-formed crystal surrounded by a later one, a cavity in a crystal filled with gas and/or liquid, a piece of older rock included in a younger rock. ($\Diamond\!\!\!\Diamond$ *Xenolith*.)

Incompetent bed. A rock layer that, during folding, flexes with appreciable flow and shear, and hence develops slaty $\Diamond$ *cleavage*. ($\Diamond\!\!\!\Diamond$ *Competent bed*.)

Inconsequent drainage. A drainage system which does not conform to the structural pattern of the region, although it may show local small adjustments to it. Two types may be recognised:

(1) SUPERIMPOSED DRAINAGE. A drainage system established upon a series of younger rocks lying with marked angular ◊ *unconformity* upon an older series. The drainage will become adjusted to the structural pattern of the younger series. In course of time, river erosion will remove the younger layers, leaving the established drainage pattern cutting down into the older rocks, the structure of which bears no relation to the drainage system. Eventually a superimposed system may slowly adjust to the structural control of the older set of rocks. An angular unconformity of some kind must exist for superimposed drainage to develop.

(2) ANTECEDENT DRAINAGE. An antecedent stream is one which has maintained itself across a geological structure developing athwart its course. Antecedent streams are usually related to rising structures, e.g. the Brahmaputra where it cuts across the Himalayas; the Rhine where it cuts into the Rhine Plateau; and in Tennessee, U.S.A., where rivers have cut into the rising Nashville Dome.

(◊ *Drainage pattern.*)

Incretion. A hollow cylindrical ◊ *concretion*, like a pipe or tube.

Index fossil. A fossil species which characterises a specific ◊ *horizon* by its abundance, although not restricted to it in time.

Index mineral. A mineral, occurring in a series of metamorphic rocks, whose first occurrence marks the outer limit of metamorphic ◊ *zone*, as one moves in the direction of increasing metamorphic ◊ *grade*.

Indicatrix. A solid geometrical figure used to represent the different vibration directions in a mineral. It is an ◊ *ellipsoid* with three rectangular axes, the length of these axes being proportional to the refractive index of any beam vibrating at right angles to it (i.e. to beams travelling parallel to it). These axes are commonly termed X, Y, Z, or a, b, c, or α, β, γ. In this publication X, Y, Z, are used, X for the smallest refractive index and the 'fast ray'; Z, for the largest refractive index and the 'slow ray'; and Y for the intermediate ray, perpendicular to the X–Z plane.

If the light passes through a mineral perpendicular to the plane containing X and Z then, because X and Z show the maximum difference in refractive index, the maximum birefringence (or maximum relative retardation) is observed.

If the mineral is isotropic then the ellipsoid degenerates into a sphere, since $X = Y = Z$. If the mineral is anisotropic, two cases are possible: (1) $X < Y < Z$ (using refractive indices); the mineral is then said to be biaxial; (2) $X = Y < Z$ or $X < Y = Z$ (using refractive indices); the mineral is then said to be uniaxial.

(◊ *Axes, optic; Interference figure.*)

Indices, crystallographic. A system of notation of crystallographic ◊ *forms* and ◊ *faces* based upon the ◊ *intercept* ratios. By taking the intercept ratios of the selected unit form (◊ *Parameters*) and using them to reduce the intercept ratios of the other forms to integral values, indices result which are known as Weiss indices. In this system, faces parallel to an axis have infinity as their index for this axis, e.g. $1 \infty 2$ implies that the face is parallel to the b-axis. This introduces serious complications when indices have to be manipulated, and consequently nowadays Weiss indices have been superseded by Miller indices. These latter are derived from Weiss indices by taking reciprocals and clearing fractions, e.g. from the example above:

Weiss $1 \infty 2$ becomes

$1 \, 0 \, \frac{1}{2}$ and, hence

$2 \, 0 \, 1$ Miller (read 'two-nought-one') (see the worked example below).

The letters h, k, l, are used as general symbols for indices (h, j, k, l, in the hexagonal system).

Indices of faces which cut negative ends of axes are indicated by means of a bar above the appropriate numeral, e.g. $\bar{2} \, 0 \, \bar{1}$ is the corresponding negative face to our previous example. If an index is enclosed within braces {h k l} this is an indication that it is the index of a form, and should be interpreted as face h k l together with all the other faces of the form required by the symmetry class involved. A symbol [h k l] in square brackets is a zone symbol which is strictly a set of coordinates.

In the hexagonal system the algebraic sum of h, j, and k is necessarily zero.

WORKED EXAMPLE OF THE DERIVATION OF CRYSTALLO-GRAPHIC INDICES. From the ◊ *interfacial angles* of the faces, as measured by a goniometer, it is possible to calculate the ◊ *intercept* ratios:

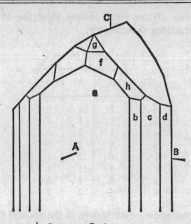

	Intercept Ratios		
Face	a	b	c
a	0·46575	∞	∞
b	0·46575	1	∞
c	0·93150	1	∞
d	1·39725	1	∞
e	∞	1	1·17302
*f	0·46575	1	0·58651
g	0·46575	1	0·29325
h	0·93150	1	1·17302

One face, marked *, is selected as the unit form and its intercept ratios are used as the ◊ *parameters*. Dividing the intercept ratios for the faces by the parameters yields the Weiss indices:

	Weiss Indices		
Face	a	b	c
a	1	∞	∞
b	1	1	∞
c	2	1	∞
d	3	1	∞
e	∞	1	2
*f	1	1	1
g	1	1	½
h	2	1	2

The reciprocals of the Weiss indices yield the Miller indices, clearing of fractions as necessary:

Face	Miller Indices		
	a	b	c
a	1	0	0
b	1	1	0
c	1	2	0
d	1	3	0
e	0	2	1
*f	1	1	1
g	1	1	2
h	1	2	1

Indicolite. A blue sodium-rich variety of the mineral ◊ *tourmaline*.

Induan. A stratigraphic stage name for the base of the European Lower ◊ *Triassic*.

Induration. A process by which soft sediment becomes hard rock. Originally the term implied hardening due to baking, i.e. proximity to an igneous rock, but nowadays it is used more generally to include hardening by pressure and cementation. (◊ *Diagenesis*.)

Inertinite. ◊ *Coal*.

Inferior. Almost always used in geology in the sense of 'lower', 'below', or 'under', and not with the meaning 'of poor quality', e.g. Inferior Oolite = lower oolite.

Infra- (prefix). Below.

Infrastructure. A name given to a ◊ *migmatite* zone showing complex tectonics, lying below a more simply folded, non-migmatitic zone (◊ *Superstructure*). The original use of the term envisaged the migmatite ◊ *front* of a granite mass moving upwards into the sedimentary and meta-sedimentary cover rocks.

Injection complex. An association of igneous rocks and, very commonly, regionally metamorphosed rocks, in which the relationships between the various rock bodies are intricately intermixed. It is commonly found that more than one period of igneous activity is involved, and several stages of metamorphism may also be detected.

Injection gneiss. ◊ *Migmatite*.

Injection structures. Under certain special circumstances of particle size, water content, etc., it is possible for one layer of sediment to force its way upward into an overlying layer and even occasionally to pierce it and flow out at the surface. Various terms for such structures have been proposed. The piercement structures have been called sand or mud volcanoes. (Mud volcanoes of enormous size have been produced by earthquake shocks.) Non-piercing structures have been referred to as flame structures, sandstone dykes (cf. ◊ *Neptunean dyke*), and streamers. (◊ Fig. 85.)

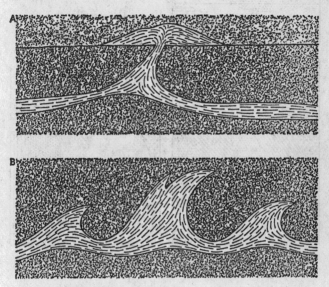

85. Two injection structures. (A) A mud 'volcano', approximately 2.5 cm. high; (B) A flame structure, approximately 10 cm. high.

Inlier. A limited area of older rocks completely surrounded by younger rocks. It may be produced by ◊ *erosion*, ◊ *faulting*, or ◊ *folding* (◊ Fig. 86), or a combination of any two, or all three, of these agencies. (Cf. ◊ *Window*; ◊◊ *Outlier*.)

Inosilicates. ◊ *Silicates*.

Inselberg (German, 'island mountain'). A steep-sided, round-topped mound, generally of granite or gneiss, which occurs in isolation, or as one of a group, rising above the general level of a pediment (◊ *Cycle of erosion*). It is the equivalent of the ◊ *monadnock* in the Davisian cycle of erosion. Inselbergs are typical of arid or

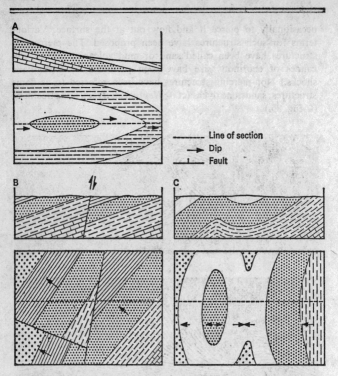

86. The formation of an inlier. (A) By erosion; (B) By faulting; (C) By folding.

semi-arid regions, a product of solar weathering, although they are also found in areas with quite different present-day conditions. They are common in tropical Africa. The Afrikaans name kopje (koppie) is a synonym. (◇ *Butte*; *Mesa*.)

Insequent. ◇ *Drainage pattern.*

Inset. A synonym of ◇ *phenocryst.*

In situ (Latin, 'in place'). Geologists use the term frequently to distinguish material, e.g. rocks, minerals, fossils, etc., found in their original position of formation, deposition or growth, as opposed to loose, disconnected, or derived material.

Insolation. The radiant heat received by rocks from the sun. (◇ *Weathering.*)

Inter- (prefix). Between.

Interambulacral areas. ◊ *Echinodermata* (Echinoidea) and Fig. 44.

Interarea (Brachiopoda). ◊ Fig. 11.

Interbedded. Literally 'between two layers', e.g. a lava flow may occur interbedded between two layers of sediments, a limestone may be interbedded between two layers of shale.

Intercept (crystallography). The position of a given plane in space may be defined in terms of the distances from the origin along three arbitrarily selected axes which the plane intersects. By suitable use of + or − signs, planes having similar intercepts but different directions of inclination may be distinguished. (◊ Fig. 4.) In crystallography the absolute intercepts are meaningless, since they depend on the size of the crystal. Since, however, it is only the direction of each plane (in morphological crystallography parallel faces in the same ◊ *octant* are the same face) in which we are interested, this direction may be defined in terms of the ratios of the intercepts, and it is simply a matter of spherical trigonometry to calculate intercept ratios from interfacial angles. Conventionally, the measurement along the b-axis is taken as unity. (For conversion of intercept ratios to indices, ◊ *Indices, crystallographic*.)

Interfacial angle. In crystallography, the angle between the normals to two crystal faces. (◊ Fig. 87.)

  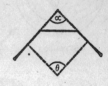

θ Angle between normals to faces   ∝ Angle between faces   θ + ∝ = 180°

87. Three examples of interfacial angle.

Interference figure. If a ◊ *thin section* of a crystal, cut perpendicular to the optic ◊ *axis* (uniaxial case) or to the acute bisectrix (biaxial case), is observed with a microscope under crossed Nicols and using convergent light, the pattern of coloured curves or rings and black areas which may be seen, if the correct technique is employed, is called an interference figure. For the uniaxial case, the pattern takes the form of concentric coloured rings and a black cross, the isogyre. (◊ Fig. 88.) For the biaxial case, the

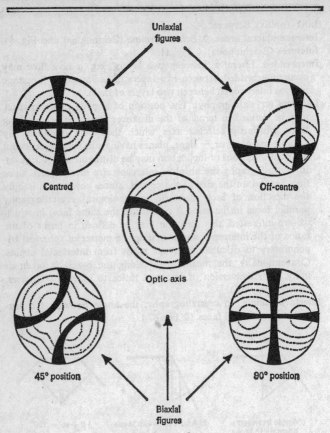

88. Interference figures.

pattern of coloured zones takes on a more complicated form, with two separate sets of curves centred around two 'eyes' and two hyperbolic isogyres, the whole pattern changing as the thin section is rotated.

The coloured lines represent zones of equal retardation (◊ *Polarisation colours*) arising from the varying directions in which the convergent light passes through the mineral plate.

In this way a synthesis of the way in which the optical properties of the mineral vary with direction can be obtained and, in particular, the determination of fast and slow directions is possible. (◊ *Vibration directions*.)

In a biaxial mineral a section perpendicular to an optic axis yields a circular arrangement of coloured rings, but the presence of a single isogyre, which rotates in the opposite direction to the stage rotation, distinguishes this case from the true uniaxial one ('compass needle isogyre'). Sections which are not accurately perpendicular to axes or bisectrix yield off-centre figures which are capable of providing partial data in some cases.

To view an interference figure, an auxiliary lens is used for best results: a ◊ *Becke lens* above the eyepiece (an ordinary low-powered hand lens may be used as a temporary expedient) or a ◊ *Bertrand lens* between the eyepiece and the objective. An unmagnified image of the interference figure may be observed by removing the eyepiece, or replacing the eyepiece by a pin hole. (◊ *Indicatrix*.)

Interfluve. The high land between two streams belonging to the same drainage system.

Interglacial period. ◊ *Ice ages* are often divisible into periods of maximum ice advance and periods of ice retreat (glacial maxima and glacial minima). A period of retreat is referred to as an interglacial. (◊ *Interstadial*.)

Intergranular. An approximate synonym of ◊ *granulitic*.

Intergrowth. A textural term applied to interlocking arrangements of two minerals, usually due to simultaneous crystallisation of the phases or the exsolution of one phase from another (◊ *Graphic texture*; *Perthite*). The term used to describe all intergrown structures is ◊ *symplectic*.

Interlimb angle. The angle between the limbs of a ◊ *fold*.

Intermediate rocks. A rock which contains less than 10% of quartz together with either a plagioclase (◊ *Feldspar*) in the andesine-oligoclase range, or an alkali feldspar, or both the feldspars. Feldspathoids may or may not be present. When plagioclase is present, the definition of the group is straightforward, but when only alkali feldspar is present the boundary between intermediate and basic is much less satisfactorily established (cf. *Acid rock*; *Basic rocks*). (For typical intermediate rocks, ◊ *Alkali syenite*; *Andesite*; *Diorite*; *Lamprophyre*; *Monzonites*; *Syenite*; *Trachyte*.)

Intermontane. Literally 'between mountain ranges'.

Internormal angle. ◊ *Interfacial angle*.

Intersertal. A ◊ *textural* term applied to an unorientated meshwork of feldspar laths in which the gaps are filled by glass or very fine crystals, the feldspar crystals forming the major part of the rock. With a decrease in the proportion of feldspar the texture becomes a variety of ◊ *ophitic texture.*

Interstadial. A pause in the advance of a ◊ *glacier* or ice sheet. (Cf. ◊ *Interglacial.*)

Intervallum (Archaeocyatha). ◊ Fig. 1.

Intra- (prefix). Within, inside.

Intraclast. A sedimentary particle composed of calcium carbonate derived by local ◊ *penecontemporaneous* erosion of the floor of the sedimentary basin. The particles may be fragments of partially consolidated calcium carbonate mud, or composite particles consisting of aggregates of smaller intraclasts, pellets, ◊ *ooliths,* skeletal material, etc. (◊ *Limestone.*)

Intraformational. A term applied to rocks, e.g. ◊ *conglomerates* and ◊ *breccias,* or structural features, which occur between two sets of defined strata. An intraformational conglomerate or breccia is one formed by the contemporaneous erosion and redeposition of a bed of sediment previously deposited. It implies temporary change in the conditions of sedimentation. Intraformational slumping or folding is also possible.

Intrusion. A body of igneous rock which has forced itself into pre-existing rocks, either along some definite structural feature, e.g. ◊ *bedding planes,* ◊ *joints,* ◊ *cleavages,* etc., or by deformation and cross-cutting of the invaded rocks. (◊ *Batholith; Boss; Dyke; Laccolith; Lopolith; Phacolith; Sill; Stock.*)

Involution. The refolding, or folding together, of one or more ◊ *nappes.*

Ionic radius ratio (I.R.R.). The ionic radius ratio is a measure of the relative sizes of a cation and an associated anion. Since oxygen is by far the commonest anion, the ionic radius ratio is most commonly given in the form:

$$\frac{\text{Radius of cation}}{\text{Radius of oxygen ion}} = \text{I.R.R.}$$

The I.R.R. enables a prediction to be made of the type of co-ordination to be expected between a cation and an anion, based upon simple geometrical considerations. The following table shows the relationship between I.R.R., coordination number and geometry of anion arrangement.

Ionic radius ratio	Coordination number	Geometry of anion arrangement
0·15–0·22	3	Plane equilateral triangle
0·22–0·41	4	Corners of tetrahedron
0·41–0·73	6	Corners of octahedron
0·73–1·00	8	Corners of a cube
1·00 and above	12	Mid-points of cube edges

These predicted coordination numbers agree well with observation, although near the limits of the classes elements are found having more than one coordination number, e.g. the I.R.R. for Al and oxygen is 0·36, and Al occurs in both coordinations 4 and 6.

Ipswichian. A stratigraphic stage name for British Upper Pleistocene (interglacial between Gipping Till and Newer Drift) (◊ *Tertiary System*).

Iron glance. Haematite, Fe_2O_3. (◊ *Glance*; ◊◊ Appendix.)

Iron hat. A ◊ *gossan*.

Iron pan. ◊ *Hard-pan* in which the cementing material is ferruginous.

Iron pyrites. ◊ *Pyrite*.

Ironstone. ◊ *Clay ironstone*; *Black-band ironstone*.

Island arc (Festoon). An arcuate chain of islands, e.g. Japan, Aleutian Islands. They are found around the margins of the Pacific, Malaya, and Indonesia, and in the Caribbean. Their abundance and distribution suggest that they have a common origin. The following are some of their main characteristics: (1) They are associated with areas of strong seismic activity and deep-focus ◊ *earthquakes*. (2) Deep oceanic trenches (fore deeps) occur on the convex side and deep basins on the concave side. (3) They are regions of intense gravitational and magnetic anomalies. (4) When a double arc occurs, the inner one is volcanic, the outer one non-volcanic. (◊◊ *Plate tectonics*.)

Iso- (prefix). Equal.

Isobath. A line joining points of equal depth of a horizon below the surface or a datum plane.

Isochemical. A term used to describe metamorphic processes in which no introduction of material from an external source takes place. (◊ *Metamorphism*.)

Isochore. A line joining points of equal vertical interval between two datum planes. (Cf. ◊ *Isopach*.)

Isoclinal fold. A ◊ *fold* in which two adjacent limbs are parallel. (◊ Fig. 62.)

Isoflor. A line connecting points of similar flora. A concept used by palaeoclimatologists in an attempt to determine climatic zones.

Isograd. A line joining points where the rocks have the same ◊ *facies* – especially rocks having attained the same ◊ *grade* of metamorphism. If the grade of metamorphism is defined in terms of ◊ *index minerals*, the isograds will limit the metamorphic zones and will thus define regions of equal temperature and pressure conditions.

Isogyre. The name given to the black areas of an ◊ *interference figure.* The black areas for a uniaxial mineral form a black cross. The interference (◊ *Polarisation*) colours (each colour representing points of equal retardation), due to the use of convergent light, form concentric circles (sections of cones, in fact) round the unique optic ◊ *axis* in the centre of the field of view. The locus of points on these circles where the ◊ *vibration directions* of the mineral and the polars of the microscope coincide, thereby producing ◊ *extinction,* forms a cross.

In a biaxial mineral, there are two sets of coaxial cones of equal retardation, one round each optic axis, lying obliquely to the plane of the slide, so that sections of the cones are hyperbolae. The result is a black cross which, upon rotation of the mineral, moves apart to form two hyperbolae, the vertices of which represent the points of emergence of the optic axes. When the hyperbolae are at maximum separation, the distance between the vertices represents the 2V (◊ *Axes, optic*) of the mineral. The hyperbolae are convex towards the acute bisectrix. If a section perpendicular to one optic axis in a biaxial crystal is observed, only one isogyre is seen, which rotates in the opposite direction to the rotation of the stage ('compass needle isogyre'). (For off-centre isogyres, ◊ *Interference figure.*)

Isoline. A synonym of ◊ *isopleth.*

Isometric system. A synonym of ◊ *cubic system.*

Isomorphous. A term originally applied to two substances crystallising in the same form, but now restricted to cases where the two substances form a series of solid solutions, e.g. rock salt and fluorite both crystallise in cubes but are not said to be isomorphous because they do not form solid solutions.

Isomorphy. ◊ *Homeomorphy* when it occurs in the Foraminifera (◊ *Protozoa*).

Isopach, Isopachyte. A line joining points of equal bed thickness. (Cf. ◊ *Isochore.*)

Isophysical. A synonym of metamorphic ◊ *facies*.

Isopleth. Any line joining points of equal value, abundance, frequency, etc., on a map, e.g. contour lines, isobars, isotherms. (Isoline is a synonym.)

Isopygous. A term applied to trilobites (◊ *Arthropoda*) which have the cephalon and pygidium of a similar size.

Isoseismic line. A line joining points of equal intensity of ◊ *earthquake* shock, which is usually a closed curve around the ◊ *epicentre*. In the absence of a large number of suitably placed seismographs, isoseismic lines are commonly determined by public questionnaires.

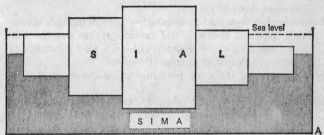

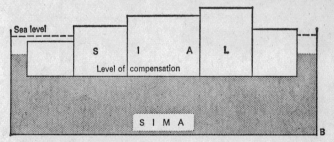

89. Two theories of isostasy. In (A), due to Airey, it is assumed that all the sialic rocks have the same density, the difference in elevation of the 'blocks' being due to their differing thickness; in (B), due to Pratt, it is assumed that the higher blocks have a lower density than the lower blocks and that consequently there was a level to which all the blocks sank – the level of compensation. It is probable that neither of these two theories is wholly correct and that a compromise between them operates.

Isostasy. The tendency of the Earth's ◊ *crust* to maintain a state of near equilibrium, i.e. if anything occurs to modify the existing state, a compensating change will occur to maintain a balance. The most obvious example of this is the deep roots of mountain chains; as erosion of the mountains occurs, reducing their height, compensation in the forms of renewed uplift occurs. Continental blocks may be depressed under the load of an ice sheet, recovering when the ice melts. In this way variations in the relative levels of land and sea may be explained. (◊ *Rejuvenation*.) The concept of isostasy depends upon the model of the Earth's crust in which lighter, continental masses 'float' on a denser sub-stratum. (◊ *Continental drift* and Fig. 89.)

Isostructural. A term used to describe two or more minerals having similar physical, chemical, and crystallographic features, but which show little or no tendency to form a solid solution, e.g. grossular and almandine varieties of ◊ *garnet*.

Isotropic. Having the same properties in all directions. (Cf. ◊ *Anisotropic*.)

Itacolumite. Sandstone which is slightly flexible in thin slabs, owing to a 'ball-and-socket' arrangement of the grains and cement.

Jacupirangite. A rock consisting mainly of titan-augite and aegirine-augite, with accessory nepheline. (A synonym is nepheline pyroxenite.) (◊ *Alkali syenite.*)

Jade. The name given to two distinct minerals: jadeite – a ◊ *pyroxene* – and nephrite – an ◊ *amphibole* – when they occur in the form of a tough compact aggregate. This material, especially the green and white varieties, has been used extensively as a gem material.

Jadeite. ◊ *Jade*; *Pyroxenes.*

Jasper. A red chert-like variety of chalcedony. (◊ *Silica group of minerals.*)

Jaspilite. A banded rock consisting of alternating layers of ◊ *jasper* and iron oxide.

Jatulian. A stratigraphic division of the Baltic ◊ *Precambrian*, the youngest ◊ *Karelian* – unmetamorphosed ◊ *Kalevian* + ◊ *Ladogian.*

Jet. A special type of compact homogeneous cannel ◊ *coal* or black lignite which can be carved, turned, and polished for jewellery and other decorative purposes.

Joint. A fracture in a rock between the sides of which there is no observable relative movement. In general, joints intersect ◊ *primary* surfaces, such as ◊ *bedding*, ◊ *cleavage*, ◊ *schistosity*, etc. They develop preferentially in the ◊ *competent* members of a series rather than in the incompetent ones. A series of parallel joints is called a 'joint set'; two or more sets intersecting produce a 'joint system'; two sets of joints nearly at right angles to one another, produced by the same ◊ *stress* system, are said to be conjugate. A master joint (a quarryman's term) is a persistent joint or set which may be horizontal or vertical. Joints may be formed as follows:

SHRINKAGE JOINTS. These are caused by tensional forces set up in a rock body as a result of cooling (in an igneous rock – ◊ *Columnar structure*) or desiccation (of a sedimentary rock).

SHEET JOINTS. A set of joints may develop which are more or less parallel to the surface of the ground, especially in plutonic igneous intrusions such as granite. They probably arise as a result of the unloading of the rock mass when the cover is eroded away.

TECTONIC JOINTS. These arise as a direct result of folding or thrusting in rocks, and in general three sets can be recognised: a strike set (longitudinal joints), parallel to the fold axes; a dip set (cross joints), perpendicular to the longitudinal joints; and a conjugate set of oblique joints, which lie at rather less than 45° to the direction of tectonic ◊ *transport*. (◊ Fig. 90.) In some cases

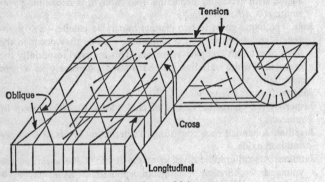

90. Types of joint.

the two sets forming the conjugate set are unequally developed. The term 'shear joints' has been used for these sets, as they correspond approximately to the theoretical directions of maximum shear. (◊ *Tension gash*.)

Joints may be open or closed, the latter being called latent, blind, or incipient joints. They may become open as a result of ◊ *weathering*, and this is especially seen in jointed limestones, where they become accentuated to form ◊ *clints* and ◊ *grikes*.

Joints are important since they commonly control drainage patterns and the shape of coastlines, and because they provide a passage whereby water may penetrate deeply into the rock mass, thus allowing weathering to take place. Jointed rocks are ◊ *pervious* to fluids and hence may act as ◊ *aquifers*, or reservoir rocks for oil or natural gas. Quarrymen and miners pay great attention to jointing, as its presence or absence in a particular direction can ease or hinder their work to a marked extent. The jointing in coal is called 'cleat', but it is probably not of tectonic origin.

Jökull (Icelandic). A small ice cap. (◊ *Glaciers and glaciation*.)
Jotnian. The youngest division of the Baltic ◊ *Precambrian*.

Jurassic System. Named from the Jura Mountains of France (by von Humboldt in 1795), the Jurassic System is the period of time extending from 195 m.y. to 135 m.y., having a duration of 60 m.y. The Jurassic is divided into the Lower Jurassic (or Lias), the Middle Jurassic (or Dogger), and the Upper Jurassic (or Malm); the boundaries are dated at 172 and 162 m.y. respectively. The Jurassic has been further sub-divided, on a very minute scale, into a number of local sub-divisions. Over 100 fossil zones have been erected, grouped into 11 stages, the lower limit being the pre-Planorbis zone, while the upper limit is marked by the zone of Cypridea punctata (an ostracod; ◊ *Arthropoda*).

All the sediments in Great Britain are of shallow-water facies and the faunas which they contain are largely derived from the Tethyan basin. ◊ *Sedimentary iron ores* are widespread at several horizons. In the Jurassic, particularly towards southern Europe, the first effects of the coming Alpine Orogeny were already being felt. In the North American area the Laramide Orogeny occurred at this time.

The fauna of the Jurassic was extremely diverse and varied, the chief members being the ammonites (◊ *Mollusca*), by which the period is zoned, the hexacorals (◊ *Coelenterata*), and the echinoids (◊ *Echinodermata*). ◊ *Brachiopoda* were abundant, represented largely by terebratulids and rhynchonellids. Lamellibranchs and gastropods (◊ *Mollusca*) were also abundant, many forms being associated with coral reefs. In addition to the ammonites, the cephalopods were also represented by the belemnites (◊ *Mollusca*). The dominant terrestrial animals were the dinosaurs, which reached their maximum size in the Jurassic and occupied most of the ecological niches. The first birds appeared in the Upper Jurassic, but the mammals, although present since the ◊ *Rhaetic*, were an insignificant element of the fauna and were rarely larger than the modern rat. The flora of the Jurassic included many forms still living at the present day, such as cycads, ginkoes, conifers, and ferns.

Juvenile. A term applied to gases and water implying 'new', i.e. arising from an underground magmatic source (e.g. volcanic emanations). (Cf. ◊ *Connate water*; ◊ *Meteoric water*.)

Kainozoic. ◊ *Cainozoic*.

Kalevian. A stratigraphic division of the Baltic ◊ *Precambrian*, the younger Karelian.

Kaliophilite. ◊ *Feldspathoids*.

Kalsilite. ◊ *Feldspathoids*.

Kame. A steep-sided alluvial cone deposited against an ice front. (◊ *Glaciers and glaciation*.)

Kaolin. The main mineral constituent of china clay. (◊ *Hydrothermal processes*; *Clay minerals*.)

Kaolinisation. ◊ *Hydrothermal processes*.

Kaolinite. ◊ *Clay minerals* and Appendix.

Karelian. A stratigraphic division of the Baltic ◊ *Precambrian* Post-◊-*Sveccofennian*.

Karnian. Stratigraphic stage name for the base of the East European and Asian Upper ◊ *Triassic*.

Karroo. (1) An arid terraced plateau in South Africa.

(2) Stratigraphic term, used for a series of sediments and lava flows of considerable extent in Southern and Central Africa. The sediments are all ◊ *continental* in character, including tillites (◊ *Drift*, 3), ◊ *evaporites*, coal seams and ◊ *red beds*. The Karroo Series is Upper Carboniferous to Lower Jurassic in age. The most important element of the Karroo fauna is the famous series of reptiles and mammal-like reptiles which have been obtained from various horizons.

Karst scenery. A landscape which shows a pattern of denudation in limestone and dolomitic rocks similar to that of the Karst region of Yugoslavia. This type of topography is produced not by normal surface run-off but by percolating ground waters and underground streams. The process may be initiated by the uplift of a limestone surface upon which normal drainage has commenced, or by downcutting of a stream, through other sediments into a limestone, or through an ◊ *unconformity*.

Limestones are normally well jointed, allowing water to follow a restricted path and not penetrating the rock as a whole. The ground water reappears at the surface as a stream issuing at the base of the limestone, having carried out its solution work. An adequate rainfall is of course necessary to provide the continual passage of water.

The leaching action of the ground water as it passes through the limestone produces a residual deposit, terra rossa, a red clay-like soil. When terra rossa does not cover the limestone surface and the rock is bare, the ground water may enlarge the joints into a conjugate pattern of clefts and ridges; this surface is called a limestone pavement or lapiés surface. The clefts in such a pavement are called grikes, and the ridges, clints.

The most characteristic feature of karst scenery is the sink (doline or swallow-hole), a conical depression in the limestone which may be several metres in diameter. Where two or more sinks coalesce a uvala is formed which may have a diameter of some 100–300 m. Water disappears underground through these hollows and by enlarging the joints and cavities in the limestone produces caverns and galleries. Occasionally the origin of uvalas has been attributed to the partial collapse of these caverns. As the water percolates downwards some of the excess calcium carbonate may be deposited on the ceilings and floors of the caverns, forming stalactites and stalagmites respectively, which often take on fantastic shapes.

Larger depressions in the landscape (covering tens of square kilometres) are known as poljes, and are characterised by steep sides and flat floors upon which small residual hills – hums or pepino hills – may remain. If the water-table is high enough lakes (polje lakes) may form. Poljes may originate from the lateral extension of uvalas, but their areal extension is such that some structural control, by faulting, etc. is generally assumed in their formation. In humid tropical regions, limestones may weather very rapidly because of the increased acidity derived from the rapid decay of overlying vegetation. Karst development is therefore very rapid and distinctive. The Cockpit Country of Jamaica is perhaps the best-known example of this type of karst. The 'cockpits' have a depth of 100–200 m. and may be up to 440 m. in diameter; they lie between small conical hills and are formed where there has been rapid solution along the joints.

The terms 'youth', 'maturity', and 'old age' have been applied to karst development in an attempt to create an analogy with the fluvial ◊ *cycle of erosion*. Youth commences with surface streams developed upon the initial limestone surface with lapiés and sinks in formation. During maturity karst development is at maximum, underground drainage is fully developed, large caverns are established and surface drainage is diminished. Old age is marked by a decrease in these features, underground streams are exposed by cavern collapse (karst windows), uvalas are formed, detached areas of original limestone form hums, and, as the land surface is worn down, surface drainage predominates once more. The net effect is that the limestone is completely removed.

Kata- (prefix). Deepest, greatest or maximum depth.

Kazanian. A stratigraphic stage name for the base of the East European Upper ◊ *Permian*.

Keel (Mollusca). ◊ Fig. 105.

Keewatin. A stratigraphic division of the Canadian ◊ *Precambrian*; a synonym of ◊ *Ontarian* or Upper Ontarian.

Kelyphitic border. A partial synonym of ◊ *corona structure*.

Kenoran. A Canadian ◊ *Precambrian* orogeny.

Kentallenite. A special variety of olivine ◊ *monzonite*.

Kenyte. An olivine ◊ *phonolite*.

Keratophyre. Trachytic and rhyolitic ◊ *spilites* containing sodipotassic feldspars and sometimes quartz. (Its use for a sodic non-feldspathoidal ◊ *trachyte* is obsolete.)

Kerogen. A solid organic material which yields petroleum-type hydrocarbons on heating and distillation. (◊ *Oil shale*.)

Kersantite. ◊ *Lamprophyre*.

Kettlehole. A depression left in the land surface when ice, formerly covered by drift, melts. (◊ *Glaciers and glaciation*.)

Keuper. A stratigraphic stage name for the West European Upper ◊ *Triassic*.

Keeweenawan. A stratigraphic division of the Canadian ◊ *Precambrian*; a synonym of Algonkian.

Kieselguhr. A synonym of ◊ *diatomite*.

Killas. The Cornish name for ◊ *slates* and low-grade ◊ *phyllites* belonging to the Devonian and Carboniferous as developed in south-west England; originally a mining term.

Kimberlite (Blue ground). A thoroughly brecciated peridotite (◊ *Ultrabasic rocks*) containing essential ◊ *mica* (phlogopite), bronzite and chrome-diopside.

Kimmeridgian. A stratigraphic stage name for the European mid Upper ◊ *Jurassic*. (◊ *Lithographic stone*.)

Kinderhookian. A stratigraphic stage name for the North American top ◊ *Devonian*/Lower ◊ *Mississippian* (Carboniferous System).

Kircher, Athanasius (1602–80). German Jesuit who produced the first attempt at a description of the Earth in physical terms in his work *Mundus subterraneus*. In ten books he covered an enormous range, from the composition of the sun, moon and Earth, to a survey of mines and minerals.

Klastic. ◊ *Clastic rocks.*

Klippe (pl. klippen). A tectonic ◊ *outlier* resulting from the erosion of a thrust sheet or ◊ *nappe*, so that an isolated mass of rock is separated from the underlying 'basement' by a thrust plane. (◊ Fig. 91.) On a map this appears as a continuous closed outcrop of the thrust plane. In the Swiss Alps klippen occur in which older rocks rest upon younger ones. (Cf. ◊ *Window.*)

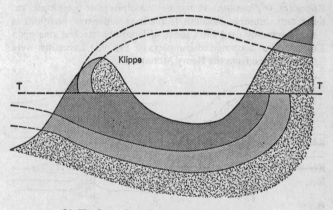

91. The formation of a klippe. T–T, thrust plane.

Knick point. The intersection between the old and new ◊ *long-profiles* of a river arising as a result of the process of ◊ *rejuvenation.* (◊ Fig. 126.)

Kopje (Koppie) (Afrikaans). An ◊ *inselberg.*

Kraton (Craton). A major structural unit of the Earth's ◊ *crust*, consisting of a large stable mass of rock, generally igneous and/or metamorphic, sometimes with a thin veneer of sediment. A typical kraton is the Canadian (Precambrian) Shield. Blocks which have been termed 'oceanic' or 'submarine' kratons probably do not conform to this definition. 'Kraton' is practically synonymous with ◊ *'shield'.*

Kungurian. A stratigraphic stage name for the top of the East European and Russian Lower ◊ *Permian.*

Kyanite. ◊ *Aluminium silicates* and Appendix.

Kylite. ◊ *Theralite.*

Labradorite. A variety of plagioclase ◊ *feldspar.*

Labrum. ◊*Echinodermata* (Echinoidea) and Fig. 45.

Laccolith. An intrusive, dome-like mass of igneous rock which arches the overlying sediment and which has a more or less flat floor (cf. ◊ *Phacolith*). It may be ◊ *composite* or ◊ *multiple*, but the term multiple laccolith (sometimes cedar-tree laccolith) is also commonly used for a series of laccoliths 'stacked' one upon the other in a connected complex. (◊ Fig. 92.) Laccoliths were first described from the Henry Mountains in Utah.

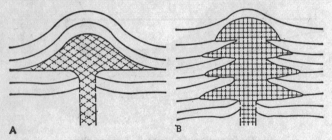

A   B

92. (A) A simple laccolith; (B) A cedar-tree laccolith.

Ladinian. A stratigraphic stage name for the top of the European Middle ◊ *Triassic.*

Ladogian. A stratigraphic division of the Baltic ◊ *Precambrian*, the older Karelian.

Lag. ◊ *Fault*; Fig. 56.

Lahar. Mud-flows and deposits resulting from torrential downpours of rain, mass melting of snow and ice, or the breaching of a crater lake acting upon loose fine-grained ◊ *pyroclastic* material (ash) on the flanks of a volcanic cone. (◊ *Gravity transport.*)

Lamarck, Jean Baptiste (1744–1829). French naturalist considered by some to have originated the science of invertebrate palaeontology. His studies led him to believe evolution to be a slow organic process much influenced by environment, as opposed to the theory of his contemporary ◊ *Cuvier*, which was much more popular at the time.

Lamellar flow. ◊ *Laminar flow.*

Lamellibranchiata. ◊ *Mollusca.*

Laminar flow. Fluid flow without turbulence, i.e. 'fluid particles' flow or move in almost straight lines, parallel to the channel walls. Laminar flow is characteristic of low velocities, and particles of sediment in the flow zones are moved by rolling or ◊ *saltation.* The term lamellar flow is effectively a synonym. ('Fluid' includes both gases and liquids; hence, both wind and water currents may develop laminar flow.)

Lamination. The development of thin, discrete layers of rock. The individual layers may differ from one another or may be homogeneous, the layers being separated by a physical discontinuity, e.g. ◊ *bedding* planes. (For igneous lamination, ◊ *Layered igneous structures.*)

Lamprophyre. Medium-grained, commonly ◊ *porphyritic,* intermediate rocks, which contain a predominance of ◊ *ferromagnesian minerals,* especially as ◊ *phenocrysts.* The common ferromagnesian minerals are associated with ◊ *orthoclase,* various ◊ *plagioclases,* and occasionally with ◊ *feldspathoids.* Quartz is an uncommon accessory mineral. ◊ *Zoning* of the ferromagnesian minerals is fairly common, especially in the ◊ *pyroxenes* and ◊ *amphiboles.*

Lamprophyres are often highly altered, the feldspars in particular being heavily decomposed. Fresh ◊ *olivine* is exceedingly rare and it is almost always altered to serpentine and carbonate. In many lamprophyres carbonate is so abundant that the rocks frequently effervesce with acid, like a limestone. This widespread alteration is presumably due to late-stage ◊ *hydrothermal* activity. Since lamprophyres are merely ◊ *melanocratic* varieties of microsyenite (◊ *Syenite*) and microdiorite (◊ *Diorite*), their chemistry is more or less identical with these types, although the high ferromagnesian mineral content causes the iron and magnesium to be higher and the silica rather lower. The calcium content should be regarded with some suspicion, owing to the previously mentioned widespread decomposition of some minerals into carbonates. This will also appear on analysis as an unusually high CO_2 content.

The classification is based upon the nature of the ferromagnesian minerals, and the type of feldspar and feldspathoid:

	Orthoclase	*Plagioclase*	*No Feldspar*
Biotite	Minette	Kersantite	Alnöite
Hornblende	Vogesite	Spessartite	
Barkevicite and/or Augite		Camptonite	Monchiquite

(Spessartite, the rock name, should not be confused with spessartite, the manganese ◊ *garnet*.)

Alnöite, which contains melilite, is one of the least silicic igneous rocks known (other than ◊ *carbonatites*, with which it is associated) – the type material from Alnö in Sweden contains only 27·3% SiO_2. Minettes and vogesites are commonly associated with granites, and granodiorites, and occasionally with syenites. Kersantites and spessartites are associated with diorites, and occasionally with granodiorites and monzonitic types, while camptonites and monchiquites are associated with syenites, especially undersaturated types and alkali gabbros.

Lamprophyres occur mainly as dykes and sills. In part, ◊ *appinites* are the coarse-grained equivalents of lamprophyres, being perhaps more accurately described as mela-diorites or mela-syenites.

Certain melanocratic ◊ *trachytes* may perhaps be regarded as volcanic equivalents of lamprophyres.

Lamp shells. A popular name for the ◊ *Brachiopoda*, especially the Terebratulid group.

Land bridge. A hypothetical transoceanic connection between continents which permitted migration of terrestial organisms. A theory of Earth history which demands fixed relative positions for the continents and ocean basins also requires ephemeral land bridges to account for the distribution of fauna and flora. Other evidence for the existence of land bridges is scanty, and the theory of ◊ *continental drift* may be preferred.

Landenian. A stratigraphic stage name for the European Upper Palaeocene (◊ *Tertiary System*).

Landscape marble. A ◊ *limestone* which shows patterns resembling a landscape scene when cut at right angles to the bedding. The most famous British landscape marble comes from the Rhaetic Beds near Bristol, and shows dark convoluted masses, resembling trees, arising from a darker layer and enclosed within a lighter sediment. It has been suggested that the dark masses are 'globules' of the darker, rather bituminous material, which have been forced into the lighter coloured calcareous sediment (while both are still soft) by the action of gas produced by decaying organisms in the bituminous layer. However, it has been pointed out that this does not explain all the features, and alternative suggestions that the masses are algal growths or some kind of exaggerated ◊ *loadcast* have been made.

Landslip, Landslide. ◊ *Gravity transport.*

Lapiés. A limestone pavement. (◊ *Karst scenery.*)

Lapilli. Fragments of ◊ *pyroclastic rocks* measuring between 4 and 32 mm.

Lapis-lazuli. The semi-precious variety of lazurite. (◊ *Feldspathoids.*)

Lappets (Mollusca). ◊ Fig. 105.

Lapworth, Charles (1842–1920). English professor who established the Ordovician System, by using parts of ◊ *Sedgwick's* Cambrian System and ◊ *Murchison*'s Silurian System, thus solving the problem of where the division lay between them. By his studies in Scotland he made important contributions to the understanding of metamorphism, and of the graptolites.

Lardalite (Laurdalite). A nepheline-bearing ◊ *larvikite.* (◊ *Syenite.*)

Larvikite (Laurvigite). A coarse-grained ◊ *syenite* consisting largely of anorthoclase ◊ *feldspar* showing blue ◊ *schillerisation.* The ferromagnesian minerals occur as clots, which consist of titanaugite, ferro-olivine, apatite, biotite and magnetite. Extensively used as an ornamental stone when cut and polished.

Laterite and bauxite. Residual deposits formed under special climatic conditions in tropical regions. Laterite consists essentially of hydrated iron oxides, bauxite of hydrated aluminium oxides, the most common impurity in both being silica; aluminous laterites and ferruginous bauxites are quite common.

The essential climatic feature is a well-marked division of the year into wet and dry seasons. During the wet season, leaching of the rock occurs, and during the dry season the solution containing the leached ions is drawn to the surface by capillary action, where it evaporates, leaving salts to be washed away during the next wet season. Thus the whole zone, from the lowest level to which the water-table falls to the highest level it reaches, is progressively depleted of the more easily leached elements, e.g. sodium, potassium, calcium, and magnesium. A solution containing these ions may have the correct pH to dissolve silica in preference to aluminium oxides or iron oxides. Thus the residuum consists mainly of these two oxides; which of them is predominant depends mainly on the characters of the original rock type; e.g.

basic igneous rocks and other iron-rich types tend to yield laterite, while granitic rocks and others low in iron yield bauxites. An essential property of the rock is that it should maintain a porous structure during the leaching process, so that fluids may circulate freely.

Fossil bauxites and laterites are known, especially interbedded between series of lava flows (bole), and this is generally taken as evidence of sub-aerial extrusion and a tropical or sub-tropical climate. Occasionally bauxite and laterite are eroded and re-deposited, although it is unlikely that they can be transported over great distances. They may become mixed with other sedimentary material, to form such rocks as bauxitic clays, etc. Spheroidal masses sometimes develop in bauxites and laterites, and the term pisolitic has been applied to these rocks, although there is no evidence that the structures have been produced in the same way as true ◊ *pisoliths*.

Latite. A synonym of ◊ *trachyandesite*.

Lattice. ◊ *Unit cell*.

Lattorfian. A stratigraphic stage name for the European Lower Oligocene (◊ *Tertiary System*).

Laurasia. The name given to the hypothetical northern hemisphere 'super-continent' of North America, Europe, and Asia north of the Himalayas, prior to its break up under the forces causing ◊ *Continental drift*. It corresponds to ◊ *Gondwanaland* in the southern hemisphere. (◊ *Pangaea*.)

Laurdalite. ◊ *Lardalite*.

Laurentian. A stratigraphic division of the Canadian ◊ *Precambrian*, the pre-Huronian granite intrusion into the ◊ *Ontarian* in Canada.

Laurvigite. ◊ *Larvikite*.

Lava. The material extruded by a ◊ *volcano* which consists of molten or part-molten silicate material (cf. ◊ *Pyroclastic rocks*; ◊ *Volatiles*). A single outpouring of lava, called a lava flow, may arise from a central-type vent or a fissure. Fissure eruptions give rise to very widespread lava-covered areas, e.g. the Deccan of India; Snake River, British Columbia; the British–Icelandic–Greenland Province. Lavas may be ◊ *vesicular* or amygdaloidal (◊ *Amygdale*), ◊ *glassy* or partially or wholly crystalline, and even-grained or ◊ *porphyritic*.

In composition they range from ◊ *acidic* to ◊ *ultrabasic*, although ◊ *basic* lavas account for more than 90% of the whole. It will be observed that all the large areas of lava mentioned are basaltic in character. Acid lavas are highly viscous and rarely cover very large areas, while basic lavas have a much lower viscosity and flow readily. Lavas which are extruded under water commonly take on the form of distorted globular masses, the characteristic so-called pillow shape. They apparently form as a result of the rapid chilling of the outer skin of the lava which forms at the liquid/liquid interface, resulting in a more or less spherical 'balloon' of lava. As this grows in size, it flattens under its own weight. The pillows have highly vesicular cores.

Lava flows interbedded with sedimentary rocks may be confused with concordant ◊ *sills*. The following criteria may be used for distinguishing purposes:

Sill	*Lava flow*
Uniform and regular	Top generally irregular and often cindery
Rarely vesicular or amygdaloidal	Commonly vesicular or amygdaloidal
Medium to fine-grained, rarely glassy, owing to slower cooling	Fine-grained or glassy owing to rapid cooling
Does not show weathering of upper surfaces	May show weathering of upper surfaces
Metamorphoses the upper and lower contact rocks	Can metamorphose the lower contact only
Stringers and veinlets of rock may penetrate both overlying and underlying rocks	Small veins in underlying rocks only. Care must be taken when flows have irregular tops
No fragments in adjoining sediments	Fragments may occur in overlying rocks, as pebbles or xenoliths

Any evidence that the rock layer transgresses the bedding planes indicates that it is a sill.

(◊ *Aa*; *Alkali basalt*; *Andesite*; *Basalt*; *Columnar structure*; *Flow structure*; *Pahoehoe*; *Pitchstone*; *Rhyolite*; *Trachyte*.)

Law of Included Fragments. The law stating that if recognisable fragments of one rock are found enclosed in another rock, then the enclosing rock must have been formed later than the rock material of the included fragments.

Law of Rational Intercepts. ◊ *Parameters*.

Law of Strata Identified by Fossils. The second of the two principles of ⟐ *stratigraphy* enunciated by William ⟐ *Smith* stating that each bed (or group of beds) contains a characteristic assemblage of fossils. (Cf. ⟐ *Superposition, Principle of*.)

Laxfordian. A stratigraphic division of the Scottish ⟐ *Precambrian*, the younger Lewisian. (Cf. ⟐ *Scourian*.)

Layer-lattice minerals. The sheet ⟐ *silicates* or phyllosilicates. The layer-lattice group of minerals has several important sub-groups, including the ⟐ *chlorite group*, the ⟐ *clay minerals*, and the ⟐ *micas*.

The basic unit is a sheet of linked $[SiO_4]$ tetrahedra, three oxygens from each tetrahedral unit being shared, thus giving the basic unit $Si_4O_{10}(OH)_2$. These are combined in two- or three-layer units with either a gibbsite layer $(Al_4(OH)_6)$ or a brucite layer $(Mg_6(OH)_6)$. Layer-lattice materials with a gibbsite layer are termed dioctahedral, and those with a brucite layer are termed trioctahedral:

	Dioctahedral	Trioctahedral
2 layer	Kaolin group (⟐ *Clay minerals*)	⟐ Serpentine
3 layer	⟐ *Pyrophyllite* Muscovite ⎫ Margarite ⎬ ⟐ *Micas* Montmorillonite ⎫ ⟐ *Clay* Illite ⎰ *minerals*	⟐*Talc* Biotite (⟐ *Micas*) ⟐ *Chlorite* ⟐ *Vermiculite*

Layered igneous structures. The term 'layering' is preferred to 'banding' so as to imply a three-dimensional rather than a two-dimensional feature. Certain basic igneous rock masses display various sorts of layered structure, in some cases amounting to a kind of pseudostratification. Examples are the Bushveld of South Africa, the Skaergaard Complex of East Greenland and the Stillwater Complex of Montana, U.S.A. The following types have been recognised:

LAYERING. The development of sheets, varying in thickness from a few centimetres to 1–2 m., of igneous rock having distinct characteristics. Contacts of layers are usually sharp, but may be gradational.

RHYTHMIC LAYERING. The occurrence at regular intervals of a layer of a characteristic rock type, or sometimes a regular repetition of a sequence of rock types. The composition of the minerals present in a single layer is effectively constant (cf. cryptic layering, below).

CRYPTIC LAYERING. The gradual change in composition of the individual minerals from the bottom to the top of an intrusion. It is not generally possible to detect this by inspection in the field or in hand specimens; a thin section, chemical analysis, or both, is essential. In some large intrusions more than one unit of cryptic layering has been recorded. The gradations found include an increase in the Fe/Mg ratio in the ◊ *mafic* minerals, and an increase in the Na/Ca ratio in the feldspars.

PHASE LAYERING (MINERAL LAYERING). The sudden appearance or disappearance of a particular mineral from one layer to the next. This is particularly striking where it involves a marked change in the appearance of the rock. Some phase layers are extremely thin, even less than a centimetre.

IGNEOUS LAMINATION. A kind of 'pseudo-schistosity' developed in layered complexes as a result of the crystallisation and deposition of platy crystals with their long axes parallel to the plane of layering. The actual distribution of the long axes is generally random, although a few cases of lineation have been recorded.

'INCH-SCALE' LAYERING. A rhythmic phase layering on a very small scale, yielding alternating bands two to three centimetres thick, of plagioclase and pyroxene.

(◊ *Cumulates.*)

Lazurite. ◊ *Feldspathoids* (Complex).

Leaching. ◊ *Weathering* (Chemical).

Lead glance. A synonym of galena, PbS. (◊ *Glance* and Appendix.)

Lechatelierite. Natural silica ◊ *glass.*

Lectotype. ◊ *Type.*

Ledian. A stratigraphic stage name for the base of the European Upper Eocene (◊ *Tertiary System*).

Leintwardinian. A stratigraphic stage name for the base of the British Upper ◊ *Ludlovian* (Silurian System).

Length, slow or fast. These terms are used to describe the relative speed (slowness or fastness) of the light vibrating parallel to the direction in which a mineral is elongated. (◊ *Vibration directions.*)

Leonardian. A stratigraphic stage name for the North American Lower ◊ *Permian.*

Leonardo da Vinci (1452–1519). He was responsible for one of the earliest recorded opinions on the organic origin of fossils – completely contrary to the accepted ideas of his time that they were the results of celestial influence. He also commented on the erosive

265

power of running water, and the presence of water-lain sediments in areas high above sea-level.

Leonhard, Carl von (1779–1862). German professor who produced one of the best macroscopic studies of the petrology of rocks, considering their modes of origin and occurrence as well as mineralogical composition.

Lepidoblastic. A texture term applied to foliations or schistosity (◊ *Cleavage, rock*) produced by planar minerals, e.g. mica. (Cf. ◊ *Nematoblastic.*)

Lepidocrosite. ◊ *Limonite.*

Lepidolite. ◊ *Micas* (Trioctahedral) and Appendix.

Leptite (Leptynite). Streaked or banded varieties of ◊ *granulite.*

Leuc-, leuco- (prefix). White, colourless.

Leucite. ◊ *Feldspathoids* and Appendix.

Leucitite. ◊ *Alkali basalt.*

Leucitophyres. ◊ *Undersaturated* ◊ *trachytes* containing dominant ◊ *leucite.*

Leucocratic. A term applied to igneous rocks consisting mainly of light-coloured minerals. 'Light minerals' are defined as quartz, feldspars, feldspathoids, white micas, topaz, etc. Many rocks consisting mainly of such minerals are dark in hand-specimen since the minerals are often dark in the mass. A thin section is essential for determining the ◊ *colour index*, which should be less than 30 on Shand's scale. (Cf. ◊ *Mesotype*; ◊ *Melanocratic*; ◊ *Hypermelanic.*)

Levantian. A stratigraphic stage name for the top of the European Pliocene (◊ *Tertiary System*).

Levée. A raised bank formed by deposition, which allows the level of the river to rise above its ◊ *flood plain.*

Lewisian. A stratigraphic division of the Scottish ◊ *Precambrian*, the oldest in Britain, covered unconformably by Torridonian.

Lherzolite. An ◊ *ultrabasic rock* consisting of ◊ *olivine* and ortho- and clino-◊*pyroxenes.*

Lias. The Lower ◊ *Jurassic.*

Ligament area (Mollusca). ◊ Fig. 101.

Ligerian. A stratigraphic stage name for the European Lower ◊ *Turonian* (Cretaceous System).

Light mineral. (1) ◊ *Leucocratic.*
(2) ◊ *Heavy mineral.*

Lignite. 'Brown ◊ *coal*'.

Limb. The part of a ◊ *fold* which lies between one hinge and the next. (◊ Fig. 60.)

Limburgite. ◊ *Basalt* glass containing olivine and augite ◊ *pheno-crysts.* Analysis suggests that it is commonly ◊ *undersaturated.*

Limestone. The term limestone is applied to any sedimentary rock consisting essentially of ◊ *carbonates.* The two most important constituents are calcite and dolomite, but small amounts of iron-bearing carbonates may also occur. (For rocks consisting essentially of siderite, ◊ *Sedimentary iron ores.*)

Limestones can be classified into three major groups: organic; chemical; and detrital or clastic. Organic and chemical limestones are known as ◊ *autochthonous* limestones, whereas the clastic limestones are known as ◊ *allochthonous.* Limestones may be freshwater or marine, and usually indicate deposition in a warm, clear-water environment.

Many of the commonly occurring limestones contain organic, detrital, and chemically precipitated material in varying propor-tions. Both calcite (hexagonal $CaCO_3$) and aragonite (orthorhom-bic $CaCO_3$) are present in modern limestone accumulations. However, since aragonite is more easily dissolved or converted to calcite, it is absent in ancient limestones. These changes, and the ease with which carbonates recrystallise during ◊ *diagenesis*, make it not uncommon for characteristic features of a rock to be lost, resulting in uncertainties as to its mode of origin. This is par-ticularly true of the ancient limestones.

ORGANIC LIMESTONES. Many plants and animals secrete $CaCO_3$, and, on the death of the organisms, their skeletons, either complete or broken up, may in the absence of any other sedimen-tary material accumulate to form a limestone. Special types of organic limestone are:

Reef limestones — biohermal limestones

Shelly limestones
Coral limestones
Algal limestones — collectively known as biostromal limestones
Crinoidal (◊ *Echinodermata*) limestones

Foraminiferal (◊ *Protozoa*) limestones — some biostromal, some ◊ *pelagic* varieties

Chalk (more or less synonymous with pelagic limestone) is the characteristic Upper Cretaceous white, fine-grained limestone, but 'chalks' of other ages have been recorded, e.g. Tertiary. (When used in the stratigraphical sense 'Chalk' must have a capital 'C'.) The type 'Chalk' from south-east England consists

267

of a small percentage of foraminiferal material, some finely comminuted shell material, and a high proportion of calcareous skeletal material from unicellular planktonic algae (◊ *coccoliths*).

CHEMICALLY PRECIPITATED LIMESTONES. There are three main types: (a) limestones belonging to an ◊ *evaporite* sequence (often dolomite); (b) oolitic and pisolitic limestones (◊ *Oolith*); (c) ◊ *calc tufa*.

CLASTIC (DETRITAL) LIMESTONES. These mechanically deposited carbonate rocks are made up essentially of fragments of organic carbonate or a pre-existing limestone rock. (Oolitic limestone is included in this group by some authors.) The classification of clastic limestones is based upon the same grain-size scale as used for the detrital ◊ *sedimentary rocks*:

Calci-rudite	2 mm. and above
Calcarenite	1/16 mm.–2 mm.
Calci-lutite	Less than 1/16 mm.

If the debris is dominantly organic in origin the prefix 'bio' may be used; bioclastic limestones are not uncommonly found interbedded with biostromal limestones. The term coquina is applied to cemented coarse shell debris. Some of the shelly 'grits' of the English Mid-Jurassic are coquinas. Note that the term 'calcite mudstone' is synonymous with calci-lutite, and does not imply an argillaceous limestone or calcareous clay or shale.

The three-fold division of limestones given, though theoretically ideal, is increasingly being replaced by a more practical rock nomenclature. This terminology is based upon the nature of the limestone matrix, and of the particle types present. Two matrix types are distinguished. Their names form the suffix of the resultant rock names:

Micrite: micro-crystalline calcite (grain size less than 0·01 mm.)

Sparite: coarse calcite (greater than 0·01 mm.) forming a mineral cement

The prefixes of the limestone names are determined by the particle types present. These may include (prefixes are in parentheses):

◊ *Intraclasts* (Intra-)

◊ *Ooliths* (Oo-)

◊ *Pellets* (Pel-)

Organic hard parts, whole or fragmented (Bio-)

Thus an oolitic rock with a sparite cement is termed an oosparite, whereas one composed of intraclasts set in micrite is an intra-micrite. A pure calci-lutite may be termed a micrite, and a limestone composed entirely of reef-building organisms, a biolithite. Rocks consisting of only micrite and sparite are termed dismicrites. This system of naming limestones has practical advantages, since it does not depend upon the determination of the origin of the sediment.

Many limestones (Chalk being a notable exception) contain a significant proportion of detrital material (sand and/or clay) as an impurity. Limestones are important as building stones and as raw materials in the manufacture of cement and as reservoir rocks for ◊ *oil accumulations*. Powdered limestone is an important agricultural commodity.

(◊ *Dolomite*; *Dolomitisation*; *Karst scenery*.)

Limnic. A descriptive term applied to the environment of freshwater lakes. Used especially of coal basins, as the opposite of ◊ *paralic*.

Limonite. The omnibus term used for a range of mixtures of hydrated iron oxides and iron hydroxides. Originally it was thought to be a mineral with a definite composition – $2Fe_2O_3,3H_2O$, or similar, but it is now known to consist of various amorphous and ◊ *cryptocrystalline* constituents. The following are the most important: goethite ($\alpha FeO . OH$), which on heating yields haematite (Fe_2O_3); lepidocrosite (γFe_2O_3); turgite – haematite with absorbed water.

Limonite also includes various amorphous iron hydroxides. It occurs widely as the weathering product of all iron-containing minerals, and is frequently intermixed with many other substances including colloidal silica, clay minerals, aluminium hydroxides, etc. The most characteristic property of the mineral is its yellow or brownish-yellow streak. Its colour ranges from yellow through various shades of orange, brown, and red, to black, and this is a function of particle size, the proportions of the various constituents, and the degree of hydration. Amongst well-known types of limonite are deposits of ◊ *gossan*, red and yellow ochres, ◊ *bog iron ore*, and ◊ *laterite*.

(◊ Appendix.)

Lineament. A large-scale linear feature which expresses itself in terms of topography, which is in itself an expression of the underlying structural features. (◊ Fig. 93.) Such features may include: valleys controlled by faulting or jointing; fronts of mountain ranges; narrow, straight mountain or hill ranges and ridges;

lines of isolated hills; straight coastlines, modern and ancient. Structural features associated with lineaments include ◊ *fault-zones*, fracture (joint) zones, ◊ *fold* axes, linear igneous intrusions, lines of volcanoes or volcanic fissure eruptions. Lineaments are

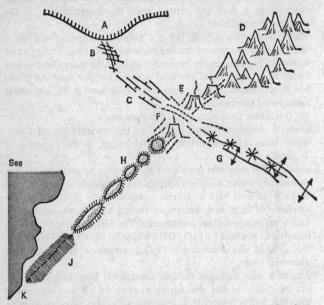

93. Types of lineament. (A) The edge of a marine transgression (unconformity); (B) Joints; (C) Faults; (D) Mountain front; (E) Volcanoes; (F) Mineral veins and dykes; (G) Fold axes; (H) Hills; (J) Ridges; (K) Straight shoreline.

often located by means of a detailed study of maps (both geological and topographic) and aerial photographs (◊ *Photogeology*). The correlation of these major structural trends is important for the elucidation of regional tectonic history.

The term 'linear' (as a noun) is not a recommended synonym.

A megalineament is a very large lineament, developed on a global scale, e.g. the ◊ *mid-oceanic ridges*, ◊ *rift valleys*, ◊ *island arcs*, and major mountain ranges.

Lineation (Linear structure). Lineation is any one-dimensional feature in a rock or shown on a rock surface. (◊ Fig. 94.) It may

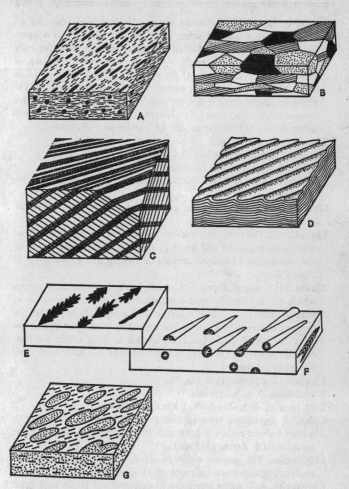

94. Types of lineation. (A) Mineral, by growth or mechanical orientation; (B) Mineral, by directed growth during recrystallisation; (C) Intersection of bedding and cleavage; (D) Phyllitic; (E) Fossil lineation (graptolites); (F) Fossil lineation (belemnites); (G) Lineation by deformation of a mineral or of rock particles.

arise in any of the following ways: (1) By a linear-parallel arrangement of minerals, either by growth or mechanical orientation; (2) By the intersection of a planar structure with the rock surface; (3) By the intersection of two planar structures, e.g. ◊ *bedding* and ◊ *cleavage*; (4) By the development of a series of small parallel puckers in a planar structure, e.g. microfolds (= phyllitic lineation); (5) By the deposition of elongate particles (sometimes fossils, especially elongate ones) in a preferred orientation during sedimentation; (6) As ◊ *striae* or ◊ *slickensides*.

Linguoid. Tongue-shaped.

Lip, inner and outer (Mollusca). ◊ Fig. 100.

Liparite. A synonym of ◊ *rhyolite*.

Liptobiolite. Peat and coal formed mainly of resins, waxes, spores and pollen.

Liptobioliths. Fossil gums and resins, including amber and copal. (◊ *Hydrocarbon minerals*.)

Liquid limit. The minimum amount of water which has to be mixed with a sediment or soil so that, under standard conditions of test, the material becomes capable of flowing like a liquid. (Cf. ◊ *Plastic limit*.)

Listric (lit. 'spoon-shaped'). A term applied to fracture planes which curve, either to near-horizontal ones which steepen, or to near-vertical ones which become less steep. Those which are concave upwards conform to the definition most strictly, and the term 'negatively listric' has been applied to those which are concave downwards. The term 'shoulder-thrust' has also been used for the latter type.

Listrium (Brachiopoda). ◊ Fig. 12.

Litchfieldite. ◊ *Alkali syenite*.

Lith- (prefix), **-lith, -lite** (suffix). Rock, rock-like.

Lithic. A descriptive term applied to rock fragments occurring in a later formed rock, e.g. lithic tuff. (◊ *Pyroclastic rocks*; for lithic sandstones, ◊ *Arenaceous rocks*.)

Lithifaction. The process which results in the formation of a massive rock from a loose sediment. (◊ *Diagenesis*.)

Lithofacies. A ◊ *facies* particularly characterised by its rock type. (Cf. ◊ *Biofacies*.)

Lithographic stone. An extremely fine-grained compact ◊ *limestone*, slabs of which were formerly used as lithographic plates. The best-known example is the ◊ *Kimmeridgian* stone from Solenhofen, Bavaria, famous for its well-preserved fauna.

Lithology. A term usually applied to sediments, referring to their general characteristics. It is used in several rather loose senses, e.g. 'lithological variation', referring to variation in composition and texture; 'lithology of a formation', meaning the rock types present in a stratigraphic sub-division. Lithology generally relates to descriptions based upon hand-specimens and outcrops rather than microscopic or chemical features.

Lithophile. Elements having a strong affinity for oxygen, concentrated in the silicate minerals. (◊ *Geochemistry*.)

Lithophysae. Hollow ◊ *spherulites*.

Lithosphere. The outer, rigid, part of the Earth's ◊ *crust*, consisting of the surface rocks, the ◊ *sial*, and the upper part of the ◊ *sima*. This part of the crust has a high strength as compared with the ◊ *asthenosphere*.

Lit-par-lit injection. The forcible intrusion of igneous material, usually granitic in character, along ◊ *bedding-planes*, foliation planes (◊ *Cleavage, rock*), etc. to give a close-spaced, more or less parallel series of veinlets. If the process is extensive and regular, a ◊ *gneiss*-like rock may be produced. Reaction between the invading magma and the country rock produces a ◊ *migmatite*. (Cf. ◊ *Stockwork*.)

Littoral. The ◊ *environment* between the highest and lowest levels of spring tides. (◊ Fig. 95.) This is practically synonymous with

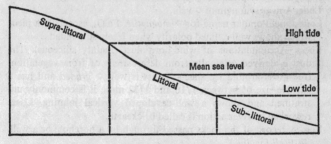

95. The littoral zone.

'beach'; more precisely with 'fore-beach'. (The area between the cliff line and high-water mark is the 'back-beach', which is usually where a 'storm-beach' is developed.) Littoral deposits are mainly sands and/or pebbles and sometimes boulders, more rarely mud. ◊ *Longshore drift* and the immediate coastal rocks will largely control the character of such deposits. Some littoral zones are bare rock, not covered by a veneer of sediment. Littoral faunas and floras have special characteristics to enable them to survive the continuously changing environment.

Llandeilian. A stratigraphic stage name for the European Lower ◊ *Ordovician.*

Llandoverian. A stratigraphic stage name for the European Lower ◊ *Silurian.*

Llanvirnian. A stratigraphic stage name for the European Lower ◊ *Ordovician.*

Loadcast. When a bed of sand is deposited on a layer of plastic mud, there is a tendency for the sand to press downwards into the mud in the form of bulbous projections, owing to unequal loading from above. These projections (which may superficially resemble ◊ *ripple marks* or ripple casts) are known as loadcasts. Occasionally they may be accentuated by later deformation (◊ *Collapse*) or by post-◊-*diagenetic* deformation.

Load of a river. The total amount of material transported by a river by ◊ *traction,* ◊ *saltation,* ◊ *suspension,* and in solution. (◊ *Transportation of sediment.*)

Load pressure. ◊ *Pressure.*

Loam. A ◊ *soil* containing sand, silt, and clay in roughly equal proportions. (◊ *Particle size.*)

Lobe, axial, occipital, and palpebral (Arthropoda). ◊ Fig. 3.

Lode. A near synonym of ◊ *vein.*

Lodestone. Popular name for ◊ *magnetite,* Fe_3O_4, especially a piece which shows well-defined polarity when freely suspended.

Loess. Accumulations of wind-born dust (mainly siliceous). The dust is derived originally from desert areas or from vegetation-free areas around ice sheets. Loess is well ◊ *graded* and has a particle size of between 1/16 and 1/32 mm.; it is commonly unstratified and shows a well-developed vertical jointing. Loess reworked by river action is called brickearth.

Log. A record of the rocks passed through by a borehole or a well. (◊ *Well-logging.*)

Logan, William (1798–1875). Canadian geologist who, while working in Wales, solved the problem of the origin of coal by his discovery of Stigmaria in rocks underlying coal seams. Later as head of the newly formed Geological Survey of Canada – then completely unknown – he set out by canoe to map the country. His most outstanding discovery was the vast extent and variation in age of the Precambrian Laurentian Shield, which he broadly mapped in its entirety during his lifetime.

Longmyndian. A stratigraphic division of the English ◊ *Precambrian*, the Shropshire Post-◊-*Uriconian* sediments.

Long-profile. The long-profile of a river is a curve which may be drawn as a graph on which height of land is plotted along the y-axis against distance from source along the x-axis. The curve can be represented by a formula of the type y = a log x+K. It is asymptotic towards the ◊ *base level*, which is usually the sea in a mature river.

The river profile of a newly emergent stream is irregular, the gradient being interrupted by waterfalls, rapids, and lakes. Each one of these irregularities acts as a temporary local base level and the long-profile appears as a series of arcs. As the river matures these irregularities are eradicated – the lakes are in-filled and the rapids and waterfalls worn back and reduced – until, in theory, the ideal curve is achieved in old age, although this condition is rarely, if ever, attained.

Should ◊ *rejuvenation* take place, the whole of the long-profile will be elevated. Adjustment will then proceed from the point of elevation towards sea level, and regress towards the source. This point of elevation will become a temporary local base level for the profile above this point. This junction of the two profiles is called the (k)nick point, which will migrate upstream as adjustment proceeds, until the ideal profile is attained or a second rejuvenation takes place. This state of affairs would involve three profiles and two knick points, and so on. Three or more rejuvenations of a river have not infrequently been recorded.

Longshore drift. The process by which material is transported along the shore. On many coasts the dominant wave-direction is oblique to the shore-line and material is thus carried obliquely up the beach, but is then moved straight down the beach by the backwash. The resulting action is that material is carried side-ways. This action is not confined to the immediate coastline, but may take place down to the lower limit of wave action. The drift may in some cases threaten to remove sand from the beaches, but this may be arrested by the erection of groynes across the beach. (Cf. ◊ *Saltation*.)

Lopolith. A saucer-shaped igneous intrusion which is concave upwards. (Cf. ◊ *Phacolith*.) Small lopoliths can be seen to have intruded into folded strata and to have their shape controlled by the folding. (◊ Fig. 96.) However, the name is also given to very large basin-shaped masses of igneous rock, e.g. the Bushveld Complex of South Africa, and the Sudbury (Ontario), and Duluth

(Minnesota) lopoliths, where the control by folding is much less obvious. These large lopoliths are amongst the greatest masses of igneous rock seen at the Earth's surface, and are to the basic rocks what ◊ *batholiths* are to the granites. The term 'mega-

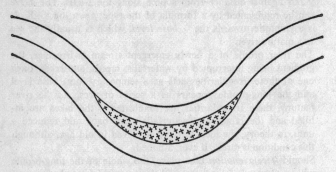

96. A lopolith.

lopolith' is here suggested for the larger kind of lopolith. They commonly display layered and banded structures, and clearly their origin must be different from that of the smaller lopoliths mentioned above, which are most nearly related to ◊ *sills*. It has been suggested that the larger lopoliths may not, in fact, be saucer-shaped but funnel-shaped.

Lowestoftian. A stratigraphic stage name for the British mid Middle Pleistocene (◊ *Tertiary System*).

Lucitanian. A synonym of ◊ *Corallian* (Jurassic System).

Ludhamian. A stratigraphic stage name for the British Lower Pleistocene (◊ *Tertiary System*).

Ludian. A stratigraphic stage name for the base of the European Lower Oligocene (◊ *Tertiary System*).

Ludlovian. A stratigraphic stage name for the European Upper ◊ *Silurian*.

Lugarite. ◊ *Theralite*. (◊ *Alkali gabbro*.)

Lumachelle. A consolidated shell bank or shell conglomerate, usually of molluscan remains. (◊ *Coquina*.)

Lunule. (1) A perforation passing through the test of some advanced echinoids. (◊ *Echinodermata*, Fig. 45.)

　　(2) ◊ *Mollusca*, Fig. 101.

Luscladite. ◊ *Theralite*. (◊ *Alkali gabbro*.)

Lustre (Luster). The character of the light reflected by minerals. This is a useful diagnostic property of minerals, generally described qualitatively and in comparative terms. The following are the most commonly used descriptive terms: Adamantine (diamond-like); Metallic; Pearly; Resinous; Silky; Splendent or specular (mirror reflection); Vitreous (glass-like); Waxy. Ideally, lustre should be described from fresh, unweathered or unoxidised, surfaces, especially freshly cleaved surfaces. The lustre of single crystals may differ from crystal aggregates, e.g. single crystals of gypsum have a pearly or vitreous lustre, while fibrous gypsum has a silky lustre.

Lustre mottling. ◊ *Poikilitic.*

Lutetian. A stratigraphic stage name for the European Middle Eocene (◊ *Tertiary System*).

Lutite. ◊ *Argillaceous rocks.*

Luxullianite. ◊ *Pneumatolysis* (Tourmalinisation).

Lyddite (Lydian stone). A dense black variety of ◊ *chert*, formerly used as a ◊ *touchstone.*

Lyell, Charles (1797–1875). Scottish geologist who won universal fame for his *Principles of Geology* which ran to 11 editions in his lifetime and had a decisive influence on the future development of the science. His main thesis, based upon the work of ◊ *Hutton*, was the uniformity of natural forces in the past and present, working in a process of slow and unending change. This brought him into opposition with the disciples of both ◊ *Werner* and ◊ *Cuvier*, but his principle of uniformitarianism has since become a basic tenet of geology. He suggested the terms Eocene, Miocene, Pliocene, and Pleistocene for the divisions of the Tertiary and attempted to define the limits of the Cambrian and Silurian (◊ *Lapworth*). As a result of his theory he also supported the fluvialist doctrine of river erosion of valleys at a time when most British geologists were diluvialists (believing that rivers had kept the same form since the Flood). His most notable failures were his inability to accept a theory of slow animal evolution, at least until he was influenced by Darwin, and his rejection of ◊ *Agassiz*'s concept of an Ice Age in favour of the Drift Theory, which became popular in Britain simply because of his influence.

Maar. The explosion vent of a ◊ *volcano.*

Maastrichtian (Maestrichtian). A stratigraphic stage name for the top of the ◊ *Cretaceous,* in Europe.

MacCulloch, John (1773–1835). Scottish geologist who worked almost exclusively in that country, and produced the first geological map of it. It is said that his clear observation and knowledge were never appreciated in his lifetime on account of his somewhat unpleasant character.

Macerals. ◊ *Coal.*

Maclure, William (1763–1840). Scottish-born American geologist who produced, single-handed, the first general geological map of the U.S.A.

Macro- (prefix). Large, great.

Maculose. Spotted or knotted; applied especially to contact ◊ *metamorphosed* rocks such as spotted ◊ *slates.*

Madreporite (Echinodermata). ◊ Fig. 44.

Mafic. A general term used to describe ◊ *ferromagnesian minerals.* It should not be applied to rocks consisting mainly of ferromagnesian minerals, for which ◊ *melanocratic* is the correct term. (Cf. ◊ *Felsic.*)

Magarian. A stratigraphic name for the North American Middle ◊ *Silurian.*

Maghemite. ◊ *Spinel group of minerals.*

Magma. A molten fluid, formed within the ◊ *crust* or upper ◊ *mantle* of the Earth, which may consolidate to form an igneous rock. Magma comprises a complex system of molten silicates with water and other gaseous material in solution. It is unlikely that any igneous rock truly represents the magma from which it was formed, as the volatile constituents are almost inevitably lost during or after consolidation. Within the molten mass various processes operate which tend to produce fractions of varying composition. (◊ *Magmatic differentiation.*) As defined above, a mixture of crystals and molten liquid, whether in equilibrium or not, is not strictly magma, but the term is loosely applied to such systems as long as they behave as a fluid rather than a solid. Magma which is extruded on to the surface of the Earth becomes ◊ *lava.*

278

Magma blister. A small ◊ *geotumor*.

Magmatic differentiation. Any process which tends to cause a magma to separate into two or more parts having different compositions is a kind of magmatic differentiation. The following are some of the mechanisms by which it has been suggested the process occurs:

(1) Liquid immiscibility: The development of two liquids which do not mix, in the same way as oil and water behave. This is regarded as unlikely to occur within the known range of igneous rock compositions.

(2) Fractional crystallisation: The separation of early formed crystals from the melt. This may occur as a result of simple gravity settling, mechanical filter-pressing, or the action of convection currents.

(3) Transport of material in solution by a gas streaming through the melt. This can be shown to be potentially important in concentrating certain elements in the upper parts of the magma chamber.

(4) Thermal diffusion: It has been suggested that if a temperature gradient is established within a mass of magma, e.g. inwards from a contact, then ions will travel along this gradient at different rates, thus producing a kind of differentiation. The more basic margins of some igneous masses may be the result of some such process.

Magmatic emanations. Fluids of magmatic origin; the term is more or less synonymous with ◊ *volatiles*.

Magmatic water. Water arising from an underground magmatic source. (◊ *Juvenile*.)

Magnesian limestone. (1) The name given to a stratigraphic formation in the ◊ *Permian* of Great Britain.

(2) A limestone containing a relatively small quantity (less than 15%) of magnesium carbonate, in which the mineral ◊ *dolomite* was not detected by the older ◊ *staining tests*. However, newer techniques can usually detect this mineral and modern usage of the term implies a limestone with a low percentage of dolomite.

Magnesite. A mineral of the ◊ *carbonate* group, $MgCO_3$, found in irregular veins in ◊ *serpentine* and formed by ◊ *replacement* of dolomite and limestone. (◊ Appendix.)

Magnetite. An important iron ore mineral, Fe_3O_4. One of the ◊ *spinels*, found in igneous rocks, ◊ *contact-metamorphic* deposits, ◊ *replacement* deposits, and ◊ *placer deposits*. (◊ Appendix.)

Malachite. A basic ◊ *carbonate* mineral, $CuCO_3.Cu(OH)_2$, a copper ore found in oxidised zones of copper deposits and, rarely, as a cementing material in sandstone. (◊ Appendix.)

Malchites. Non-porphyritic micro-◊*diorites*.

Malignite. ◊ *Alkali syenite.*

Malm. (1) Obsolete. A chalky-clayey soil or rock, in part equivalent to the term marl (◊ *Argillaceous rocks*), and used loosely of other impure calcareous rocks.

(2) Old name for the Upper ◊ *Jurassic.*

Malvernian. A stratigraphic division of the English ◊ *Precambrian*, the Herefordshire Pre-Uriconian (?).

Mamelon (Echinodermata). ◊ Fig. 44.

Mammalia. ◊ *Chordata* (Vertebrata).

Mammilated. ◊ *Botryoidal.*

Mantle. That portion of the Earth's interior lying between the Mohorovičić and Gutenburg ◊ *Discontinuities*, that is from a depth of 35 km. to 2,900 km. The density ranges from 3·3 at the Mohorovičić Discontinuity to 5·7 at the Gutenburg Discontinuity. The rate at which density changes increases markedly below 1,100 km. The mantle probably consists mainly of ◊ *olivine*, hence the term 'peridotite shell' which is sometimes used. It has been suggested that there are substantial quantities of sulphides in the upper mantle and some nickel-iron in the lower mantle. It has been claimed that minor discontinuities can be detected within the mantle. (◊ *Crust of the Earth*; *Core*.)

Mantled gneiss dome. In many of the Precambrian ◊ *shields* there occur masses of granite sheathed in gneiss, which intrude medium to low ◊ *grade* metamorphosed sediments. They were first described from Finland and have subsequently been found in other regions. The most favoured explanation is that due to Eskola. He suggests that a granite of ◊ *synkinematic* type was intruded into rocks undergoing metamorphism, thus developing a gneissose outer layer. Denudation of the area exposed the metamorphic rocks and associated granite, and a new series of sediments was then deposited ◊ *unconformably* upon this basement. A subsequent ◊ *orogeny* occurred which reactivated the ◊ *basement* and caused the gneiss-covered granite masses to become mobile, and intrude and dome-up the cover of younger rocks. (◊ Fig. 97.)

Marble. A metamorphosed ◊ *limestone*, produced by recrystallisation under conditions of thermal ◊ *metamorphism*, although marbles of regional metamorphic origin are not uncommon. (◊ *Calc-silicate hornfels* for metamorphosed limestones which contain little or no carbonate after metamorphism.)

Marbles derived from pure limestones consist simply of recrystallised calcite, but impurities in the form of ◊ *dolomite*, silica, iron compounds, ◊ *clay minerals*, etc. may give rise to other minerals which give certain marbles their characteristic appearance, e.g. a

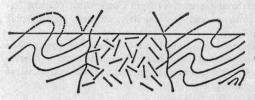

Synkinematic granite, later denuded

Subsequent sedimentation

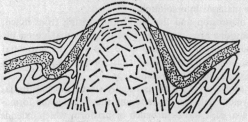

Second orogeny re-activation of basement

97. The formation of a mantled gneiss dome.

slightly dolomitic limestone containing a little sand could on metamorphism produce a forsterite marble, thus:

$$2(CaCO_3 . MgCO_3) + SiO_2 = Mg_2SiO_4 + 2CaCO_3 + 2CO_2$$
$$\text{Dolomite} + \text{Silica} = \text{Forsterite}$$

It should be noted that stone-masons apply the term marble to almost any rock which can be easily polished – more especially limestones – which have not been metamorphosed. Thus many so-called 'marbles' do not conform to the above definition, e.g. 'birds-eye marble' is a crinoidal limestone. (◊ *Landscape marble*.) On the other hand the marbles from Connemara, Western Ireland (◊ *Ophicalcite*), and Carrara, Italy, are truly metamorphic marbles.

Marcasite. A mineral, FeS_2, found in low-temperature near-surface deposits in sedimentary rocks, and in limestone often in concretions or as ◊ *replacements*. (◊ Appendix.) The 'marcasite' of jewellers is ◊ *haematite*.

Margarite. ◊ *Mica* (Dioctahedral).

Margarites. ◊ *Devitrification*.

Marine band. A layer of sediment characterised by the presence of marine fossils, interbedded with non-marine sediments. The term is especially applied to such bands in coal-bearing strata, where they constitute a major means of correlation betwen basins of deposition. (◊ *Marker bed*.)

Marine erosion. The processes responsible for the wearing-back and destruction of coastlines. Erosion is carried out by the sea itself acting as a 'hammer' on the rocks and thereby shattering them, but the process is reinforced when the waves are armed with rock fragments, and when the air contained in crevices and cavities of the rocks is suddenly compressed by the waves. In some areas (limestone), erosion may also be augmented by seawater taking material into solution.

The course of erosion and the land-forms arising from it will depend, in any one area, upon the type of rocks making up the coastline, their attitude with respect to the sea, the structural nature of the area, and the initial coast profile, depending on whether it is of highland or lowland type.

The general sequence of events is usually started by the formation of a wave-cut notch in the land and this initiates the formation, by the sea, of a wave-cut bench or platform which extends inland. The bench may consist of the bare rock or it may in time become covered with debris of all grades. This deposit is known as the beach, and is a temporary feature. In many instances it is in constant movement (◊ *Longshore drift*) and in time of severe storms it may disappear completely. If the land area is ◊ *rejuvenated* with respect to the sea, then the beach becomes 'raised' above sea level. Erosion of the land leads to the deposition of material on the seaward side and forms a marine terrace so that a profile of equilibrium is approached. A marine cycle of erosion is not recognisable in the same sense as a fluvial one (◊ *Cycle of erosion*), but analogies may be drawn. In the 'youth' stage of a marine cycle the coastline is straight with only minor indentations, wave action is strong, and hence the effect of both erosion and transportation is at a maximum. Indentations become modified into a series of bays and headlands (maturity) when the development of a beach is characteristic. Old age occurs when headlands are worn back and destroyed, producing a straighter coastline once again, wave action is minimal and the wave-cut bench is very wide. Thus the net effect is one of regression.

Mariupolite. ◊ *Alkali syenite.*

Marker bed, Marker horizon. A relatively thin layer of rock (usually a sediment but occasionally a ◊ *pyroclastic* or ◊ *lava* layer) which because of some peculiarity of lithology, structure, or faunal content is easily recognised. It is usually assumed that such a thin layer is not diachronous and hence marks a definite point in time. Marker bands are of great importance in stratigraphic correlation and elucidation of structure, particularly when they occur in thick series of unfossiliferous beds or of uniform lithology. Occasionally the term marker horizon is applied to the contact between two relatively thick layers – obviously a less satisfactory usage. The concept of marker horizons has been extended to ◊ *layered igneous structures.* (◊ *Marine band* and Fig. 35.)

Markfieldite. ◊ *Diorite.*

Marl. A calcareous mudstone. (◊ *Argillaceous rocks.*)

Marmorize. The process of recrystallisation of a limestone by heat (with or without pressure effects) to form a ◊ *marble.* The term is practically obsolete, but is occasionally used for the slight thermal effects produced by a ◊ *dyke* or ◊ *sill* intruded into limestone.

Marsh, Othniel Charles (1831–99). American palaeontologist who amassed and classified an enormous collection of dinosaurs, and over a thousand new species of Mammalia. A professor at Yale and a nephew of the philanthropist Peabody, he used his influence and his wealth in his lifelong quarrel with Edward ◊ *Cope.*

Mass wasting. A synonym of ◊ *gravity transport.*

Maturity. (1) Of rivers, ◊ *Cycle of erosion.*
(2) Of sediments (sandstones), ◊ *Arenaceous rocks.*

Maysvillian. A stratigraphic stage name for the North American mid Upper ◊ *Ordovician.*

Meander. Any naturally occurring bend which a river or stream may develop along its course, often making a series of re-entrant curves. The precise cause of meandering is unknown, but recent work suggests that meanders are produced as a result of the river taking a preferred route, determined by gradient, cross-sectional area of channel, and ratio of rate of discharge to load. Only in exceptional circumstances is a meander generated by casual obstructions. The effect of a meander is to lengthen the river over a particular reach, and thus reduce its gradient and hence its velocity. These reductions may permit the river to approach the graded state. In order to be ◊ *graded* it is necessary for a river to discharge a maximum volume of water at a minimum velocity, and if a river meanders this ideal condition can be attained.

Once a meander has been generated – by whatever means – others will start to form, and with continuous erosion of the outer banks of the curve and deposition on the inside of the curves (slip-off slopes), the meander (or meander belt) moves downstream. Such belts contribute to the formation of ◊ *flood plains.* (◊ *Ox-bow lake*; *Braided stream.*) The term is derived from the River Maiandros (now Menderes) in Turkey.

Measures. A mining term formerly applied to sequences of sedimentary rocks which could be measured in bore-holes or mine shafts. In this sense the term is now obsolete, although preserved in the name 'Coal Measures' (Upper Carboniferous; Pennsylvanian) in Western Europe. Occasionally the term is used as a synonym for 'beds' 'series' or 'division' of strata, e.g. the Barren Red Measures at the Carboniferous/Permian junction in Europe.

Medinian (Albion). A stratigraphic stage name for the North American Lower ◊ *Silurian.* Synonym of Albion.

Megacyclothem. A cycle of ◊ *cyclothems.*

Megalineament. ◊ *Lineament.*

Mela- (prefix). Dark.

Melanite. ◊ *Garnet.*

Melano- (prefix). Black or dark.

Melanocratic. A term used to describe igneous rocks which consist mainly of dark minerals, essentially the ◊ *ferromagnesian minerals.* The corresponding ◊ *colour index* ranges from 60 to 90 on Shand's scale. (Cf. ◊ *Leucocratic*; ◊ *Hypermelanic*; ◊ *Mesotype.*)

Melaphyre. An altered amygdaloidal ◊ *basalt.*

Melilite. A sorosilicate mineral. (◊ *Silicates.*)

Melilite basalt. A feldspar-free ◊ *alkali basalt* related to ◊ *alnöite.*

Melteigite. ◊ *Alkali syenite.*

Member. ◊ *Stratigraphic nomenclature.*

Menapian. A stratigraphic stage name for the top of the European Lower Pleistocene (◊ *Tertiary System*).

Menevian. A local stratigraphic stage name for the top of the Welsh Middle ◊ *Cambrian.*

Meramecian. A stratigraphic stage name for the North American Middle ◊ *Mississippian* (Carboniferous System).

Mercalli Scale, Modified. ◊ *Earthquake.*

Mero- (prefix). Part, fraction of.

Merocrystalline. A term applied to rocks which contain both crystals and ◊ *glass.* (◊ *Texture.*)

Mesa. Spanish term for an isolated table-land area with steep sides, the result of a horizontal capping of hard strata having resisted denudation. In the course of time with continual erosion of the sides a mesa is reduced to a smaller flat-topped hill – a butte. Mesas and buttes often occur in groups and may be regarded as a late stage in the dissection of a region of horizontally layered rocks. (Cf. ◊ *Inselberg.*)

Meso- (prefix). Middle of.

Mesocratic. Term originally proposed for the group of rocks now described as ◊ *mesotype.* Mesotype is to be preferred, as the suffix '-cratic' implies dominance and for the 'mean' to be dominant in this connotation is obviously absurd.

Mesostasis. In igneous rocks, the last fraction of the melt to consolidate in the spaces between the earlier-formed crystals. It may take the form of a ◊ *glass* or a eutectic, or appear as a ◊ *reaction rim* around an earlier crystal.

Mesothermal. A ◊ *hydrothermal* ore deposit formed at intermediate temperatures (200–300°C.) and depths. (Cf. ◊ *Epithermal;* ◊ *Hypothermal.*)

Mesotype. A term applied to igneous rocks which consist of approximately equal quantities of light and dark minerals (◊ *Leucocratic; Melanocratic*) with a ◊ *colour index* of between 30 and 60 on Shand's scale. (◊ *'Mesocratic'*, which was originally suggested for this group of rocks, is a barbarism.)

Mesozoic (lit. 'middle life'). An era ranging in time from 230 to 70 m.y., a duration of 160 m.y. It comprises the ◊ *Triassic,* ◊ *Jurassic,* and ◊ *Cretaceous* Systems. The Mesozoic was preceded by the Palaeozoic and followed by the Tertiary. In older works, the Mesozoic may sometimes be referred to as the 'Secondary Era'. The lower boundary of the Mesozoic is unfortunately placed between two systems (the ◊ *Permian* and the ◊ *Tri-*

assic) which in many areas consist of continental sediments, which makes accurate delimitation difficult. The beginning of the era saw the final phase of the Variscan ◊ *Orogeny*, while the close was an important event in the Alpine Orogeny (= the Laramide Orogeny of North America). There were extensive lava flows in South America, South Africa, and India.

The most spectacular elements of the Mesozoic faunas were the giant reptiles; the first mammals appeared at the commencement of the era. The chief invertebrates are ◊ *Mollusca*, ◊ *Brachiopoda*, and ◊ *Echinodermata*.

Meta- (prefix). Denotes metamorphism of the rock so qualified.

Metabasite. Any metamorphosed basic igneous rock.

Metallogenetic (lit. 'metal-forming'). The term is used with the essential implication of ◊ *mineralising*, especially ore deposition. It is mainly used in the two phrases 'metallogenetic epoch' and 'metallogenetic province'. Metallogenetic epochs are definite periods of time during which ◊ *mineral deposits* were formed. Metallogenetic provinces are regions in which a series of mineral deposits possess common characteristics.

Metamict. A term used to describe a mineral which, although possessing external crystal form, fails to yield a normal X-ray diffraction pattern. It can be shown that this is due to an imperfect development of the atomic lattice (◊ *Unit cell*) which destroys the basic pattern upon which the X-ray diffraction depends. In a number of cases the metamict state can be converted to the regular crystalline state by prolonging heating at a temperature below the melting or decomposition point of the substance. The development of metamict structures is now believed to be brought about by radioactive bombardment by atoms of uranium and/or thorium contained within the mineral, or, more rarely, in minerals in contact with it.

Metamorphic aureole. The zone around an igneous mass within which metamorphic changes, mainly thermal effects, have occurred. (◊ Fig. 98.) The outer limit is drawn at the first detectable change in texture or mineralogy of the rocks. The width of the aureole will vary (a) with different rock types, being widest in ◊ *argillaceous rocks* and narrowest in ◊ *arenaceous rocks*, and (b) with the size, temperature of intrusion, and volatile content of the igneous rock mass. (◊ *Hornfels*; *Marble*; *Calc-silicate hornfels*; *Epidiorite*; *Quartzite*.)

Metamorphism. The processes by which changes are brought about in rocks within the Earth's crust by the agencies of heat, ◊

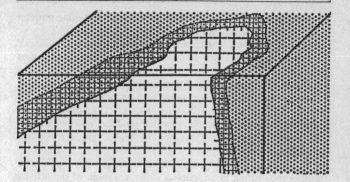

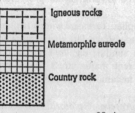

Igneous rocks

Metamorphic aureole

Country rock

98. A metamorphic aureole.

pressure, and chemically active fluids. The processes of ◊ *weathering* and ◊ *diagenesis* are excluded from metamorphism, since they involve external agencies. Metamorphic processes are considered as taking place in the solid state, and hence processes involving melting or solution of the rocks, either wholly or in part, are also excluded. The end-product of metamorphism, the metamorphic rock, is the result of the interaction of the metamorphic agencies and the parent rock. Metamorphic processes are classified as isochemical when no introduction of material from an external source takes place, or metasomatic (◊ *Metasomatism*) when alteration of the bulk composition of the rock is involved. (◊◊ *Migmatite*.) Five types of metamorphism can be recognised:

(1) THERMAL METAMORPHISM. This involves heat alone, without significant pressure effects. It is normally associated with igneous intrusions, when the term contact metamorphism may be used. (◊◊ *Metamorphic aureole*.)

(2) DYNAMIC METAMORPHISM (DISLOCATION METAMORPHISM). This involves intense localised stresses which tend

287

to break up the rocks, sometimes reducing them to fine powder. (◊ *Cataclasis; Cataclastic rock; Mylonite; Rock flour.*)

(3) REGIONAL METAMORPHISM (DYNAMOTHERMAL METAMORPHISM). This is the large-scale action of both heat and 'pressure', producing a wide range of new minerals and a widespread development of ◊ *tectonites*. Regional metamorphism is always associated with orogenies and igneous intrusions.

(4) RETROGRADE METAMORPHISM (DIAPHTHORESIS). Any reversal of metamorphism which produces rocks of a lower metamorphic ◊ *grade* from one of higher grade (excluding weathering processes).

(5) AUTOMETAMORPHISM. This involves changes which occur during the cooling of an igneous mass as a result of the activity of residual fluids within the mass. (◊ *Hydrothermal processes; Pneumatolysis; Serpentinisation.*)

The term polymetamorphism is applied to cases where two or more successive metamorphic episodes have left their traces in the rocks. These episodes may be similar, e.g. two successive regional metamorphisms with different major stress directions, or of different kinds, e.g. a thermal metamorphism superimposed upon a regional metamorphism.

(◊◊ *Amphibolite; Calc-silicate hornfels; Epidiorite; Gneiss; Granulite; Hornblende schist; Hornfels; Marble; Mylonite; Phyllite; Quartzite; Replacement* (7); *Schist; Skarn; Slates; Facies, metamorphic; Zone, metamorphic.*)

Metasomatism. A metamorphic change which involves the introduction of material from an external source. Whether or not there is a corresponding amount of material expelled from the system will depend upon the conditions. Under some circumstances metasomatism may result in the complete replacement of one mineral by another without loss of the original ◊ *texture.* If the metasomatic fluids are energetic enough, recrystallisation may accompany metasomatism, and this may conceal some of the more obvious effects of the process. (◊◊ *Ectinite; Migmatite; Replacement; Granitisation; Albitisation.*)

Meteoric water. A term applied to water which penetrates the rocks from above, i.e. rain, dew, hail, snow, and also the water of rivers and streams. (Cf. ◊ *Connate water;* ◊ *Juvenile;* ◊ *Ground water.*)

Meteorite. Extra-terrestrial material which falls to the surface of the Earth when captured by the Earth's gravitational field. Much of the material that reaches the surface of the Earth does so in the form of a fine dust and only exceptionally do large fragments survive the passage through the atmosphere. Four classes of meteorite may be recognised:

(1) Those composed entirely of metal (nickel-iron) – siderites.

(2) Those consisting of both metal and silicate – stony-irons, siderolites, pallasites.

(3) Those composed dominantly of silicate material – stones, aerolites.

(4) Glassy meteorites – tektites.

Meteorites belonging to classes 1 and 2 may also contain Fe,Co, and Ni sulphides, carbides, phosphides, and sometimes graphite. The stony meteorites are divided into the chondrites (the commonest kind) and the achondrites, depending upon whether chondrules (small, globular masses of pyroxene, olivine, and sometimes glass) are present or not. Tektites include forms which have been given specific names such as Australites, Moldavites, etc. They are very rich in silica and superficially resemble obsidian. Their meteoric origin has been challenged and it has been suggested that they are globules 'splashed' by meteoric impact on the Earth or even the Moon.

Meteorites are important to the geologist because they may well represent a sample of planetary material, and it is on this assumption that the average composition of meteorites is used as an estimate of the composition of the planets as a whole. A comparison of the figures for the number of meteorites found but not seen to fall with those for meteorites actually seen to fall reveals a remarkable discrepancy:

Type	Found: not seen to fall	Seen to fall
Irons	66%	5%
Stony-irons	7½%	1½%
Stones	26½%	93½%

This is presumably because a stony meteorite is extremely easily overlooked, while an iron meteorite stands out as an unusual object.

There is evidence that, in the past, giant meteorites have struck the Earth forming enormous craters. Ages of 7×10^9 years have been assigned to certain meteorites, but doubts have been expressed as to the validity of these high figures: ages up to $5 \cdot 5 \times 10^9$ seem authentic. The occurrence of micro-organisms in some meteorites has been suggested, but the evidence seems equivocal and is not generally accepted.

Miarolitic rocks. Plutonic rocks in which small cavities occur into which crystals project. (Cf. ◊ *Druse*; ◊ *Vugh*.)

Mica-plate. A quarter-wave plate (◊ *Accessory plate*).

Micas. The micas are ◊ *layer-lattice minerals* of the three-layer type, and may be divided into the dioctahedral muscovite group and the trioctahedral phlogopite-biotite group. An essential feature of the mica group is the replacement of one Si by one Al in the silicon-oxygen lattice to give a structural unit $(Si,Al)_4O_{10}(OH,F)_2$. A characteristic feature of the micas is a perfect basal cleavage. Cleavage flakes are flexible and elastic. (Cf. ◊ *Chlorites*.)

DIOCTAHEDRAL. Muscovite – $K_2Al_4(Si_6Al_2)O_{20}(OH,F)_4$ – is the commonest dioctahedral mica. It is transparent and this, coupled with its resistance to heat, has made it useful for furnace windows etc. Muscovite occurs primarily in acid plutonic rocks including ◊ *pegmatites*. It also occurs in schists and gneisses and as a detrital mineral in sediments. Chromium-bearing muscovite is called fuchsite. Sodium-bearing muscovite is called paragonite, which is usually regarded as a rare mineral, but because of the difficulties of distinguishing it from muscovite it may well be commoner than is generally thought. Fluorine-rich muscovite is called gilbertite. Glauconite may well be a mica, possibly in part dioctahedral and in part trioctahedral, as it contains both Al and $Fe''Mg$ in the lattice. It is a common authigenic constituent of marine sediments, and is characteristically bright green. Glauconite as a cavity filling is called celadonite. Margarite is a member of the group of brittle micas in which the cleavage flakes are neither flexible nor elastic. Its composition is $Ca_2Al_4(Si_4Al_4)O_{20}$ $(OH,F)_4$ and it occurs essentially in metamorphic rocks.

TRIOCTAHEDRAL. Phlogopite – $K_2(Mg,Fe'')_6(Si_6Al_2)O_{20}(OH,F)_4$ – and biotite – $K_2(Mg,Fe''Fe''',Al)_6(Si_{6-5}Al_{2-3})O_{20}(OH,F)_4$ – are characteristically various shades of brown to black. Minerals in this group are exceptionally widespread, occurring in almost all types of igneous and metamorphic rocks. They are, however, rather uncommon in sediments, as they tend to be unstable in sea water. It has been observed that calc-silicate metamorphic rocks and basic and ultrabasic rocks tend to contain Mg-phlogopites, while acid igneous rocks, schists, and gneisses tend to contain the more iron-rich biotites. Lithium-bearing biotites occur and are known as zinnwaldite. Lepidolite is a more or less iron-free trioctahedral mica with a composition $K_2(Li,Al)_{5-6}(Si_{6-5},Al_{2-3})O_{20}$ $(OH,F)_4$. It was thought to be dioctahedral, but the lithium must be regarded as replacing the Mg of the brucite layer. It is characteristically pale violet in colour and is an important source of lithium. Its main occurrence is in pegmatites. Clintonite is a brittle mica which is the trioctahedral analogue of margarite – $Ca_2(Mg,Al)_6(Si_{2.5},Al_{5.5})O_{20}(OH,F)_4$ – occurring in talc schists and in metamorphosed limestones.

Sub-micas is a convenient term for certain mica-like substances possibly intermediate between the micas and the ◊ *clay minerals*. They are characterised by (1) a deficiency of aluminium replacing Si in 4 coordination compared with the micas, (2) an excess of hydroxyl (hence the name hydromica), and (3) a deficiency of potassium or sodium in the structure. The best known is perhaps sericite, although this name has been given to many micaceous substances, e.g. fine-grained muscovite aggregates, kaolin-muscovite aggregates, and in fact almost any fine-grained aggregate of a colourless layer-lattice mineral.

Micrite. ◊ *Limestones.*

Micro-. A prefix meaning 'small', 'below the limit of vision of the naked eye'. It is generally assumed, however, that it does not imply that the phenomenon is below the limits of visibility using an optical microscope. 'Micro-' prefixing a rock name (e.g. ◊ *microgranite*) implies a rock having a grain size too small for the individual grains to be visible to the naked eye. Occasionally 'micro-' is used in the sense of 'very small', and this is sometimes applied in a relative way, e.g. a micro-fold may be clearly visible with the naked eye, but will have an amplitude measured in centimetres or millimetres, compared with the kilometres or tens of kilometres of a major fold.

Microadamellite. ◊ *Microgranite.*

Microcline. A triclinic alkali ◊ *feldspar* mineral, $KAlSi_3O_8$, found in acid igneous rocks, pegmatites, and metamorphic rocks. (◊ Appendix.)

Microcrystalline. A term used to describe a texture which is so fine-grained that a microscope is needed to resolve the individual crystals. (Cf. ◊ *Cryptocrystalline.*)

Microgabbro. ◊ *Dolerite.*

Microgranite. A medium-grained acid igneous rock having very similar mineralogical and chemical properties to ◊ *granites* (or ◊ *adamellites* in the case of microadamellites, and ◊ *granodiorite* in the case of microgranodiorite). It is often difficult to distinguish these three types. Riebeckite and aegirine microgranites are less-common soda-rich varieties called paisanite and grorudite respectively. ◊ *Porphyritic* microgranites have long been recognised under the name quartz porphyry (or less happily granite porphyry); these are the rocks termed elvans by the Cornish miners Microgranites displaying granophyric textures (◊ *Graphic texture*) are termed granophyres.

Microgranites and their associates occur in ◊*dykes*, ◊ *sills*, and small ◊ *plugs*, but much larger bodies of granophyre occasionally occur; the 'granites' of the British Tertiary Volcanic Province are commonly granophyric.

Microgranodiorite. ◊ *Microgranite*.

Microlite. A very small crystal, but not so small as to be completely indeterminate. Usually found embedded in glassy material. (Cf. ◊ *Crystallite*.)

Mid-oceanic ridge. Detailed oceanographic work has established the existence of a narrow belt of submarine mountains approximately midway between Europe and North America, and Africa and South America. This Mid-Atlantic Ridge, as it was named, is typical of mid-oceanic ridges such as were subsequently discovered in other oceans. Occasionally the ridge reaches the surface as islands or reefs, such as Jan Mayen Island, St Paul Rocks, Bouvet Island, and Rodriguez Island. Mid-oceanic ridges are the sites of intense volcanic and earthquake activity, and there is considerable evidence to suggest that they may have a form analogous to that of ◊ *rift valleys*. Most authorities believe that they are closely related to ◊ *continental drift*. A number of tear ◊ *faults* have been located striking at right-angles to the line of the ridges. The theory of ◊ *plate tectonics* suggests that oceanic ridges are the boundaries between plates which are moving apart, with new oceanic crust being created at the same time. The vulcanicity, earthquakes, rift valley structures and the number of tear ◊ *faults* which have been located striking at right-angles to the line of the ridges support this theory.

Migma- (prefix). Mixed.

Migmatite. A term more or less synonymous with injection ◊ *gneiss*. Modern usage preserves the idea that a migmatite is a mixed rock, i.e. made up from two different sources. Generally, the two materials are a pre-existing host rock, which is inevitably a metamorphic rock of some sort, and an invading granitic material which may take the form of (1) a ◊ *magma*, (2) a ◊ *hydrothermal* solution or (3) a rather indefinite diffusion of material ('emanations') through the host rock.

In all migmatites there is clearly a reaction between the host rock and the invading material, and the migmatite may develop a totally new structure of its own, destroying in the process all traces of the original ◊ *fabric*. Migmatites generally pass through a plastic stage, with the development of flow ◊ *folds* (ptygmatic folding). When traces and patches of the original fabric remain, the rock is termed a nebulite, while if all traces of the original texture and structure are destroyed the term anatexite is used. Agmatites are produced when the introduced material appears to form a network of veins throughout the host rock. (◊ *Lit-par-lit*.) Migmatites occur mainly in the highest ◊ *grade* of regional metamorphism, but contact migmatites have been described at the immediate contacts of granite and country rock. It should be noted that regional migmatites need not be associated with identifiable granite bodies. (◊ *Front*.)

Miller, Hugh (1802–56). Scottish geologist and writer – formerly a stone-mason – who discovered a number of fossil fish in the barely fossiliferous Old Red Sandstone of northern Scotland. He is remembered chiefly for the collection of geological essays he published under the title *The Old Red Sandstone*.

Miller indices. ◊ *Indices, crystallographic*.

Milli- (prefix). Thousandth part.

Mimetic. Imitative; e.g. mimetic crystallisation – the fabric of a recrystallised metamorphic rock which imitates a pre-existing ◊ *fabric*.

Mineral. A structurally homogeneous solid of definite chemical composition, formed by the inorganic processes of nature. This definition includes ice as a mineral, but excludes coal, natural oil and gas. The only allowable exception to the rule that a mineral must be solid is native mercury (quicksilver), which is a liquid. The term 'definite chemical composition' is not synonymous with 'fixed or constant composition', since many minerals have compositions which are variable between certain limits, which are defined in terms of end members; e.g. the composition of the common ◊ *olivines* is expressible in terms of the two compounds, Mg_2SiO_4 (forsterite) and Fe_2SiO_4 (fayalite). The general rule is that minor variations of composition which do not markedly alter fundamental properties are discounted. 'Structurally homogeneous' implies that the fundamental atomic structure is continuous and constant through the mineral unit, e.g. in ◊ *silicates* the silicon-oxygen lattice (◊ *Unit cell*) will be constant in character, although the interstitial cations may vary in different parts of the lattice.

Although strictly of organic origin, the constituents of many limestones, siliceous rocks, and bedded phosphate deposits are treated as though they were true mineral species.

Mineral deposit. The term is used for any naturally occurring body of minerals which is wholly or partly of economic value. It is usually reserved for cases where the value lies in the individual minerals, and not for cases where the assemblage of minerals is valuable as an entity (i.e. a rock unit), e.g. a building stone. (◊ *Ore*; *Ore mineral*.) Mineral deposits are classified as follows:

(1) Magmatic concentrates, e.g. diamonds, ◊ *chromite*, ◊ *magnetite*. Pegmatitic deposits (e.g. ◊ *mica*, ◊ *corundum*, ◊ *cassiterite*) may be included here (◊ *Pegmatite*).

(2) Products of contact ◊ *metasomatism*, e.g. some ◊ *pyrite* deposits and iron ores.

(3) ◊ *Hydrothermal* deposits: (a) Cavity fillings (◊ *Vein*), including ◊ *lodes*, ◊ *reefs*, and ◊ *stockworks*; (b) ◊ *Replacement* deposits. Most of the major copper, lead, and zinc deposits come under these two headings.

(4) Products of ◊ *pneumatolysis* – tin and tungsten deposits.

(5) Sedimentary deposits: (a) Primary sediments, e.g. ◊ *sedimentary iron ores*; (b) ◊ *Evaporites*, e.g. ◊ *gypsum*, salt, and potash minerals; (c) Residual and mechanical deposits, ◊ *placer deposits*, e.g. tin ore (cassiterite), gold, and diamonds, and ◊ *bauxite*.

(6) Metamorphic deposits, e.g. ◊ *asbestos*, ◊ *graphite*, and ◊ *talc*. Many of these may be subjected to the processes of oxidation and ◊ *secondary enrichment*.

(For zonation of mineral deposits, ◊ *Hydrothermal processes*.)

Mineralisation. This term is used almost exclusively for the introduction of ◊ *ore minerals* and ◊ *gangue* minerals into pre-existing rocks, whether by ◊ *veins*, ◊ *replacement* (5 and 6), or in a disseminated fashion. Mineralisation can be considered on any scale, from a hand specimen to a region of several thousand square miles. The term can also be used in a stratigraphic sense; e.g. one can refer to a 'post-Carboniferous mineralisation', meaning the formation of the mineral deposits associated with the Hercynian (Armorican) granites. (Cf. ◊ *Metallogenetic*. ◊ *Mineral deposit*; *Ore body*.)

Mineralography. The microscopic study of polished surfaces of opaque minerals, especially ore minerals, by reflected light.

Minette. (1) ◊ *Lamprophyre*.

(2) Sedimentary ironstones of Jurassic Age in the Lorraine area

of France. Sometimes applied to any sedimentary iron ore, especially if it consists mainly of ◊ *limonite*.

Minnesotaite. A rare iron analogue of ◊ *talc*.

Minor intrusions. ◊ *Dyke*; *Sill*.

Minverite. Albite gabbro or dolerite containing barkevicite but no feldspathoids. (◊ *Alkali gabbro*.)

Miocene. The epoch of the ◊ *Tertiary* period between the Oligocene and the Pliocene epochs.

Miotian. A stratigraphic stage name for the base of the North American Upper Miocene (◊ *Tertiary System*).

Misfit stream. Under normal, uninterrupted conditions, a stream cuts a valley into which it fits precisely. A misfit stream or river is one which flows in a valley which is too large for it, i.e. a meandering stream flowing in a meandering valley. Such streams arise as a result of a reduction in the total amount of water available. It has been suggested that one possible cause is a reduction in the average annual rainfall. Alternative explanations have been put forward, e.g. ◊ *river capture*, but this cannot always be proved to have taken place.

Mispickel. Old name for the mineral arsenopyrite, FeAsS (◊ Appendix).

Mississippian. The older of the two sub-systems into which the ◊ *Carboniferous* period in North America is divided.

Missourian. A stratigraphic stage name for the North American Middle ◊ *Pennsylvanian* (Carboniferous System).

Missourite. ◊ *Alkali syenite*.

Mobile belt. An elongated zone of the Earth's crust, in which major deformation, igneous activity, ◊ *metamorphism* and ◊ *migmatisation* occur. Commonly it is the earliest stage in the development of a ◊ *geosyncline*, and seismic and volcanic activity are frequent phenomena in such zones.

Mode, Modal analysis. The mode is the percentage (by weight) of the individual minerals which make up a rock (cf. ◊ *Norm*). The term is not normally applied to sedimentary rocks, and only occasionally to metamorphic ones. Modal analysis is the process by which the volumes of the individual minerals are obtained (which can be converted to weight per cent by multiplying by the density and recalculating to 100%). Determination of the mode may be carried out by one of the following methods:

(1) Direct determination: by crushing the rock and handpicking or otherwise separating the minerals.

(2) Measuring the total area of each mineral exposed on a flat

surface, e.g. a polished slab or thin section. It can be shown that the area of each mineral is proportional to the volume of each mineral present, providing a large enough total area is measured. (3) The method of Delesse and Rosiwal. This involves making a closely spaced series of linear traverses across a thin section, recording the total length of each mineral which is intercepted by the traverse line. If a sufficient length is measured, it can be shown that the total length of intercept of a particular mineral is proportional to the area and hence to the volume of the mineral present.

(4) Point-counting technique. This is a development of method 3, in which the intercept length for a particular mineral is measured in a series of discrete steps instead of continuously. The steps are counted, and the number of counts (points) for a particular mineral is recorded. The number of points is proportional to the linear intercept, and hence the area and volume of the mineral present.

Mofette. A type of ◊ *fumarole*.

Mohawkian. A stratigraphic stage name for the North American Middle and Upper ◊ *Ordovician*.

Moho. A colloquial abbreviation for the Mohorovičić ◊ *Discontinuity*.

Mohorovičić Discontinuity. ◊ *Discontinuity*.

Mohs' Scale. ◊ *Hardness*.

Moine. A stratigraphic division of the Scottish ◊ *Precambrian*, the metamorphic equivalent of the Torridonian in Scotland, which passes up (conformably?) into the Lower Dalradian.

Molasse. The term originated in Switzerland, and is used to describe sediments produced by the erosion of mountain ranges after the final phase of an ◊ *orogeny* (post-tectonic) (cf. ◊ *Flysch*). In Switzerland, these post-tectonic sediments of the Alpine Orogeny include arkoses (◊ *Arenaceous rocks*), polymict ◊ *conglomerates* and ◊ *breccias*, and reddish-brown shales (◊ *Argillaceous rocks*). They appear to have developed in ◊ *intermontane* basins, which are commonly non-marine. Other deposits which have been claimed as molasse are the Newark Sandstone (Triassic) of the eastern United States, derived from the Appalachians, and the Old Red Sandstone (Devonian) of Britain, derived from the Caledonian Mountains.

Mollusca. A group of invertebrate animals which may be terrestrial, freshwater, or marine in habitat, commonly having a shell secreted by the mantle. The shell may be either external univalve, external bivalve, internal or missing. The molluscs possess a muscular 'foot' which is modified in the various groups. They are divided into a number of important classes:

AMPHINEURA (PLACOPHORA). Normally called chitons, these are unspecialised molluscs usually having a segmented univalve shell which may become reduced to a number of spicules situated in a leathery mantle. They range in size from 1 to 3 cm. A flat creeping foot is present and the mouth is armed with a chitinous radula with which they obtain their algal food. ◊ *Palaeozoic* chitons differ from ◊ *Mesozoic* to Recent forms in that the plates making up the shell do not articulate with one another. The range is from ◊ *Ordovician* to Recent. This group is of little geological importance.

SCAPHOPODA (◊ Fig. 99). Molluscs with a tubular mantle

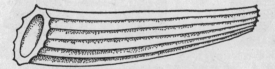

99. A scaphopod (*Dentallium*).

which secretes a tubular shell open at both ends. They range in length from 2 to 20 cm. They are all burrowing forms equipped with a flattened digging foot. Their range is from (? ◊ *Ordovician*) ◊ *Silurian* to Recent. They are locally abundant in ◊ *Mesozoic* and ◊ *Tertiary* clays, but are of little geological importance.

GASTROPODA (◊ Fig. 100). A group of marine, freshwater, or terrestrial molluscs ranging in size from 1 mm. to 10 cm., although exceptionally they may reach 50 cm. in length. The shell is calcareous and univalve, and is usually coiled, the coiling (either dextral or sinistral – ◊ Fig. 100) being typically helical spirals, although plane spirals do occur. Growth proceeds from a smooth chamber, a protoconch secreted by the planktonic larva, even where the adult form is ornamented. The palaeontological classification is based on the shape of the shell and aperture, whereas the zoological classification is based on the nature of the soft parts; there is little agreement between the two. The following terms are used to describe the shape of shell:

Biconical Body whorl is usually larger than the spire and conical in shape. Spire angle ± 90°

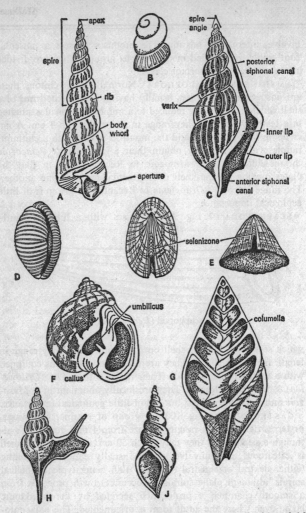

100. Morphology displayed by various gastropod shells. (A) and (C) General descriptive terms; (B) Enlargement of protoconch; (D) *Trivia;* (E) Patella-type gastropod showing selenizone; (F) *Natica;* (G) Longitudinal section of a gastropod showing the columella, internal structure, and coiling; (H) *Apporrhais,* showing apertural spine and digitation; (J) Sinistrally coiled shell – the aperture appears on the left (cf. dextral coiling in A, C, F, and H).

Discoidal   Flat or plane spiral coiling

Extraconical   Spire concave

Fusiform   A long anterior siphonal canal, more or less equal in length to the spire

Patellate   Shell uncoiled, depressed and conical, like a limpet

Pupaeform   Spire convex instead of being typically straight-sided

Trochiform   Flat base to body whorl. Spire angle 70° ±

Turbinate   Body whorl as long as spire. Spire angle 70° ±

Turreted   Spire elongated. Spire angle 20° ±

The aperture may be either:

Holostomatous   (Aperture entire) or

Siphonostomatous   (Aperture margin broken by a siphonal canal or canals)

Gastropods are most abundant in the Tertiary, but ranges from Cambrian to Recent. They have been used as zonal indices in the Tertiary.

LAMELLIBRANCHIATA (PELECYPODA, BIVALVIA) (◊ Figs. 101–103). A group of marine, brackish, or freshwater molluscs having a bivalve calcareous (usually mainly aragonite) shell which is secreted by a bivalve mantle. The valves are often equal in size (equivalve lamellibranchs – the opposite is inequivalve) and are situated laterally about the animal, rather than dorsally and ventrally as in the Brachiopoda. The lamellibranchs adopt a number of modes of life which give rise to various modifications in the morphology of the shell (◊ Fig. 102). Early classifications were based on these morphological variations, but they have now been superseded by a classification based upon the hinge teeth. (◊ Fig. 103.) Early lamellibranchs have a large number of hinge teeth which become progressively reduced in number until the position is reached of the advanced forms in which either four or six teeth may be present. Shell size ranges from 0·5 to 10 cm., with occasional examples with a length of 150 cm. The following terms are used to describe their muscle scars:

Anisomyarian Two muscle scars of unequal size in each valve

Dimyarian Two muscle scars present in each valve

Isomyarian Muscle scars of equal size

Monomyarian One muscle scar to each valve

The range is from Lower ◊ *Cambrian* to Recent. Lamellibranchs have been used successfully as zonal indices in the Coal Measures, and have also been used in the ◊ *Tertiary* and ◊ *Rhaetic*.

CEPHALOPODA. A group of entirely marine molluscs in which

the foot is modified to a ring of tentacles around the mouth. The univalve shell may be either external, internal or missing, and is usually composed mainly of aragonite. It may either be wholly chambered when external, or partially chambered when internal. These chambers are gas filled and may be used to make the

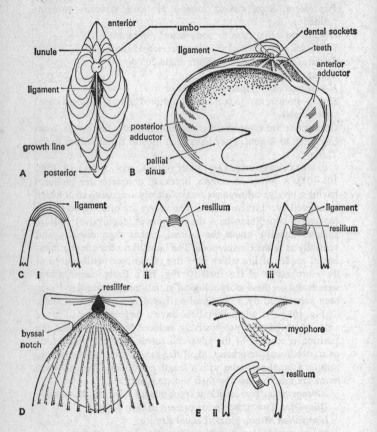

101. Lamellibranchs. (A) Dorsal view of the two valves; (B) Interior of the left valve; (C) The three main types of ligament structure: i, external; ii, internal; iii, combined internal and external; (D) Internal view of a *Pecten*; (E) Specialised ligament structures (e.g. Mya): i, internal view; ii, sectional view.

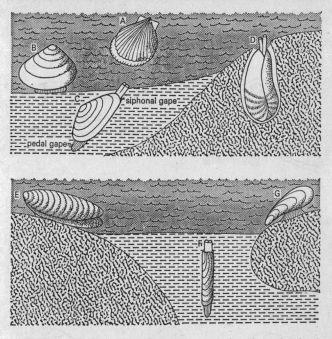

102. Modes of life of lamellibranchs. (A) Active, free-swimming (*Pecten*); (B) Active, crawling (*Venus*); (C) Burrowing, shallow (*Mactra*); (D) Burrowing, rock-boring (*Pholas*); (E) Fixed, cemented (*Ostraea*); (F) Burrowing, deep (*Ensis*); (G) Fixed, byssal attachment (*Mytilus*).

animal buoyant. In those forms having an external shell, the animal lives in the most recently formed chamber. The cephalopods range in size from 1 cm. to 10 m., and rarely 20 m. They are divided into a number of groups, which include the modern squids, octopuses, and argonauts. Important fossil groups include the Nautilioidea (◊ Fig. 104), Ammonoidea (◊ Fig. 105) (the body chamber is connected with the protoconch by a thin thread of body tissue situated within a tube which perforates the septa – the siphuncle, in these two groups), and the Belemnoidea (◊ Fig. 106). The first two groups have external chambered shells, while the last one has an internal chambered shell. (Belemnoidea –

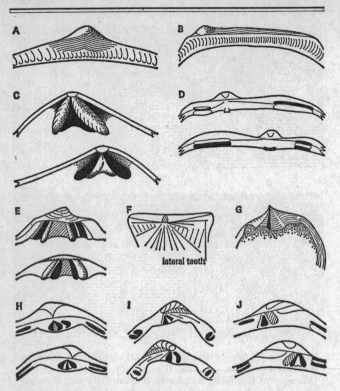

103. Examples of the main dentition types of lamellibranchs. (The black shading indicates the sockets, unshaded areas the teeth, and cross-shading the resilifer.) (A) Taxodont, right valve (*Glycimeris*); (B) Taxodont, right valve (*Arca*); (C) Schizodont: i, right valve; ii, left valve (*Trigonia*); (D) Schizodont: i, right valve; ii, left valve (*Unio*); (E) Isodont: i, right valve; ii, left valve (*Plicatula*); (F) Dysodont, right valve (*Pteria*); (G) Dysodont, right valve (*Ostraea*); (H) Heterodont Cyrenoid: i, right valve; ii, left valve (*Cyrena*); (I) Heterodont Lucinoid: i, right valve; ii, left valve (*Lucina*); (J) Desmodont: i, left valve; ii, right valve (*Mactra*).

(? ◊ *Cambrian*) Upper ◊ *Carboniferous* to Lower ◊ *Tertiary*.)

(a) Nautiloidea. Straight, curved or coiled forms, in which the chamber walls (septa) have a sinuous boundary (the suture line) where they intersect the outer shell. The siphuncle is in a central

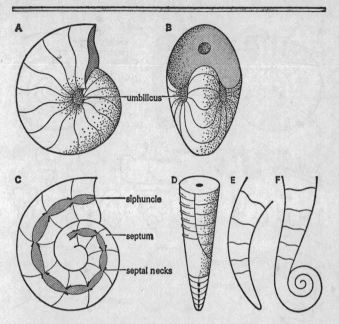

104. Nautiloids. (A) Equatorial view; (B) Apertural view; (C) Equatorial section; the septal necks are shown here backward-facing (retrosiphonate); (D) *Orthoceras*; (E) Uncoiled nautiloid (*Lytoceras*); (F) Initiation of coiling (*Cyrtoceras*).

position. The range is from (? Lower ◊ *Cambrian*) Upper Cambrian to Recent.

(b) Ammonoidea. A group which evolved from the nautiloids in ◊ *Silurian* times, which show an increase in the complexity of the suture line and in which the siphuncle migrates to the outer margin of the shell. In ◊ *Palaeozoic* times diverse simple forms of ammonoid occur such as the goniatites (zone fossils in the ◊ *Devonian* and ◊ *Carboniferous*), globose forms with sharp-angled suture lines. In the ◊ *Mesozoic*, the advanced ammonites appear with the complex ornament and suturing for which they are well known, and some partly coiled and uncoiled forms appear. Their range is Lower Devonian to Upper ◊ *Cretaceous*. The ammonites have been used extensively as zonal indices for the Mesozoic systems.

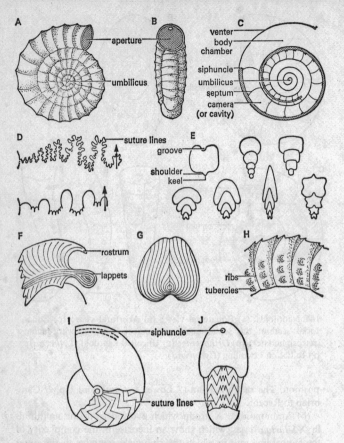

105. Ammonites. (A) External view of an evolute ammonite; (B) Apertural view of (A) showing whorl section; (C) Equatorial section through (A); (D) Detail of sutures: phylloceratid suture (top) and ceratitid suture; (E) Transverse section through some typical ammonites from venter to umbilicus showing the whorl section; (F) Detail of aperture; (G) The aptychus – the plate closing the aperture of a cephalopod; (H) Detail of venter whorl ornament; (I) A goniatite, with part of the shell omitted to show the suture lines; (J) Apertural view of (I).

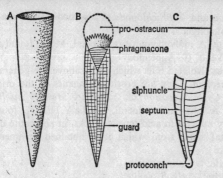

106. Belemnites. (A) Guard; (B) Vertical section; (C) Detail of phrag-macone.

The following terms are used for the description of the shell shape of cephalopods:

Ascocone   Shell cyrtoconic in the early stage becoming inflated in the mature stage

Brevicone   Compressed orthocone

Cadicone   Evolute form with a broad venter angular shoulder, and deep umbilicus

Conispiral   Coiled like a screw

Cyrtocone   Curved elongated cone

Evolute   Shell coiled with the whorls in contact and whorls visible

Gyrocone   A loosely coiled shell in which the whorls are not in contact

Involute   Shell coiled with the last whorl covering earlier whorls

Lituitcone   Early shell coiled, mature part orthoconic

Orthocone   Straight elongated cone

Oxycone   Shell laterally flattened. Venter is acutely angled

Trochiform (Trochoid)   See Gastropods, above.

Cephalopod ribbing is described by the following terms:

Bifurcation   Ribs divide into two before crossing venter

Trifurcation   Ribs divide into three before crossing venter

Modification of the aperture:

Lappets   Lateral spoon-shaped extension of the apertural margin

Rostrum   Ventral extension of the apertural margin

These are considered by some to be sexual characters.

Molluscoidea. Obsolete name for supposed phyla containing the

◊ *Polyzoa* and ◊ *Brachiopoda.* These two groups are regarded as separate phyla.

Molybdenite. A molybdenum ore mineral, MoS_2, found in hydrothermal veins. (◊ Appendix.)

Monadnocks. Isolated hills which stand above the general level of a ◊ *peneplain.* They are erosion remnants of the original surface. (◊ *Cycle of erosion;* cf. ◊ *Butte;* ◊ *Mesa;* ◊ *Inselberg.*)

Monazite. A rare-earth mineral, $(Ce,La,Y,Th)PO_4$, found as an accessory mineral in acid igneous rocks, in pegmatite dykes, and in detrital sand, from which it is obtained commercially. (◊ Appendix.)

Monchiquite. ◊ *Lamprophyre.*

Monmouthite. ◊ *Alkali syenite.*

Mono- (prefix). Single, one.

Monocline. ◊ *Fold.* Monoclinal is not synonymous with ◊ *uniclinal.*

Monoclinic system. A ◊ *crystal system* divided into three symmetry classes, as follows:

	International symbol	Centre	Planes of symmetry	Axes of rotation symmetry	Axes of rotary inversion	Examples
1	2/m	C	1	1 ii	—	Orthoclase, epidote, augite, hornblende, micas, chlorite
2	m or $\overline{2}$*	—	1	—	(1 ii)*	Scolecite (◊ *Zeolite*), kaolinite
3	2	—	—	1 ii	—	Rare

* A unique plane of symmetry is equivalent to a 2-fold inversion axis.

It includes all those crystals that can be referred to three axes, two of which are at right angles (vertical and horizontal), with the third making an angle, other than a right angle, with the vertical axis, the plane defined by the vertical axis and the inclined axis being perpendicular to the horizontal axis. The three parameters are all unequal. Elements of the monoclinic system will include therefore the three parameters and the angle between the vertical and inclined axes (β). (For international symbols, ◊ *Symmetry*; ◊ *Axes, crystallographic.*)

Monomineralic rocks. Rocks composed of one mineral. For example, dunite is made up of pure olivine, anorthosite of pure feldspar, and white marble is pure recrystallised calcite. Even the 'purest' monomineralic rock is likely to be found to contain some other minerals if a sufficiently large sample is examined in detail.

Montian. A stratigraphic stage name for the West European Middle Palaeocene (◊ *Tertiary System*).

Monticellite. A mineral of the ◊ *olivine* group.

Montmorillonite. ◊ *Clay minerals.*

Monzonites. Coarse-grained igneous rocks, ranging from ◊ *acid* quartz-bearing types to ◊ *basic* olivine-bearing varieties with the essential feature of the presence of approximately equal amounts of alkali and calc-alkali ◊ *feldspar*. Unmodified, the term monzonite implies a ◊ *saturated* rock. The chemistry of the monzonites is variable between the normal limits for acid and basic rocks, and between alkaline and calc-alkaline rocks.

Acid	Adamellite (◊ *Granite*)
	Quartz monzonite (banatite)
	Monzonite (s.s.) (syenodiorite is a synonym)
Basic	Olivine monzonite (kentallenite is a special variety of olivine monzonite). Syenogabbro is a partial synonym.

Monzonites rarely form large homogeneous bodies of rock, but are more commonly found as members of igneous rock complexes or as small individual ◊ *bosses* or ◊ *plugs* associated with major granitic masses. The more acid members tend to occur in association with granites, granodiorites, diorites, etc., while the more basic members tend to occur with alkali gabbros and syenites. It is probable that much monzonite occurs unrecognised in many 'granite' ◊ *batholiths*. Micromonzonites occur, but often pass unrecognised or are combined with other types.

The volcanic equivalent of monzonite (s.s.) is ◊ *trachyandesite*; latite is a synonym. Trachyandesites occur in association with andesites more commonly than with trachytes and may also be more widespread than records suggest, for similar reasons to those mentioned above. (For the volcanic equivalent of adamellite, ◊ *rhyolite*, and of olivine monzonite ◊ *trachybasalt*.)

The porphyry copper ores of the United States are associated with porphyritic quartz monzonites.

Moraine. An accumulation of material which has been transported or deposited by ice. Transportation of material may take place in three ways: (1) On the surface of the ice. Rock debris composed of weathered material and scree falls on to the ◊ *glacier* to form a lateral morain. When two glaciers merge, two lateral moraines will coalesce into a median moraine or moraines. It should be noted that lateral and median moraines are only transient features of the ice surface. (2) In the ice; material carried within the ice is termed englacial. (3) Beneath the ice; material which is carried at the base of the glacier and which performs much of the scouring action is termed subglacial moraine.

Rock fragments carried by the ice are deposited when the ice melts and by subglacial streams; such deposition occurs during a period when the ice is neither advancing nor retreating, and results in a terminal or end-moraine. A series of recessional moraines are formed as the ice retreats, each moraine representing a temporary halt in the retreat. Should the ice advance across a moraine, the sediments may become contorted and folded, often producing structures resembling tectonic deformation. Such a feature is known as a push moraine.

Ground moraine is the sheet of debris left after a steady retreat of the ice (◊ *Drift* (3)). Moraine material is usually an ungraded mass of sediment ranging in size from clay-grade to boulders, and is subject to subsequent modification and destruction by melt waters.

Morgannian. A stratigraphic stage name for the European Upper ◊ *Westphalian* (Carboniferous System).

Moro, Anton Lazzaro (1687–1740). Venetian abbot who was one of the earliest advocates of the theory of the volcanic origin of mountains. He was the first to suggest a distinction in structure between Primary (unstratified) and Secondary (stratified) rocks.

Morrowan. A stratigraphic stage name for the base of the North American ◊ *Pennsylvanian* (Carboniferous System).

Mortar structure. An early stage in the development of a ◊ *mylonite*, in which the larger crystals in a rock are surrounded by finely granular material formed by the crushing of interstitial crystals. It is most often observed in quartzo-feldspathic rocks.

Mortlake. A synonym of ◊ *ox-bow lake*.

Mosaic. In aerial photography, an overall picture of an area built up by fitting together (and overlapping where necessary) a series of individual air photographs.

Mosaic texture. A synonym of ◊ *saccharoidal* texture.

Moscovian. A stratigraphic stage name for the Russian Upper ◊ *Westphalian* (Carboniferous System).

Mould. The impression obtained from an original form. (◊ Fig. 15.)

Mountain building. ◊ *Orogeny.*

Mouth. (Arthropoda) ◊ Fig. 3; (Echinodermata) ◊ Figs. 41–45, 47.

Mud. A wet mass of clayey material in a liquid or semi-liquid state (◊ *Liquid limit*). The term is also used for the artificial 'slurries' used when drilling bore holes.

Mud flow. ◊ *Gravity transport* (1).

Mudstone. ◊ *Argillaceous rocks.*

Mud volcano. (1) A hot spring in a volcanic region in which is incorporated sufficient ◊ *argillaceous* material or volcanic ash to produce a boiling mass of mud.

(2) A conical mound, which mimics a normal volcano, formed as a result of earthquake activity during which masses of semi-liquid mud are forced up through fissures. (◊ *Injection structures.*)

Mugearite. ◊ *Andesite.*

Mullion structure, Mullions. Linear structures developed by the *compression* of ◊ *competent beds,* consisting of elongated fluted columns of rock, varying in diameter from 5–6 cm. up to 50 cm. or more. They may be produced: (1) by the development of a series of tight folds which 'pinch-off' into isolated units (fold mullions); (2) as a result of the intersection of ◊ *bedding* and ◊ *cleavage* (cleavage mullions); and (3) by the operation of both the above processes, giving irregular mullions. Mullions develop parallel to the plunge of the fold axis. A very similar type of structure is termed 'rodding'. This consists of a series of elongated rod-like masses of quartz, with a cross-section which is more or less circular. These are derived from quartz veins in the original rock which become dislocated and 'rolled-up' during the process of folding. Wilson (see reference below) has shown that the original quartz veins may occupy cleavage or schistosity planes, original bedding planes or fissures (e.g. tension gashes) oblique to both bedding and cleavage. During folding there is a tendency for the quartz veins to break up and segregate into the noses of the folds, where further deformation tends to roll them into rods. Thus quartz rods are parallel to the plunge of the fold in precisely the same way as are mullions. (◊ Fig. 107.)

G. Wilson, 'Mullion and Rodding Structures in the Moine Series of Scotland', *Proceedings of the Geologists' Association,* Vol. 64, 1953.

Mullite. ◊ *Aluminium silicates.*

Multi- (prefix). Many.

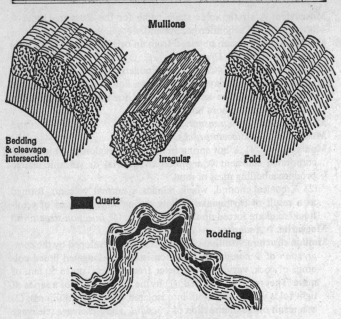

107. Mullions and rodding.

Multiple. A term applied to igneous ◊ *intrusions* where several emplacements of more or less identical material have occurred to make up the whole mass (cf. ◊ *Composite*). ◊ *Dykes* and ◊ *sills* are most commonly multiple, but other bodies of rock sometimes show this phenomenon.

Murchison, Roderick Impey (1792–1871). Distinguished Scottish geologist who was responsible for the first efforts to determine the Palaeozoic succession. The result of his work was an entirely new stratigraphic system which he named ◊ *Silurian*, an achievement marred only by his failure to recognise the need to distinguish it from the Cambrian System being investigated by his former friend ◊ *Sedgwick*. He also established the ◊ *Permian System* during a visit to Russia, at the invitation of the Tsar, in 1841. He succeeded de la ◊ *Beche* as Director General of the Geological Survey of Great Britain in 1855.

Muschelkalk. A stratigraphic stage name for the West European Middle ◊ *Triassic*.

Muscles, adductor. ◊ *Adductor muscles.*

Muscovite. ◊ *Micas* (Dioctahedral) and Appendix.

Mylonite. If the shearing and dislocation process causing ◊ *cataclasis* is prolonged and intense, the individual crystals in the rock become fractured and the whole rock becomes more and more fine-grained, sometimes developing a crude foliation. This fine-grained rock is known as mylonite; it is typically found associated with the great zones of tectonic dislocation.

Myrmekite, Myrmekitic texture. An intergrowth of quartz and plagioclase feldspar in which the quartz occurs as worm-like rods within the feldspar. It is generally the result of the replacement of potash feldspar by a soda-rich plagioclase feldspar with separation of excess quartz. It is a typical ◊ *symplektite.*

Nacrite. A ◊ *clay mineral* of the kaolinite group.

Nail-head spar. A variety of ◊ *calcite.*

Namurian. A stratigraphic stage name for the base of the European Upper ◊ *Carboniferous.*

Nanoplankton. The name given to fossil material of ultra-microscopic size; in general it is so small that it can only be seen by using the highest power of a phase-contrast microscope, or an electron microscope. Because the organisms are ◊ *planktonic* they are found over enormous geographical ranges and are practically independent of ◊ *facies* (excluding continental and fresh-water ones).

These fossils have proved useful in correlating both short range and long range, and a particular point in their favour is that they occur in rocks which are barren of other fossils. They have also been recorded from unmetamorphosed ◊ *Precambrian* sediments and the possibility exists that they will provide a means of correlation for these beds. It is not always possible to assign the fossils to individual organisms but there is no doubt that both zooplankton and phytoplankton are involved.

Napoleonite. An ◊ *orbicular* ◊ *diorite* containing large layered spherical arrangements of crystals.

Nappe. A ◊ *fold* in which the axial plane is horizontal or sub-horizontal.

Nassauian. A stratigraphic stage name for the European middle ◊ *Dinantian* (Carboniferous System).

Native element. An element which occurs in the free state as a mineral, e.g. copper, silver, gold, and carbon (graphite and diamond).

Natural gas. The use of this term is generally restricted to gaseous hydrocarbons, which may occur in association with oil accumulations. Because of its economic value, it has become particularly important in recent years. (◊ *Oil.*)

Nautiloidea. ◊ *Mollusca* (Cephalopoda).

Nebulite. A ◊ *migmatite* in which traces and patches of the original fabric remain.

Neck. A volcanic ◊ *plug.*

312

Needian. A stratigraphic stage name for the European Elster/Saale Pleistocene interglacial (◊ *Tertiary System*).

Negative, optically. ◊ *Axes, optic.*

Negative area. An area where deposition is taking place. The opposite of a ◊ *positive area.*

Negative crystal. A ◊ *pseudomorph* consisting of a hollow mould of a crystal shape, the original crystal having been removed from it by solution. Also applied to cavities having the form of crystals enclosed within a crystal.

Nektonic. Free-swimming, not bottom-living.

Nema (Chordata). ◊ Fig. 18.

Nematoblastic. A texture term applied to a foliation (◊ *Cleavage, rock*) which is produced by prismatic minerals, e.g. hornblende and augite. (Cf. ◊ *Lepidoblastic.*)

Neocomian. A stratigraphic stage name for the base of the British Lower ◊ *Cretaceous.* Wealden is a synonym.

Neogene. The name given to the Miocene and Pliocene periods of the ◊ *Tertiary,* when grouped together. It is almost obsolete.

Neotony. The retention of larval characters in the adult form. A common synonym is paedomorphosis.

Neotype. ◊ *Type.*

Nepheline. ◊ *Feldspathoids* and Appendix.

Nephelinite. ◊ *Alkali basalt.*

Nephrite. ◊ *Jade.*

Neptunean dyke. A cross-cutting, more or less vertical sheet of sediment, commonly a sandstone or breccia, contained within a contrasting rock type, often igneous or metamorphic. They appear to be formed by the infilling of cracks or fissures by sedimentary material. (◊ Fig. 108.) In certain circumstances neptunean dykes

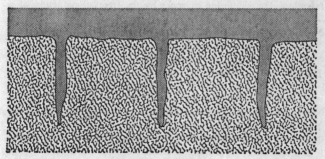

108. Neptunean dykes.

preserve the only evidence that deposits of a particular age were laid down in a particular area. Sandstone dyke is a term used, but it is better applied to sand injected into an overlying layer. (For sedimentary dykes caused by injection from below, ◊ *Diapir*; *Injection structures*.)

Neptunic. The term suggested by H. H. Read to embrace the sedimentary rocks. (Cf. ◊ *Volcanic*; *Plutonic*.)

Neritic. ◊ *Sub-littoral*. (◊ *Environment* and Fig. 49.)

Nesosilicates. ◊ *Silicates*.

Névé. Compacted snow formed during the formation of a ◊ *glacier*.

Newbournian. Obsolete stratigraphic stage name for the (post- ◊ *Ludhamian*) base of the British Pleistocene (◊ *Tertiary System*).

New Red Sandstone. The name given, especially in Europe, to the red beds occurring above the ◊ *Carboniferous* and below the ◊ *Jurassic* (cf. ◊ *Old Red Sandstone*). The name is effectively synonymous with Permo-Triassic or Permian + Trias.

Newton's Scale. ◊ *Polarisation colours*.

Niagaran. A stratigraphic stage name for the North American middle ◊ *Silurian*.

Nick point. An American spelling of ◊ *knick point*.

Nicol, William (1810–79). Scottish professor who first perfected the preparation of thin sections of petrified wood and bones for microscopic examination, later more widely applied by ◊ *Sorby*. He is chiefly remembered for his invention of the calcite prism for the production of polarised light which bears his name.

Nicol prism. ◊ *Polariser*.

NiFe. ◊ *Core of the Earth*.

Niggli number. A variant of the ◊ *norm* classification developed by P. Niggli.

Nivation. The shattering of rocks by the freeze-thaw action of a patchy snow cover. (◊ *Cirque*.)

Nodule. A spherical, oval, or similarly rounded ◊ *concretion* (of either type), here defined as not exceeding 256 mm. in diameter. The terms nodule and concretion are often used synonymously, nodule being preferred for more rounded types. (Cf. ◊ *Dogger*; ◊ *Armoured clay balls*.) Nodular ◊ *rhyolites* contain rather large ◊ *spherulites* which weather out from the rest of the rock.

Nomen ambiguum, Nomen confusum, Nomen dubium. Terms applied to names of organisms or groups which are invalid.

NOMEN AMBIGUUM: A name which is ambiguous by virtue of usage with two or more different meanings (cf. ◊ *Homonym*).

NOMEN CONFUSUM: A name established as a result of confusion of features belonging to two separate entities which are in fact discrete, e.g. the association of the skull of one animal with the limb bones of another in the belief that they form a single organism.

NOMEN DUBIUM: A name which is invalid because its use is uncertain, generally owing to inadequate description.

Nomen nudum (pl. nomina nuda). A name given to a new or supposedly new organism (or group) and published without a technical description of the organism or group. It is therefore impossible for other workers to make use of the name in the correct sense and hence the name is automatically invalid. Where it is desired to record the use of a name of this kind the abbreviation 'Nom. nud.' is commonly placed in brackets after the name.

Nomina conservanda. Names of organisms which although technically invalid are retained, as it is thought that their replacement by other names would cause confusion. Names are declared 'nomina conservanda' by the International Commission on Zoological Nomenclature.

Non-conformity. An obsolete synonym of ◊ unconformity.

Non-sequence. A synonym of non-depositional ◊ unconformity, where the break in sedimentation is short and often localised.

Nontronite. A ◊ clay mineral of the montmorillonite group.

Nordmarkite. An ◊ oversaturated ◊ syenite.

Norian. A stratigraphic stage name for the European mid Upper ◊ Triassic.

Norite. ◊ Gabbro.

Norm. A method of expressing the chemical composition of an igneous rock in terms of a series of arbitrarily selected minerals, according to a prescribed set of rules. It was originally devised as an essential first step in the classification of igneous rocks according to the system of Cross, Iddings, Pirsson, and Washington (the ◊ CIPW Classification). In general the norm has little relationship to the ◊ mode, although in certain respects partial agreement is possible, e.g. free quartz in the norm usually means free quartz in the mode. Since all the normative minerals are anhydrous, the norm of rocks containing notable amounts of hydrated or hydroxyl-bearing minerals will show significant deviations from their mode. Although the classification scheme is more or less obsolete, norms are still calculated as a device for presenting chemical data. (Rules for calculating the Norm may be found in *Descriptive Petrography of the Igneous Rocks*, A. Johanssen, Vol. 1, 1931, and elsewhere.)

Nosean. ◊ *Feldspathoids.*

Notothyrium (Brachiopoda). ◊ Fig. 12.

Novaculite. A Palaeozoic ◊ *chert*-like material from Arkansas, used as a honing stone.

Nuée ardente. An incandescent cloud of gas and volcanic ash, violently emitted during the eruption of certain types of ◊ *volcano.* The most famous example of one occurring in recent times is that of Mont Pelée, which in 1902 erupted and overwhelmed the town of Saint-Pierre, the capital of Martinique. Within a few minutes of the eruption the whole city and its 30,000 inhabitants were completely annihilated. Deposits formed by nuées ardentes are welded tuffs (ignimbrites). (◊ *Pyroclastic rocks.*)

Nummulites. ◊ *Protozoa.*

Nummulitic. An obsolete synonym of Eocene (◊ *Tertiary System*).

Nunatak (Eskimo). A rock mass which projects through an ice sheet, generally found at the margins of a sheet where the ice is thinnest.

Ob- (prefix). Inverted.

Obsequent. ◊ *Drainage pattern.*

Obsidian. ◊ *Rhyolite.*

Ocean floor spreading. ◊ *Mid-oceanic ridge*; *Plate tectonics.*

Oceanite. An olivine-rich mela-◊*basalt.*

Occult minerals. Minerals which, on the basis of chemical analysis, might reasonably have been expected to form from a ◊ *glass* if it had crystallised out completely.

Ochoan. A stratigraphic stage name for the North American Upper ◊ *Permian.*

Ochres. Natural pigments. (◊ *Limonite.*)

Octant. A sector of space defined by three intersecting planes, no two of which are parallel, and all of which pass through a common point, the origin. (◊ *Axes, crystallographic.*)

Oddo and Harkins Rule. ◊ *Geochemistry.*

Offlap. A term used to describe the disposition of beds in a marine regression where the lowest beds extend further than the youngest ones. (◊ *Unconformity.*)

Offshore. The zone extending from the lowest limit of low spring tides to the edge of the ◊ *continental shelf.*

Oil, Oil accumulations. The term oil is used both for naturally occurring liquid hydrocarbons and for certain refined products obtained from these. We are concerned here only with the natural occurrences and associations. Oil is generally found in association with ◊ *natural gas,* salt water and sometimes solid ◊ *hydrocarbon minerals.* The term petroleum is often used as a synonym for oil. The origin of oil is still disputed, but in broad terms it is most probably of organic origin, i.e. produced by anaerobic decay of plants and/or animals under special conditions. Most oil appears to have formed under marine conditions, but some may have been formed under estuarine or deltaic conditions. It is important to notice that oil is rarely found at its point of formation, since it is a fluid and migrates quite readily through any openings (e.g. pores, joints, cleavage, bedding planes; ◊ *Porosity*; *Pervious*; *Permeability*). During this migration the oil–water system tends to break up into two separate layers, the oil floating on the water. Subsequently the lower boiling-point hydro-

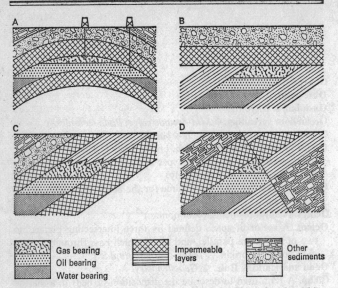

109. Oil accumulation. (A) Anticlinal fold; (B) Unconformity; (C) Stratigraphic, by change of facies; (D) Fault.

Gas bearing
Oil bearing
Water bearing
Impermeable layers
Other sediments

carbons may 'boil' off from the oil to give a gaseous layer. The liquid oil usually contains some solid hydrocarbons (waxes) in solution, and, if these are present to any great extent, they may separate 'out when oil is brought to the surface, owing to the fall in temperature.

Unless the migration of the oil fluid is prevented, it will flow upwards to the surface, where the liquid portions will evaporate, leaving behind the solid waxes and ◊ *asphalts*. For oil to be commercially exploitable a 'trap' must exist which prevents upward movement of the fluid. The essential feature of a trap is an impermeable cap-rock so arranged that the oil cannot by-pass it. Two main sorts of trap are known: (a) the structural trap, formed by faulting or folding; (b) the stratigraphic or lithological trap, e.g. ◊ *unconformity*, ◊ *facies* variation, localised lithological variation. (◊ Fig. 109; ◊ *Salt domes*.)

Oil-bearing strata may be pervious or porous. The term 'oil-pool' is used for an area of oil-bearing rocks below a cap-rock; it does not imply an open cavity in the rocks in which a 'lake' of oil exists. The pressure of the gas-cap and the upward pressure of the underlying water are used to drive the oil towards the surface, thus eliminating or reducing the need for pumping. In the search for oil, most efforts are directed towards determination of structural conditions which produce a possible trap.

Natural gas may occur without associated oil, but the converse is practically unknown.

Oil shale. A fine-grained black or dark grey ◊ *argillaceous rock* containing ◊ *kerogen*, which yields liquid hydrocarbons, i.e. oil, on distillation. Oil shales do not contain free oil. They appear to form in a special kind of anaerobic environment and are not uncommonly found in association with coal-bearing strata. (◊ *Boghead coal*.)

Old Red Sandstone. The name given to the continental facies of the ◊ *Devonian* in Britain – commonly abbreviated to O.R.S.

Olenekian. A stratigraphical stage name for the base of the East European Lower ◊ *Triassic*.

Oligo- (prefix). Small.

Oligocene. The epoch of the ◊ *Tertiary* period between the Eocene and Miocene epochs.

Oligoclase. A variety of plagioclase ◊ *feldspar*.

Oligoclasite. A ◊ *diorite* containing 5% or less of ◊ *ferromagnesian minerals*, the remainder being ◊ *oligoclase*.

Oligomict. A term describing detrital rock containing fragments of only one kind of material (or with one kind in great predominance). It is applied especially to ◊ *rudaceous rocks*. (Cf. ◊ *Polymict*.)

Olivines. A group of rock-forming neso◊*silicates* having a general composition of $R_2''SiO_4$ (where R'' may be Mg,Fe$''$,Mn,Ca (part)). The essential structure consists of a series of isolated SiO_4 tetrahedra which are linked by means of metal cations. The olivines are orthorhombic, but well-formed crystals are rare. They show no cleavage but display conchoidal fracture. Common olivine (◊ Appendix), which is usually green or brownish-green in colour, is a member of a continuous series of solid solutions between Mg_2SiO_4 (forsterite, abbreviated Fo) and Fe_2SiO_4 (fayalite, abbreviated Fa). Various names have been given to the intermediate compounds between these pure end members:

Forsterite	0 – 10% Fa
Chrysolite	10 – 30% Fa
Hyalosiderite	30 – 50% Fa
Hortonolite	50 – 70% Fa
Ferrohortonolite	70 – 90% Fa
Fayalite	90 – 100% Fa

Refractive indices and density increase from forsterite to fayalite, while the melting points decrease from forsterite to fayalite.
The term 'olivine' used without qualification may usually be
taken to mean chrysolite. Olivine is an unsaturated mineral (i.e.
it will not normally occur in association with free silica) and is
rather easily altered by ◊ *weathering* or ◊ *hydrothermal processes*
to serpentine (◊ *Ophicalcite*) or ◊ *chlorite* material. Gem-quality
olivines are referred to as peridot.

The majority of the olivine minerals described occur in ◊ *basic*
and ◊ *ultrabasic* igneous rocks (◊ *Peridotite*), but iron-rich olivines
may sometimes occur in acid rocks, either as a metastable mineral
(in certain ◊ *pitchstones*) or because iron-rich olivines may co-
exist with free silica. Olivine is rare as a detrital mineral owing to
its low stability in sedimentary environments. Forsterite occurs in
thermally metamorphosed dolomites containing a limited amount
of silica (◊ *Dedolomitisation*):

$$2CaMg(CO_3)_2 + SiO_2 = Mg_2SiO_4 + 2CaCo_3 + 2CO_2$$
$$\text{Dolomite} \qquad\qquad \text{Forsterite}$$

Other minerals having the olivine structure are tephroite
(Mn_2SiO_4) and monticellite ($CaMgSiO_4$). Minerals which somewhat resemble olivine but contain hydroxyl and fluorine ions
belong to the humite group of minerals: humite has the composition $3Mg_2SiO_4.Mg(OH,F)_2$, but chondrodite ($2Mg_2SiO_4.Mg(OH,F)_2$)
is perhaps the commonest member of the group.

It is believed that olivine may be a common constituent of the
◊ *sima* and the ◊ *mantle*.

Olivinite. A synonym of ◊ *dunite*, but its use is not recommended
owing to possible confusion with olivenite – a copper arsenate
mineral.

Omphacite. ◊ *Pyroxenes*.

Onesquethawian. A stratigraphic stage name for the top of the
North American Lower ◊ *Devonian*.

Onlap. ◊ *Unconformity*.

Ontarian. A stratigraphic stage name for the oldest Canadian ◊
Precambrian.

Onyx. A banded chalcedonic silica. (◊ *Silica group of minerals*.)

Oolite. (1) A synonym of oolitic ◊ *limestone*.

(2) An old name for the Upper ◊ *Jurassic* in Britain and Europe.

Oolith. A spherical or subspherical rock particle which has grown
by accretion around a nucleus. The commonest type of oolith is
calcareous, and there is some evidence that non-calcareous ooliths
are derived by replacement of calcareous ones. However some

oolitic iron ores may contain ◊ *primary* ooliths of iron minerals. Ooliths often show a concentrically banded structure, but radial and combined radial and concentric developments are recorded (◊ Fig. 110). The nucleus may be inorganic (e.g. a sand grain), or organic (e.g. a shell fragment). It has been suggested that algal filaments may promote the deposition of calcareous layers. Ooliths tend to be concentrated in beds to the exclusion of non-calcareous material, forming oolitic ◊ *limestones*, oolitic iron-stones, etc. Their formation depends upon the nuclei being constantly agitated so that the calcareous material is uniformly deposited. This agitation acts as a sorting mechanism, and con-

| Concentric | Combined | Radial |

110. Types of oolith.

sequently beds of ooliths are often well graded. Current action often results in oolitic limestones displaying current-bedding (◊ *Cross-bedding*).

Pisoliths are large ooliths about the size of a garden pea – 3 to 6 mm. in diameter, which occasionally form beds (pisolite). Algae commonly play a significant part in their formation.

Ooliths are occasionally found which consist of a thin skin of calcite around a perfectly spherical quartz grain.

Oolitic ironstone. ◊ *Sedimentary iron ores.*

Ooze. ◊ *Abyssal deposits.*

Opal. ◊ *Silica group of minerals* and Appendix.

Ophi- (prefix). Serpentine, i.e. snake-like.

Ophicalcite. A variety of ◊ *marble* in which forsterite (Mg_2SiO_4), formed by thermal ◊ *metamorphism*, has been altered to ◊ *serpentine*. Ophicalcites are commonly greenish streaked or mottled rocks and are much used in decorative work. Connemara marble is an ophicalcite with a high proportion of dark green serpentine.

Ophiolites. ◊ *Basic* and ◊ *ultrabasic* lavas and minor intrusions associated with the infilling of a ◊ *geosyncline*. The term is also applied to their metamorphic equivalents, which are usually rich in ◊ *albite* and ◊ *amphiboles*. (◊ *Spilites*.)

Ophitic texture. A texture characteristic of some basic igneous rocks, especially ◊ *dolerites*, consisting of large ◊ *augite* crystals enclosing, either partially or wholly, laths of ◊ *plagioclase*. The term is rarely applied to the similar occurrence of other pairs of minerals, although in certain metamorphosed dolerites this texture is preserved with ◊ *hornblende* replacing augite. It is very similar to ◊ *poikilitic* texture.

Ophiuroidea. ◊ *Echinodermata.*

Oppel, Albert (1831–65). German palaeontologist who suggested the term 'zone' for the horizons delimited by significant fossil types. He concentrated on the Jurassic, and his methods were very influential generally in later German research.

Optical properties of minerals. ◊ *Analyser; Axes, optic; Becke lens; Becke line and test; Birefringence; Dispersion; Ellipsoid; Extinction; Indicatrix; Interference figure; Optic orientation; Pleochroism; Polarisation colours; Polarised light; Polariser; Shadow test; Thin section; Vibration directions.*

Optic axes. ◊ *Axes, optic.*

Optic orientation. The spatial relationships between the optical elements and the crystallographic elements of a mineral. The usual method of expressing it is to relate the optic ◊ *axes* X, Y, Z to the crystallographic ◊ *axes* a, b, c. Optic orientation is often shown by means of a block diagram on which other features, such as cleavage, may also be shown. (◊ Fig. 112.)

Orbicular. A term used to describe a structure of large spherical or sub-spherical masses, made up of concentric shells of different mineral composition, which occur within certain plutonic igneous rocks, mainly ◊ *granites* and ◊ *diorites*. The names napoleonite and corsite are given to orbicular diorites. The orbicular units range in size from 2 cm. to 15 cm. in diameter, and often show traces of a ◊ *xenolith* at their centres, which appears to have acted as a nucleus.

Ordinary ray. ◊ *Vibration directions.*

Ordovician System. Named from the Ordovices – an ancient Celtic tribe of Central Wales. The period extended from 500 to 435 m.y., a duration of 65 m.y. The lower limit of the Ordovician is the base of the Arenig Series, which is defined on the first appearance of two-stiped extensiform graptolites. The upper limit is the top of the Ashgill Series, which is overlain by the Lower Llandovery of the Silurian Series. The top of the Ashgillian is the zone of Dicellograptus anceps: the last of the Trinucleid trilobites occurs at this horizon.

In old works (and especially old maps of the Geological Survey of Great Britain) the Ordovician is referred to as Lower Silurian. This obsolete terminology appears to have survived also amongst certain European authors.

Widespread vulcanicity characterised the period in the type area, and the onset of the Caledonian ◊ *orogeny* is evident.

Compared with the Cambrian, more advanced trilobites (◊ *Arthropoda*) became abundant and the ◊ *Brachiopoda* are represented by articulate forms. Crinoids (◊ *Echinodermata*) became abundant at some horizons and the first tabulate and rugose corals (◊ *Coelenterata*) appeared. The first vertebrates (fish) appeared in North America, but have not yet been recorded in Europe. The most important fossils are the graptolites (◊ *Chordata*) by means of which the system is zoned.

In the north-west of Scotland only the lower part of the Ordovician is represented, and this purely by limestones which are conformable upon the Cambrian succession. The beds contain a fauna of trilobites, cephalopods and gasteropods (◊ *Mollusca*) having strong North American affinities.

Ore. The sum total of ◊ *ore minerals*, ◊ *gangue* minerals, and ◊ *country rock* which constitute the material worked for the purpose of extracting a metal from the ore mineral. Technically, for material to be called ore, the economic factor must be taken into account, and it must be possible to extract metals profitably from the ore. The term is often used loosely, as a synonym of ore mineral.

Ore body. A mass of ◊ *ore* which is economically capable of being worked. The following is an abbreviated classification of ore bodies:

(1) Intrusive and cavity-filling bodies, e.g. ◊ *veins* (which includes lodes and stockworks); pipes – more or less cylindrical ore bodies; ◊ *breccia* fillings; pore (◊ *Impregnation*) and vesicle fillings; ◊ *pegmatite* dykes; and various irregular bodies of ore which are intrusive, filling cavities or ◊ *disseminations*. (◊ Fig. 111.)

(2) ◊ *Replacement* bodies and ◊ *metasomatic* bodies.

(3) Sedimentary ore bodies, including those produced by normal deposition, evaporation (◊ *Evaporites*), mechanical concentrates (◊ *Placer deposits*), and leached or residual deposits (bauxite – ◊ *Laterite and bauxite*).

The disposition of ore bodies is largely controlled by structural considerations and to a lesser extent by lithology.

(⟡ *Hydrothermal processes*; *Mineral deposit*.)

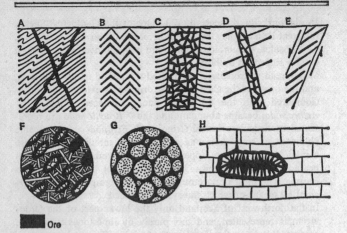

Ore

111. Cavity-filling ore bodies. (A) Veins; (B) Saddle reefs; (C) Pipe breccia; (D) Fault breccia; (E) Tension gashes in shear zone; (F) Vesicle fillings; (G) Pore fillings; (H) Solution cavity in limestone.

Orellan. A stratigraphic stage name for the North American Upper Oligocene (◊ *Tertiary System*).

Ore microscopy. Most ore minerals are opaque, even in thin section (◊ *sphalerite*, ◊ *cassiterite*, and ◊ *haematite* are exceptional in being transparent). This is especially true of most sulphides. These minerals are, however, capable of being polished to a high degree, and a technique of examining them by reflected light has been evolved to which the name 'ore microscopy' has been given. The apparatus used is more or less identical with the standard metallurgical microscope, used for examining polished samples of metal. The major technical problem which had to be overcome before ore microscopy could be fully utilised was the difficulty of producing a highly polished surface which was flat and did not show differential abrasion of harder and softer minerals.

Considerable information has been obtained using polarised light and a range of micro-chemical tests on the polished surface. Etching and micro-hardness tests are also valuable.

Ore microscopy has yielded a great deal of information about the mineralogy of ores and has been responsible for the discovery of several new minerals. Elucidation of orders of crystallisation, mineral reactions, and intergrowths have also resulted from such studies.

Ore mineral. A mineral from which a useful metal may be extracted, which may serve as a source of this metal if it can be profitably worked. (◊ *Ore.*)

Orenburgian. A stratigraphic stage name for the Russian Upper ◊ *Stephanian* (Carboniferous System).

Orientation diagram. A diagram which relates the crystallographic data to the optical data for a mineral (◊ Fig. 112). (◊ *Optic orientation.*)

Origin. Of a crystal, the point at which the crystallographic ◊ *axes* intersect; the geometric centre of the crystal.

Orogeny, Orogenesis. An orogeny is a period of mountain building. Orogenesis is the process of mountain building, leading to the formation of the intensely deformed belts which constitute mountain ranges. Most orogenic belts arise on the sites of ◊ *geosynclines* and the resulting mountains therefore consist of sediments and volcanic rocks deformed and metamorphosed to a greater or lesser extent according to their position and depth in the orogenic belt. Regional metamorphism and granite emplacement are always found in the deeper levels of an orogenic belt.

An orogeny extends in time for some tens of millions of years, and it is not surprising therefore that most orogenic periods display several maxima which often develop in different parts of the orogenic belt.

Correlation of orogenies in the absence of reliable ◊ *radioactive dating* is a difficult process. On the whole orogenies appear to be more or less world-wide events, although individual maxima are often highly localised. There appears to be an average interval of 200–300 m.y. between orogenies, but there is little evidence for a marked periodicity in their occurrence. Owing to erosion processes the older an orogenic belt is, the deeper the level which is exposed at the surface at the present day. (See table overleaf.)

Orpiment. A mineral, As_2S_3, found in the oxidised zones of arsenic minerals, in veins, and in hot spring deposits. (◊ Appendix.)

O.R.S. An abbreviation for ◊ *Old Red Sandstone.*

Ortho- (prefix). Straight, rectangular, regular.

Orthoclase. ◊ *Feldspars* and Appendix.

Orthogneiss. A ◊ *gneiss* presumed to have been formed from an original igneous rock.

Orthomagmatic. The name given to the stage in the crystallisation of an igneous mass during which the main mass of silicates

Periods	Britain	N.W. Europe	N. America	Africa	Elsewhere
			OROGENIES		
Tertiary					
Pliocene					
Miocene			Pasadenian		
Oligocene	Alpine	Alpine		Alpine	
Eocene					Andean (S. America)
Cretaceous			Laramide	Cape Folding	
Jurassic		Kimmerian (Cimeric)			
Permo-Triassic		Variscan	Appalachian		
Carboniferous	Variscan				
Devonian			Acadian		
Silurian	Caledonian				
Ordovician			Taconic		Tasman (Australia)
Cambrian					

Precambrian	Early Caledonian (Charnian?)	Wichita	Mozambiquian / Damaran / Late Katangan	Adelaide (Australia)
	Gothic	Grenville	Karagwe-Ankola	Satpura (India)
		Beltian	Ubendian	
Laxfordian		Gt Bear Lake	Tarkwaran	
		Karelian / Svecofennian		
		Hudsonian	Toro	
	Marealbian	Huronian		
	Saamian	Older Laurentian (Yellowknife)	Younger Swaziland	Upper Dharwar / Lower Dharwar (India)
Scourian	Ukrainian	Dahomeyan		
(Pre-Scourian?)		Older Swaziland		

Intercontinental equivalence is only approximate. 'Peaks' of activity are shown; all orogenies have earlier and later minor activity, e.g. the Alpine in Europe probably commenced during the Jurassic and is still not completely dead today.

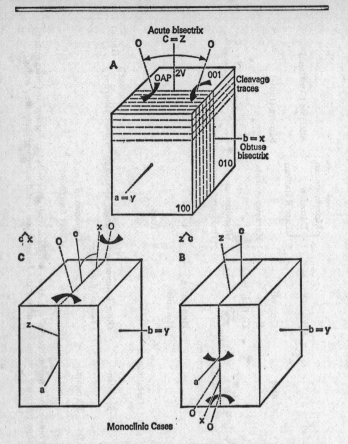

112. Orientation diagrams. (A) Orthorhombic case. OAP, optic axial plane; 2V, optic axial angle; O, optic axes; (B, C) Monoclinic cases. Two possible cases are shown, z making an acute angle with c (B), and x making an acute angle with c (C). Note that b is always equal to y in these cases.

crystallises. It is followed by the postmagmatic phase, during which such processes as ◊ *pneumatolysis*, ◊ *hydrothermy*, and ◊ *pegmatite* formation occur. The orthomagmatic phase is sometimes divided into a first or early magmatic phase, during which

328

anhydrous silicates crystallise, and a second or late magmatic phase, during which both anhydrous and hydroxyl-bearing silicates crystallise. The temperature and other conditions at which the orthomagmatic stage passes into the postmagmatic stage are rarely definable accurately, and much depends upon the amount of volatiles present.

Orthorhombic system. A ◊ *crystal system* divided into three symmetry classes, as in the following table:

	International symbol	*Centre*	*Planes of symmetry*	*Axes of rotation symmetry*	*Axes of rotary inversion*	*Example*
1	mmm (2/mm)	C	3	3 ii	—	Barytes, sulphur, olivine and certain pyroxenes
2	mm (2m)	—	2	1 ii	—	Hemimorphite, natrolite
3	222	—	—	3 ii	—	Epsomite ($MgSO_4 . 7H_2O$)

This system includes all crystals referred to three axes mutually at right angles, employing different parameters on each axis. The name should never be abbreviated to rhombic. (For international symbols, ◊ *Symmetry*. ◊◊ *Axes, crystallographic*.)

Osagean. A stratigraphic stage name for the North American Lower ◊ *Mississippian* (Carboniferous System).

Osar. A synonym of ◊ *esker*.

Oscillatory zoning. ◊ *Zoned crystal*.

Oslo essexite. ◊ *Essexite*. (◊◊ *Alkali gabbro*.)

Ossicle. A single echinoderm plate. They are often so abundant that they may comprise the greater part of a ◊ *limestone*. (◊ *Echinodermata*.)

Ostracoda. ◊ *Arthropoda*.

Otolith. The ear-bone of a fish; occasionally found fossil (or as a microfossil) in the absence of other vertebrate remains.

Outcrop. (1) Often used as a synonym of ◊ *exposure*, but should strictly be used only with the sense of (2) below:

(2) The total area over which a particular rock unit occurs at the surface – whether visibly exposed or not. The term is especially used for the delineation of such an area on a geological map.

Strictly speaking, the verb derived from the noun 'outcrop' is to 'crop-out'.

Outlier. A limited area of younger rocks completely surrounded by older rocks. It may be produced by ◊ *erosion*, ◊ *faulting*, or ◊ *folding*, or a combination of any two, or all three, of these agencies (◊ Fig. 113). (Cf. ◊ *Inlier*.)

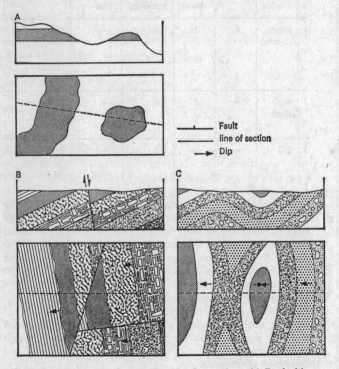

113. The formation of an outlier. (A) By erosion; (B) By faulting; (c) By folding.

Overburden. (1) Useless material which overlies a bed of useful material, e.g. shale overlying a coal seam.
(2) Any loose unconsolidated material which rests upon solid rock.

Overgrowth. The deposition of secondary material in optical continuity with an original growth, as found for example in certain quartz-cemented sandstones (◊ *Arenaceous rocks*).

Overlap. A term used to describe the relationship of the beds in an ◊ *unconformity* where progressively younger members of an upper series rest upon an older series.

Overprinting. ◊ *Replacement* (7).

Oversaturated. A rock containing free silica (in the form of quartz, tridymite, cristobalite, or silica glass; ◊ *Silica group of minerals*) is said to be oversaturated. (Cf. ◊ *Undersaturated*; ◊ *Saturation*.)

Overstep. An ◊ *unconformity* which develops during a marine transgression when the younger series rests upon progressively older members of the underlying series.

Overthrust. A low-angle reverse ◊ *fault*.

Owen, Richard (1804–92). Probably the greatest of the early British anatomists, he produced a system of classification for the fossil-reptiles after some fifty years' work, and originated the term 'dinosaur'.

Ox-bow lake. ◊ *Meander* loops tend to enlarge themselves by the erosion of their outer banks, which gradually narrows the neck between successive loops. In time of flood the neck may be breached and the ends of the loop become silted up, the river taking the new, 'preferred', straighter course. The cut-off loop is variously called an ox-bow lake, a mortlake, a billabong, or a bayou. The lake subsequently becomes a marsh and in time dries out.

Oxfordian. A stratigraphic stage name for the base of the European Upper ◊ *Jurassic*.

Oxidates. A geochemical division of the ◊ *sedimentary rocks* containing the sedimentary iron and manganese ores.

Oxyhornblende. ◊ *Amphiboles*.

Ozokerite. A ◊ *hydrocarbon mineral*.

Paedomorphosis. A synonym of ◊ *neotony*.

Pahoehoe (Hawaiian). A type of ◊ *lava* having a festooned, ropey surface structure. (Cf. ◊ *Aa*.)

Paisanite. ◊ *Microgranite*.

Palaeo-. A prefix common in geological terminology, meaning 'ancient, of past times', and sometimes suggesting an early or primitive nature.

Palaeobotany. The plant kingdom is divided into a number of groups, as is the animal kingdom. These groups, however, are not given the prefix 'phylum' and the names are used simply as labels to identify the various groups. The major groups are: algae, liverworts and mosses, which are not vascular and rare as fossils (although certain Ca-secreting algae occur in rock-forming abundance, e.g. algal ◊ *limestones*); and the vascular plants, divided into two major groups, (a) the spore-bearing (pteri-dophyte) and (b) the seed-bearing (spermatophyte) forms. The spermatophytes can be further sub-divided into two groups, the Gymnosperms or conifers, and the Angiosperms or flowering plants.

Non-vascular plants are represented as fossils from Precambrian times onwards but they are rare. Vascular plants were the first forms to colonise land because of their additional rigidity and their ability to translocate water. They first appeared in the Upper Silurian and were abundant from Lower Devonian times onwards. These forms continued through the Devonian and Carboniferous where they reached considerable size. The first true seed-bearing plants appeared in the Permian and were represented by primitive conifers; the flowering plants did not appear until the Cretaceous. (◊ *Coal*.)

Palaeocene. Lowermost division of the ◊ *Tertiary System*.

Palaeoclimatology. The study of the climatic conditions obtaining in past geological eras. Detailed study of sedimentary rocks and their enclosed fossils has made possible estimates of such climatic factors as wind direction, rainfall, atmospheric and oceanic temperatures, and the effects of atmospheric changes. The most obvious palaeoclimatic determinations are the recognition of ◊ *ice ages*, and the hot, dry periods in which ◊ *evaporites* form.

Estimates of climatic conditions become less and less reliable as they are projected further and further back in time. Thus, Pleistocene climates are relatively well known whereas climates for Lower Palaeozoic periods are probably little better than intelligent guesswork. In exceptional circumstances, e.g. deposition of ◊ *varved* clays, seasonal variations from year to year can be detected. (◊ *Palaeoecology.*)

Palaeocurrent. A current which existed during the deposition of a sediment at some period of geological history. Palaeocurrent analysis is the process of attempting to determine the direction of such currents by means of ◊ *sedimentary structures* which display current or current-orientated features, e.g. ◊ *cross-bedding*, ◊ *ripple marks*, ◊ *imbricate structure*, etc. The determination of major current direction depends upon a statistical analysis of many local observations, and great care has to be taken that minor local currents are not misinterpreted as major ones.

Palaeoecology. Fossils were once animals and plants, and were therefore governed by precisely the same ecological factors as now govern living animals and plants. Palaeoecology is therefore a study of these factors and how they affected the mode of life of organisms in the past. A palaeoecological study is designed to obtain as much information as possible about the ◊ *environment* of the fossil or fossils, particularly their relation to the sediment. This is necessary, since many of the factors affecting living forms, such as temperature, salinity, and depth of water, can rarely be determined for fossils in any other way. Much palaeoecological work depends upon comparative studies of ancient and modern forms, e.g. adaptive morphology, associations of organisms, etc. (◊ *Trace fossils.*)

Palaeogene. The name given to the Eocene and Oligocene periods of the ◊ *Tertiary*, when grouped together. The term is almost obsolete.

Palaeogeography. A reconstruction of the presumed geography, especially the relative positions of land and water, at some particular period in the past. It is generally more difficult to reconstruct the geographies of the older periods and in many cases published maps are no more than approximations. It should be noted that features, e.g. coastlines, shown in a particular position superimposed on the present geography probably did not occupy this position relative to other modern features and to other palaeogeographical features, owing to the fact that palaeogeographical maps are rarely ◊ *palinspastic*. (◊ *Continental drift.*)

333

Palaeomagnetism. It is well known to physicists that the positions of the Earth's magnetic poles change over a period of time. The investigation of such changes during geological time is the field of palaeomagnetic studies. Some rocks acquire a permanent record of the Earth's magnetic field by virtue of certain iron-bearing minerals which they contain. The most important groups are:

(a) Lavas containing ◊ *magnetite*, Fe_3O_4 (or certain related minerals – ◊ *Spinels*). Crystals of this mineral orientate themselves in the magnetic field once the temperature of the lava has fallen below the critical point for magnetisation to take place – the Curie point.

(b) Sediments containing detrital magnetite, in which the grains orientate themselves when allowed to settle freely. Sediments which have been deposited in the presence of strong current are generally unsuitable, as the force of the current is many times the relatively weak magnetic forces.

(c) Red sandstones and shales, in which the colouring is due to ◊ *haematite*, Fe_2O_3, show a strong effect, probably due to crystallisation of the haematite cement under the influence of the magnetic field.

The superimposition of later (or recent) magnetisations has to be allowed for, and the residual original magnetisation is often very weak.

Collection of orientated samples is required for measurement of the direction of orientation of the magnetic particles, and structural complications, such as folding or tilting, must be removed. The direction of magnetisation preserved is expressed as a bearing to the North (or South) Magnetic Pole. By taking a sufficiently widespread series of samples from the same horizon, statistical methods yield a position for the magnetic poles at that particular time. Successive determinations yield a picture of movements of the poles through geological time.

Determinations of the positions of the magnetic poles in different parts of the globe has yielded a variety of results and it has been suggested that the only reasonable explanation is that continents have moved relative to one another (◊ *Continental drift*). Reversals of the magnetic field are now generally thought to be world-wide and a sequence of reversals has been established, back to the Cretaceous. A study of the distribution of normal and reversed magnetism on either side of ◊ *mid-oceanic ridges* has supported the concept of ocean floor spreading. (◊ *Plate tectonics*.)

Palaeomagnetic properties have been suggested as a means of correlation of horizons over wide areas and even inter-continent-ally.

Palaeontology. The study of ancient life. Strictly speaking it includes palaeozoology and ◊ *palaeobotany*, but the term is not infrequently used as a synonym of the former. It was introduced by de Blanville and von Waldheim in 1834. Palaeobiology is a synonym. Palaeontology is the essential tool of the stratigrapher for purposes of correlation, strata identification, establishment of sequences, and determination of environments (◊ *Palaeoecology*). From a biological point of view palaeontology yields important evidence for evolution and adaption of organisms to different environments. (◊ *Fossil*; *Fossil Man*; *Annelida*; *Archaeocyatha*; *Arthropoda*; *Brachiopoda*; *Chordata*; *Coelenterata*; *Echinodermata*; *Mollusca*; *Polyzoa* (*Bryozoa*); *Porifera*; *Protozoa*.)

Palaeoslope. The inclination of an ancient land surface as existing in some period of geological history.

Palaeozoic. The era ranging in time from 600–230 m.y., a duration of 370 m.y. It comprises the ◊ *Cambrian*, ◊ *Ordovician*, and ◊ *Silurian* Systems in the older or Lower Palaeozoic sub-era, and the ◊ *Devonian*, ◊ *Carboniferous*, and ◊ *Permian* Systems in the newer or Upper Palaeozoic sub-era. The boundary between the Lower and Upper Palaeozoic is drawn at 400 m.y. The Palaeozoic was preceeded by the ◊ *Precambrian* and followed by the ◊ *Mesozoic*. In older works, it may sometimes be referred to as the 'Primary Era'. (For its upper limit, ◊ *Mesozoic*.) Its lower boundary is marked by the appearance of the first fossil trilobites (◊ *Arthropoda*), but this is unsatisfactory in some areas (◊ *Cambrian*).

Two major orogenies occur, the Caledonian in the Lower Palaeozoic and the Variscan at the top of the Upper Palaeozoic, together with their accompanying granite intrusions and metamorphism. (Caledonian = Taconic+Acadian, and Appalachian = Variscan in North America.)

The Lower Palaeozoic faunas are characterised mainly by invertebrates, including trilobites, graptolites (◊ *Chordata*), and brachiopods (◊ *Brachiopoda*). The earliest fish remains occur in the Ordovician. In the Upper Palaeozoic, graptolites and trilobites became extinct, while corals (◊ *Coelenterata*) and crinoids (◊ *Echinodermata*) increased in abundance. Fish were important in the Devonian; Amphibia and reptiles developed in the Carboniferous. The first terrestrial floras appeared in the Devonian.

Palagonite. ◊ *Pyroclastic rocks.*

Palimpsest structure. Relics of an original texture or structure which is visible through, and which may even exert a modifying influence on, a superimposed texture or structure. The superimposed structure must of necessity be a metamorphic one. Practically synonymous with relict structure.

Palin- (prefix). Again.

Palingenesis. (1) The melting or partial melting of pre-existing rocks to form new magma. (◊ *Anatexis*; *Syntexis.*)

(2) The recapitulation of morphological and other features of adult ancestors at earlier stages during the life of the descendants.

Palinspastic map. A geological map on which is restored a set of conditions which existed prior to certain events. Typical examples are maps which unfold folded strata, replace thrusts, and restore sedimentary strata to their former extent. (◊ *Palaeogeography.*)

Palissy, Bernard (c. 1510–90). A Frenchman, chiefly remembered for his ceramics, whose depiction of fossils on his work led him to take sufficient interest to write – far ahead of his time – of their possible organic origin. He was also one of the first to suggest that the only source of water for springs and rivers was rain and melting snow.

Pallas, Peter Simon (1741–1811). German naturalist who undertook a six-year exploration into the entire natural history of Asiatic Russia. His most important geological contributions arising from this trip were his account of the large mammal remains of Siberia, and his observations on the apparent three-fold structure of mountains – with a central 'Primary' core, a calcareous 'Secondary' layer and an arenaceous 'Tertiary' layer.

Pallasite. A ◊ *meteorite* consisting of both metals and silicates.

Pallial sinus (Mollusca). ◊ Fig. 101.

Palygorskite group of minerals. ◊ *Clay minerals.*

Palynology. The study of fossil spores. This definition is taken to include the study of the pollen from higher plants, but not the seeds. Palynology is a subject which is becoming increasingly important, since correlation of strata by pollen analysis has advantages over other methods, the main one being that spores and pollen are extremely resistant to destruction and, being light, they may be carried for some considerable distance out to sea; thus they can be present in both marine and freshwater sediments, a fact of some importance in correlation, since few other fossils are common to the two environments. Palynology has also been used to determine climatic changes in the ◊ *Quaternary*.

Pan- (prefix). All.

Pangaea. ◊ *Laurasia* and ◊ *Gondwanaland* together. A hypothetical super-continent which was supposed by Wegener to have existed before being fragmented by ◊ *continental drift*.

Pandiomorphic. A texture term applied to a rock made up of well-developed crystals.

Pannonian. A stratigraphic stage name for the European Upper Miocene (*Tertiary System*).

Pantellerite. A ◊ *rhyolite* containing a soda-potash feldspar, quartz, aegirine, and sometimes sodic ◊ *amphibole*. Some pantellerites are low enough in quartz to be considered varieties of ◊ *trachyte*.

Paper shale. A shale (◊ *Argillaceous rocks*) which splits uniformly into thin, slightly flexible laminae. Commonly, paper shales are carbonaceous, and are usually regarded as representing deposition under very quiet (lagoonal or lacustrine) conditions.

Par-, para- (prefix). Besides.

Paragenesis. The relationship of minerals expressed in terms of a time sequence, i.e. if conditions are changing progressively along a time-axis, then a particular mineral will appear as a derivative of an earlier mineral, and may ultimately pass into yet another mineral, the process taking place being recrystallisation. The term is sometimes used with the specific meaning of the order of crystallisation of the minerals forming a rock, and it is also used loosely in the sense of the mode of origin of a rock or mineral.

Paragneiss. A ◊ *gneiss* presumed to have been formed from an original sedimentary rock.

Paragonite. ◊ *Micas* (Dioctahedral).

Paralic. A term applied to sedimentary basins developed in marginal marine environments, e.g. lagoons, littoral basins, etc. (Cf. ◊ *Limnic.*)

Parallel growth. During the formation of crystals it commonly happens that, owing to local controls, all the crystals of a particular substance develop the same ◊ *habit* and have one crystallographic direction in common. Parallel growth should be carefully distinguished from twinning (◊ *Twinned crystal*); it is an accidental feature of crystal growth, with an element of randomness, compared with the strictly regular laws governing twinning. The commonest type of parallel growth involves crystals having a prismatic habit, the axes of which are all parallel. (◊ Fig. 114.)

Parameters. When the ◊ *intercept* ratios for a number of forms making up a crystal are calculated, it will be observed that for each face they are either multiples or sub-multiples of the corresponding intercept ratios of every other face. It is therefore clear that the intercept ratio can be simplified by selecting one

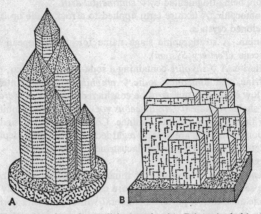

114. Two examples of parallel growth. (A) Prismatic habit (e.g. quartz); (B) Tabular habit (e.g. barytes).

form – the unit form – to provide a unit of measurement in which all the other intercepts may be expressed. The intercept ratios of this unit form are called parameters. The term axial ratio is also used, although the implication that axes have a finite length is undesirable.

It is found that for all crystals a set of parameters may be selected in such a way that all faces of the crystal yield intercepts which are integral multiples or sub-multiples of them. This is the Law of Rational Intercepts.

When selecting a form to be a unit form the following criteria are usually adopted: (1) The face must cut all the axes. (2) It should be a relatively commonly occurring form. (3) The form selected should make other commonly occurring forms have the simplest indices possible. If for example two otherwise equally suitable forms are being considered and it is found that one makes a common prism 110 while the other makes the same prism 230, then the first should be preferred.

It is significant that parameters derived from the examination of crystals are commonly identical with, or simple multiples or sub-multiples of, the ratios of the absolute lengths of the edges of the ◊ *unit cell*. (◊ *Axes, crystallographic.*)

Paramorph. A type of ◊ *pseudomorph* formed by one ◊ *polymorph* converting to another.

Parasitic cone. A subsidiary conelet on the slopes of a ◊ *volcano*.

Parataxitic. A term used to describe the streaky appearance of some ◊ *pyroclastic rocks*. (◊ *Eutaxitic.*)

Paratype. A specimen selected to demonstrate characters additional to those of the original holotype. (◊ *Type.*)

Parautochthonous. A term used to describe a feature formed at great distance from the place in which it is found. It is most commonly used to describe ◊ *nappe* structure in the Alps. (◊ *Granite series.*)

Pargasite. A sodium ◊ *amphibole*.

Parieties. ◊ *Archaeocyatha* and Fig. 1.

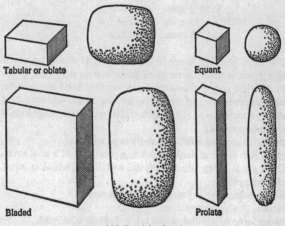

Tabular or oblate   Equant

Bladed   Prolate

115. Particle shape.

Particle shape. The shape of ◊ *clastic* fragments is defined in terms of the ratio of the dimensions of the fragment. The three measurements necessary are length, breadth, and thickness. Four classes have been established by Zingg as follows: (a) tabular, oblate or discoidal; (b) equant, equiaxial, or spherical; (c) bladed or triaxial; (d) prolate or rod-shaped. (◊ Fig. 115.) It is not uncommon for pebbles to have characteristic shapes which are not adequately described by these terms, e.g. ◊ *dreikanter*, wedge-shaped, or tetrahedral forms (◊ *Ventifact*). (◊ *Roundness*; *Sphericity*.)

Particle size. In dealing with sediments and ◊ *sedimentary rocks* it is necessary that precise dimensions should be applied to such terms as clay, sand, pebble, etc. Numerous scales have been suggested, but in this work the Wentworth-Udden scale is used, as it is widely accepted as an international standard. In the table which follows, particle size limits are shown, but within most groups further subdivision is possible; for example, sand may be described as very fine, medium, coarse, very coarse.

Size Range	Particle
>256 mm.	Boulder
64–256 mm.	Cobble
4–64 mm.	Pebble
2–4 mm.	'Granule', gravel
1/16–2 mm.	Sand
1/256–1/16 mm.	Silt
<1/256 mm.	Clay

The term 'granule' is not in common use. It should be noted that in a given sediment a range of particle sizes may exist (◊ *Grade*, 3). Particle size is normally determined by hand measurement of pebbles, cobbles and boulders, sieving of gravel, sand and silt, and ◊ *elutriation* of silt and clay. (For grain size in igneous rocks, ◊ *Igneous rocks*.)

Parting. (1) A small joint, especially in a coal seam.

(2) A thin layer of shale separating two parts of a coal seam.

(3) A plane along which a crystal will divide, which is not a true cleavage plane.

Passage beds. A synonym of ◊ *transition series*.

Patellate. ◊ *Coelenterata* and Fig. 23.

Peacock ore. A popular name for ◊ *bornite* (Cu_5FeS_4).

Pearlspar. A name given to the rhombohedral form of the mineral ◊ *dolomite* or ◊ *ankerite* ($(Ca,Mg,Fe'')CO_3$).

Peat. A partially decomposed mass of vegetation which has grown in a shallow lake or marsh. Generally peat is dark brown or black, and contains recognisable vegetable fragments, but very little mineral material (◊ *Coal*). The main plants which make up peat are the 'peat-mosses' Sphagnum and Hypnum, with some contributions from rushes, sedges, horse-tails, etc. Peat deposits may be some tens of metres thick and cover large areas. For large-scale peat formation, rapid plant growth, high moisture content, and the development of anaerobic conditions are essential. Highland and lowland (or fen) peats are recognised, the former being spongy and mainly formed from mosses, the latter being more compact and homogeneous, mainly formed from sedges, rushes etc. ◊ *Diagenesis* of peat leads to lignite and coal formation, under the right conditions.

Pebble. A rock fragment with a diameter of between 4 and 64 mm., i.e. larger than gravel and smaller than cobble. (◊ *Particle size*.)

Pebble phosphate. ◊ *Phosphatic deposits*.

Pectolite. ◊ *Pyroxenoids*.

Pedalfer. A leached ◊ *soil* in a region of high rainfall.

Pedal gape (Mollusca). ◊ Fig. 102.

Pedicle foramen (Brachiopoda). ◊ Figs. 11 and 12.

Pedicle valve. ◊ *Brachiopoda* and Figs. 11, 12, and 13.

Pediment. A plain of eroded bedrock (which may or may not be covered by a thin veneer of alluvium) in an arid region developed between mountain and basin areas. (◊ *Cycle of erosion*.)

Pedion. A crystal ◊ *form* consisting of only one face.

Pediplain. The end-product of the ◊ *cycle of erosion*, according to the scheme of W. Penck and L. C. King.

Pedocal. An unleached ◊ *soil* in a region of low rainfall.

Pedology. The study of ◊ *soil*.

Pegmatite. A very coarse-grained igneous rock having a grain size of 3 cm. or larger, with crystals occasionally reaching a metre or so in length; very rarely, lengths of some ten metres or so have been recorded. Conventionally, the word pegmatite, unqualified, refers to a rock of ◊ *granite* composition, but the term is also used in combination with the appropriate plutonic rock name to refer to very coarse-grained varieties of these rocks, e.g. gabbro-pegmatite, syenite-pegmatite. Although many granite-pegmatites consist more or less exclusively of the common granite minerals, there are also many examples of pegmatites which contain suites of minerals which are otherwise rare, e.g. minerals containing lithium, beryllium, and rare earths. Pegmatites also commonly contain typical ◊ *pneumatolytic* and ◊ *hydrothermal* minerals, e.g. tourmaline, topaz, cassiterite, fluorite, apatite, etc. They are economically important as a source of many rare elements, including radioactive ones, together with tin and tungsten.

It is generally agreed that the ultra-coarse grain size is a result of slow crystallisation from a volatile-rich melt. (Cf. ◊ *aplites*, with which pegmatites are often associated, sometimes as lenses within aplites.)

In metamorphic terrains it is not uncommon for pegmatites to show evidence of having assimilated the ◊ *country rock*, and cases where metamorphic minerals, such as garnet, occur in pegmatites are usually due to contamination in this way.

Pelagic. A term used to describe the mode of life of those animals which live in the open sea but not on the sea floor, including forms which are free-swimming (nektonic), e.g. fish, and forms which float passively in the surface waters (planktonic), such as jellyfish.

Pelecypoda. A synonym of Lamellibranchiata (◊ *Mollusca*).

Pelées hair. A fine mass of hair-like glass which is formed as a result of lava being exuded through a small orifice and blown about by the wind. (◊ *Pyroclastic rocks*.)

Pelitic. ◊ *Argillaceous*, but now used almost entirely in relation to metamorphosed argillaceous rocks (pelites).

Pellets. Ovoid particles of sediment, commonly composed only of calcium carbonate, which range in size from 0·25 mm. to 5 mm. in length. They are usually considered to be faecal pellets which may have been excreted by molluscs, echinoderms, and possibly some other groups of invertebrate animals. Pellets consisting of phosphate and/or other materials have been recorded, and cases of replacement of pellets by ◊ *glauconite*, ◊ *pyrite*, etc. are known. It is probable that many occurrences of faecal pellets have been overlooked, owing to their similarity to inorganic particles. Larger faecal pellets are termed ◊ *coprolites*.

Pelmatozoa. ◊ *Echinodermata*.

Penck, Albrecht (1858–1945). German geologist who gave the first scientific account of the Ice Age, proving that it consisted of alternating cold and warm periods.

Penecontemporaneous. Almost at the same time.

Peneplain. The end-product of the ◊ *cycle of erosion* in humid climates, according to the scheme of W. M. Davis.

Penninite. A mineral of the ◊ *chlorite group*.

Pennsylvanian. The younger of the two sub-systems into which the ◊ *Carboniferous* period in North America is divided.

Peperino. A rock of mixed ◊ *pyroclastic* and ◊ *sedimentary* origin, including pyroclastic material, and weathered and eroded volcanic material (including scoriae, cinders, etc.) cemented together.

Pepino. ◊ *Karst scenery*.

Per- (prefix). Throughout, completely, very.

Percussion figure. A figure produced when a ⬦ *layer-lattice mineral* is struck sharply with a pointed instrument. It consists of lines radiating from the point of impact, which bear a constant relationship to the symmetry elements of the crystal.

Percussion mark. A crescentic chip in a pebble, formed by the impact of one pebble on another.

Peri- (prefix). Around, beyond.

Periclinal structure. A term applied collectively to domes and basins (⬦ *Folds*), or, by some authors, to domes alone.

Peridot. ⬦ *Olivine* of gem quality. (Cf. ⬦ *Peridotite*.)

Peridotite. A class of ⬦ *ultrabasic rocks* consisting predominantly of olivine with or without other ⬦ *ferromagnesian minerals*.

Peridotite shell. The ⬦ *mantle* of the Earth.

Periglacial ('around a glacier'). A term applied to a region adjacent to an ice sheet. It is suggested that temporary snow caps and ⬦ *permafrost* may have developed in these areas. Characteristic deposits of periglacial areas include brickearth, ⬦ *loess*, ⬦ *combe rock*, and other ⬦ *solifluxion* deposits.

Period. ⬦ *Stratigraphic nomenclature*.

Periproct (Echinodermata). ⬦ Fig. 44.

Peristome. ⬦ *Echinodermata* (Echinoidea) and Fig. 44.

Perknite. A class of ⬦ *ultrabasic rocks* consisting predominantly of ⬦ *ferromagnesian minerals* other than olivine.

Perlite. A glassy rock of rhyolitic composition displaying ⬦ *perlitic texture*.

Perlitic texture. A texture found in glassy and devitrified igneous rocks, consisting of curved or spherical to sub-spherical cracks. These are produced by contraction during cooling and occasionally are so strongly developed that the rocks break into a series of spherical units. (⬦ *Perlite*.)

Permafrost. Permanent frost or permanently frozen ground. It extends from 2,000 to 3,500 km. south of the North Pole. It can be divided into three classes: (1) Continuous permafrost, in which the subsoil never thaws; from 0·5–500 m. (maximum) below the surface. (2) Discontinuous permafrost, in which the frozen ground is interrupted by unfrozen patches. Freezing takes place up to 20 m. below the surface and part usually melts in summer. (3) Sporadic permafrost, where the areas of unfrozen ground exceed the frozen areas.

For permafrost to form, the heat loss from the ground must counterbalance insolation and the interior heat of the Earth. It is encouraged by low humidity and clear skies, and is found in those regions of the tundra with least snow and with mean annual temperatures below 0°C. with rainfall of 25 cm. or less.

Permafrost was not extensive in the areas which were covered by the Pleistocene ice sheet, but its effects can be observed in areas which were adjacent to the ice sheet. At the present time it is thought to be receding.

Two processes are associated with permafrost: frost-shattering or riving, which tends to reduce larger fragments ultimately to a silt grade, such material becoming mixed with soil; and frost-stirring, producing polygonal nets (called stone-stripes) of angular material, especially on slopes.

(◊ *Fold*, Cambering.)

Permeability. A rock is said to be permeable if water or other liquid in contact with its upper surface tends to pass through the rock more or less freely to the lower surface. Permeability may be achieved by the rock being either ◊ *porous* or ◊ *pervious*. The essential feature of a bed of permeable rock is that the liquid it contains may be extracted by pumping. The opposite of permeable is impermeable. Permeability in a rock is measured in ◊ *darcies*.

Permian System (named at the suggestion of ◊ *Murchison* in 1841 from the province of Perm in Russia). The period of time from 280 to 225 m.y., a duration of 55 m.y. It marks the end of the ◊ *Palaeozoic Era*. Because of the widespread occurrence of continental conditions during late ◊ *Carboniferous*, Permian and ◊ *Triassic* times, defining the lower and upper limits is often difficult. Where marine deposits occur, the incoming of Pseudoschwagerina (a large Foraminifera – ◊ *Protozoa*) marks the base. The upper limit of the continental facies has been defined as the top of the 'zone' of Cisticephalus (a reptile), but this is not of very great general use. Local limits have been used in many localities which may not define the same time plane.

The continental facies of the Permian is represented by red marls and arkosic sandstones, dolomitic ◊ *limestones*, and ◊ *evaporites*, which are important economic deposits. The period saw limited vulcanicity, the continuation of the Variscan Orogeny, and the climax of the Southern Hemisphere glaciation.

The period marked the extinction of a number of fossil groups, the most important being the trilobites (◊ *Arthropoda*) and the tabulate and rugose corals (◊ *Coelenterata*). The only new group to become widely established was the reptiles, which were the first vertebrates to sever their association with water. Floras also showed a marked change in the Permian, the large, primitive forms of the Carboniferous being largely replaced by the more advanced conifers. Fossils are extremely rare in the British Permian, the most common being tracks and trails of the early reptiles. (◊ *Trias*.) Some authorities deny the existence of a separate Permian System, believing that the strata assigned to the period are in fact equivalent to the top part of the Carboniferous and the Lower Trias.

It is now fairly common practice to unite the Permian and Triassic in a single Permo-Triassic System. 'New Red Sandstone' is often used in Europe as a synonym for this grouping.

Permo-Trias. The ◊ *Permian System* and the ◊ *Triassic System* considered together.

Perthite. The name given to an intergrowth of two ◊ *feldspar* minerals (microperthite is properly the term used for intergrowths

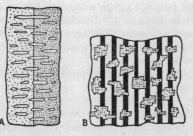

116. (A) Perthite; albite rods in orthoclase; (B) Anti-perthite; microcline 'patches' in albite.

visible under a microscope). Strictly used, the term implies a potash-rich feldspar containing strings, patches, or lenticles of sodic-plagioclase. If the reverse relationship exists, i.e. a plagioclase containing similar inclusions of orthoclase or microcline, the term anti-perthite is used. (◊ Fig. 116.) Two origins are suggested: (1) Exsolution, i.e. a feldspar, homogeneous at high temperatures, becomes unstable at lower temperatures and one component appears as a separate phase. (2) Replacement perthite, produced by the reaction of soda-rich liquids upon early-formed potash-feldspar, or vice-versa. (Cf. ◊ *Graphic texture*; ◊ *Symplektite texture*.)

Perthosite. A ◊ *syenite* consisting almost entirely of alkali feldspar, but containing small amounts of ◊ *aegirine* and other ◊ *ferromagnesian minerals*, and often ◊ *nepheline*.

Pervious. A rock is said to be pervious if it is permeable by virtue of mechanical discontinuities such as joints, bedding planes, fissures, etc. (e.g. some limestones and some igneous rocks). The opposite is impervious. (Cf. ◊ *Porosity*; *Permeability*.)

Petrifaction. ◊ *Fossils*.

Petro-, petr- (prefix). Rock.

Petrofabric analysis. ◊ *Fabric*.

Petrogenesis. A comprehensive term covering all aspects of the formation of rocks, i.e. processes, mechanisms, reactions, sequences of events, later modifications, etc. which result in the final rock.

Petrographic province. A region characterised by the occurrence of a group of igneous rocks which are genetically related, and which belong to a similar period of igneous activity. It is usually found that such associations of rocks can be characterised by some special feature, e.g. their mineral content, or a characteristic suite of trace elements (◊ *Geochemistry*). Examples are the Brito–Icelandic–Greenland Tertiary Province, the Andes, and the Roman Province (especially characterised by the prevalence of leucite-bearing rocks).

Petrography. The systematic description of rocks in hand specimen and thin section.

Petroleum. ◊ *Oil, Oil accumulations*.

Petrology. The general term for the study of rocks in all their aspects, including their mineralogies, textures, and structures (◊ *petrography*), their origins (◊ *petrogenesis*), field occurrences, alterations (◊ *diagenesis*, ◊ *metamorphism*, etc.), and their relationships to other rocks.

Petromict. A synonym of ◊ *polymict*.

Phacelloid. Synonym of fasciculate (◊ *Coelenterata* and Fig. 23).

Phacolith. A concavo-convex body of igneous rock which is anticlinal in form, i.e. the concave surface faces downwards. (◊ Fig. 117.) It is generally assumed that the igneous rocks were intruded

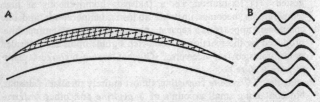

117. (A) A simple phacolith; (B) Saddle reefs.

into position during folding and were deformed along with the enclosing sedimentary rocks. Phacoliths sometimes occur in vertical sets occupying the crestal zone of an anticlinal fold. They range in size from a few metres to a few kilometres in width. Occasionally ore deposits, e.g. gold-quartz veins, occur in this form and are called saddle reefs (e.g. at Bendigo, Australia). (Cf. ◊ *Laccolith*.)

-phainos (suffix). Showing.

Phanero- (prefix). Visible.

Phanerocrystalline, Phaneritic. A textural term applied to igneous rock in which the crystals comprising it can be separately distinguished with the naked eye. (Cf. ◊ *Aphanitic*.)

Phanerozoic (literally 'obvious life'). That period of time during which sediments have accumulated containing obvious and abundant remains of animals and plants. It is a convenient term for the stratigraphic systems from the Cambrian to the Quaternary. (Cf. ◊ *Precambrian*.)

Phenoclast. A general term for a large fragment (larger than 4 mm.) in a sedimentary rock, i.e. a pebble, cobble, or boulder. The term should be used with care, since it is easily capable of confusion with the igneous term ◊ *phenocryst*.

Phenocrysts. The relatively large crystals which are found set in a finer-grained groundmass, constituting the texture termed ◊ *porphyritic*; 'inset' is a synonym. Phenocrysts are generally idiomorphic (euhedral) (◊ *Texture*).

Phillips, William. ◊ *Conybeare*.

Phlogopite. ◊ *Micas* (Trioctahedral) and Appendix.

Phonolite. ◊ *Undersaturated* ◊ *trachytes* containing dominant ◊ *nepheline*.

Phosphatic deposits. Under this heading are included those sedimentary phosphatic deposits commonly included under the headings Phosphorite (Rock Phosphate) and Pebble Phosphate. The mineralogy of phosphate deposits is usually complex, as they consist of a fine-grained mixture of various calcium phosphates, of which hydroxyl-apatite, carbonate-apatite, fluor-apatite, and solid solutions of these end-members are perhaps the most important. Collophane is apparently a ◊ *cryptocrystalline* or amorphous calcium phosphate complex. Three main types of phosphate deposit may be recognised:

(1) PRIMARY MARINE PHOSPHATES. All marine sediments, especially limestones, contain a proportion of phosphate and under certain conditions this proportion may rise to a notable level

(phosphatic limestone), and more rarely reach an economically useful concentration. The phosphate deposits of commerce have usually arisen as a result either of selective leaching of $CaCO_3$, or of extraction of phosphate from higher levels followed by concentration from downward-percolating groundwaters. In the case of pebble phosphates, mechanical concentration has also taken place. The presence of hydrocarbons and ◊ *glauconite* in many primary phosphates suggests that they were formed under anaerobic conditions. The conditions for preferential deposition of phosphates are obviously uncommon, since they are rare as sediments.

(2) BONE BEDS. Local accumulations of bone, teeth, scales, and ◊ *coprolites* occur at a number of horizons. They are often residual deposits of condensed deposits and are rarely of great thickness. It is possible that leaching of these horizons may have occasionally resulted in phosphatisation of limestone beds at a lower level.

(3) GUANO. The accumulated excreta of seabirds or bats, generally found on oceanic islands, or, in the case of bat guano, in very large cave systems. It requires for its maximum development a dry climate, but even so, underlying limestones are occasionally phosphatised, probably owing to the slightly acid nature of the excreta.

Phosphorite (Rock phosphate). ◊ *Phosphatic deposits.*

Photic. An environmental term meaning 'light' (sunlight). The zone in the sea down to 200 m. (◊ Fig. 49.)

Photogeology. Strictly speaking, the term may be used for the geological interpretation of any photograph. In practice it is used almost exclusively for the geological interpretation of aerial photographs, mainly vertical photographs, although 'high obliques' can also be used. It is usual to study vertical photographs in stereoscopic pairs, using a stereoscope, so that relief may be fully appreciated.

Detailed study under the stereoscope in many cases provides almost enough data to plot a geological map, especially one showing structural features. It is not, however, usually possible to identify rock types from air photographs; an exception is limestone, which can often be recognised by virtue of the ◊ *karst* topography displayed. Air photographs commonly reveal details of the geology which are not detectable on the ground. Even in heavily forested regions, slight changes in slope, soil, type of vegetation, etc. can be detected, and provide valuable evidence of geological features. In regions where there is only a thin soil and/or vegetation cover, considerable detail can be recognised, e.g. small faults, fine-scale stratigraphy, minor folding, etc.

The use of photogeology for the rapid mapping of large areas of undeveloped and inaccessible territory has revolutionised the work of geological surveys in the underdeveloped countries. A series of flown surveys together with a limited number of ground traverses can result in a map showing a surprising amount of detail and of sufficient accuracy for many purposes. Excellent examples are to be found in the maps produced by the Geological Surveys of many African states, notably Kenya, Uganda, Tanzania, Rhodesia, Zambia, Ghana, Sierra Leone, and Malawi.

A future development of photogeology may lie in the study of satellite photographs, which are likely to be especially suitable for the study of large-scale geological phenomena.

Phragmacone (Mollusca). ◊ Fig. 106.

Phreatic eruption. A volcanic eruption caused by the rapid and violent conversion of ◊ *ground water* to steam, possibly as a result of contact between such water and a source of magmatic heat, e.g. lava. Such eruptions are usually low temperature and are not accompanied by the usual products of a volcanic eruption.

Phreatic water. Generally regarded as meaning ◊ *ground water*, particularly below the water table (cf. ◊ *Vadose water*). The possibility that the term is synonymous with ◊ *juvenile* water arises from the same source as the confusion between vadose and ◊ *meteoric water*.

Phyllite. A cleaved metamorphic rock having affinities with both ◊ *slates* and mica ◊ *schists*. The term is rather loosely used for rocks which are coarser-grained and less perfectly cleaved than slates, but which are finer-grained and better cleaved than mica schists. They are formed by low-temperature regional metamorphism.

Phyllitic lineation. ◊ *Lineation*.

Phyllonite. A ◊ *mylonite* in which recrystallisation or growth of new minerals has taken place.

Phyllosilicates. The ◊ *layer-lattice minerals*.

-phyre, -phyric. A suffix denoting a ◊ *porphyritic* rock.

Physiography. Synonym of ◊ *geomorphology* or geomorphology plus climatology and oceanography; not now in common use.

Piacenzian. A stratigraphic stage name for the European Upper Pliocene (◊ *Tertiary System*).

Picotite. ◊ *Spinel group of minerals*.

Picrite. A class of ◊ *ultrabasic rocks* consisting of 90% or more ◊ *ferromagnesian minerals* and up to 10% feldspars.

Piemontite. ◊ *Epidotes.*

Piezoelectric. If, when a crystal is very slightly distorted by applying a stress to one end of it, an electric charge develops at the other end, the crystal is said to be piezoelectric. Only crystals belonging to classes which lack a centre of ◊ *symmetry* can be piezoelectric. (Cf. ◊ *Pyroelectric.*) The property is made use of in the detection of shock waves, utilising discs of tourmaline. Quartz is notably piezoelectric and by means of various refined techniques plates of it may be used as oscillators in electronic circuits.

Pigeonite. ◊ *Pyroxenes.*

Pillow lava. ◊ *Lava* in the form of distorted globular masses which has been extruded under water.

Pilo- (prefix). Felt-like.

Pilotaxitic. A texture consisting of a felted mass of acicular or lath-like crystals, sometimes showing ◊ *flow structure* (cf. trachytic, ◊ *Trachyte*). Hyalopilitic is a variety of pilotaxitic texture in which the felt of crystals or crystallites is embedded in a glassy base.

Pinnule. ◊ *Echinodermata* (Crinoidea) and Fig. 43.

Pipe amygdale. ◊ *Amygdale.*

Pipe clay (Ball clay). A deposit of reworked ◊ *china clay* (kaolin).

Pisces. ◊ *Chordata* (Vertebrata).

Pisolite. Beds of ◊ *pisolitic* ◊ *limestone.*

Pisolith. Large ◊ *ooliths* about the size of a garden pea (3 to 6 mm. in diameter) which occasionally form rocks (pisolite).

Pitch. (1) In a ◊ *fold* the angle between the horizontal and the axis measured on the axial plane. (Cf. ◊ *Plunge.*)

(2) A synonym of ◊ *Asphalt.*

Pitchblende. Massive encrusting ◊ *uraninite*, oxidised wholly or in part to U_3O_8; theoretically its composition is UO_2.

Pitchstone. A glassy igneous rock often somewhat ◊ *devitrified*, which is characterised by a dull, 'pitchy' lustre and a rather flat fracture, as opposed to the ◊ *conchoidal* fracture of obsidian (◊ *Rhyolite*). Pitchstones vary in composition from ◊ *acidic* to near-◊ *basic*. Sometimes they contain traces of crystals and are occasionally ◊ *spherulitic*, ◊ *porphyritic*, etc. (cf. tachylite – ◊ *Basalt*). Pitchstones occur as ◊ *dykes*, ◊ *sills*, ◊ *lavas*, and chilled margins.

Placer deposits. ◊ *Alluvial*, ◊ *eluvial*, and ◊ *colluvial* material which contains economical quantities of some valuable mineral – those containing gold, platinum, diamonds, and tin (cassiterite) are the most important types. For a mineral to occur as a placer mineral it must be highly resistant to water and abrasive action, and be a ◊ *heavy mineral*. Eluvial placer minerals are termed 'float'. Concentration of the ore mineral usually takes place, and many placer deposits are derived from ◊ *primary* deposits which are too poor or too small to be worked directly. During transport of the material local concentrations of the ore mineral may be expected at certain points, e.g. below rapids or waterfalls, in pools or lakes along the course of a river, in flood-plain deposits, deltas, etc. Marine placer deposits are known, and fossil placer deposits also occur. (◊ *Banket*.)

Placophora. ◊ *Mollusca*.

Plagioclase. A series of sodium/calcium ◊ *feldspars*. (◊ Appendix.)

Plaisancian. A stratigraphic stage name for the European Upper Pliocene (◊ *Tertiary System*).

Planktonic. ◊ *Pelagic* and Fig. 49.

Plastic limit. The minimum amount of water which has to be mixed with a sediment or soil so that, under standard conditions of test, the material becomes capable of being permanently deformed without breaking. (Cf. ◊ *Liquid limit*.)

Plastron (Echinodermata). ◊ Fig. 45.

Plate. ◊ *Echinodermata* and Figs. 42–46.

Plate tectonics. Modern study of the major architectural features of the Earth's crust – ◊ *mid-oceanic ridges*, major tear faults, oceanic trenches, continental blocks, mountain ranges, earthquake belts, volcanic zones, etc. – suggests that these may be used to define a series of major regions of the Earth's crust (termed plates). The typical plate includes the continental shelf, sea, and oceanic areas. The theory of ◊ *continental drift* requires relative movement between continental blocks: it is suggested that in fact the movement is between plates, and features such as tear-faulting (transcurrent faulting) represent lateral movement of plates, mid-oceanic ridges result from plates moving apart (with creation of crust at the 'join'), and trenches (◊ *island arcs*) occur where one plate moves under another. Other features, such as mountain ranges, earthquakes, etc., can also be related to plate movements. The whole concept seems to unify a number of major observations, and there is little doubt that plate tectonics will be much used to explain both recent and more remote events in Earth history.

Plateau basalt. Large and extensive eruptions of basalt lava resulting from fissure eruptions are termed plateau basalts. The term has also been used as a 'magma-type', originally for olivine basalts, later for tholeiites. In view of the confusion existing the term should be abandoned in its latter sense.

Playfair, John (1748–1819). Scottish mathematics professor who produced a lucid account of ◊ *Hutton*'s theory of the Earth. He also gave clear descriptions of the fluvial erosion of valleys and the transport of erratics by former glaciers, but these explanations were overshadowed in his lifetime by the controversy aroused by Hutton's theory.

Pleistocene. ◊ *Tertiary*; *Quaternary*.

Pleochroism (Dichroism). A term used in the description of minerals in thin-section. Certain minerals which are coloured in thin-section exhibit a variation in colour when rotated in plane polarised light (◊ *Analyser*). This is the phenomenon of pleochroism. The colour shown by a mineral in thin section results from the absorption of certain 'colours' (wavelengths) from the incident white light, the resulting transmitted light being complementary in colour to that absorbed. Minerals which exhibit pleochroism do so by virtue of the fact that they absorb different wavelengths in different directions. Pleochroism is defined in terms of the ◊ *vibration directions*. The colours given represent the extremes observed, e.g. for a particular hornblende: $X =$ blue-green; $Y =$ yellow-green; $Z =$ greenish-brown. Note that ◊ *isotropic* substances and sections do not display pleochroism.

Pleuron, Pleura (Arthropoda). ◊ Fig. 3.

Pliensbachian. A stratigraphic stage name for the top of the European Lower ◊ *Jurassic*.

Pliocene. ◊ *Tertiary*.

Plug. (1) A volcanic plug (or neck) is the nearly circular vertical feed channel of a ◊ *volcano* which has been filled with solidified lava and/or ◊ *pyroclastic* material, and has subsequently been exposed by erosion of the volcanic cone. (◊ Fig. 118.) Often it is the only remaining trace of an ancient volcanic vent. (◊ *Puy*.)

(2) An intrusive plug is any rather small more or less vertical cylindrical mass of igneous rock, usually of ◊ *plutonic* or ◊ *hypabyssal rock*. (◊ Fig. 118.) The term bysmalith is approximately synonymous. (Cf. ◊ *Stock*.)

Plumasite. An undersaturated oligoclase-corundum ◊ *diorite*.

Plumbago. An obsolete name for the mineral ◊ *graphite*.

Plunge. A ◊ *fold* is said to plunge if the axis is not horizontal, the amount of plunge being the angle between the axis and the horizontal line lying in a common vertical plane. (Cf. ◊ *Pitch*.)

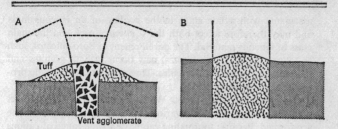

118. Plugs. (A) Volcanic; (B) Intrusive.

Pluton. Originally a general term for any large-scale mass of igneous rock either consolidated from a magma or formed as a ◊ *metasomatic* replacement. However, H. H. Read used it in the sense of a high-level, cylindrical mass of granitic rock which was emplaced at low temperature in a near-solid state. This usage is now rather widespread.

Plutonic. A word in common use, the precise meaning of which requires the context to be taken into account. The following are some of the variations in use: (1) To describe a body of igneous material of presumed deep-seated origin. (2) To describe a body of igneous rock implying a large intrusion formed at depth. (3) Loosely, as a synonym for coarse-grained (also plutonite). (4) By H. H. Read for a division of rocks comprising granites and their associated ◊ *migmatites*, and regionally metamorphosed rocks. (Cf. ◊ *Neptunic*; ◊ *Volcanic*.)

Pluvial period. The rainy period in non-glaciated regions which corresponds to the glacial maxima of the ice-covered regions, e.g. four pluvial periods are recognised in Africa, corresponding to the four glacial maxima in the Northern Hemisphere during the Pleistocene (◊ *Tertiary*); 'pluvial' and 'inter-pluvial' are sometimes used in the same way as 'glacial' and 'inter-glacial', as nouns rather than adjectives.

Pneumatolysis. Those changes brought about by the action of hot gaseous substances (other than water) associated with igneous activity. The distinction between this process and ◊ *hydrothermal processes* is largely arbitrary, but the distinction is perhaps useful in terms of the special minerals and rock types produced. The commonest ◊ *volatiles* involved in pneumatolysis are fluorine, hydrofluoric acid, and boron fluorides; other substances may be present to a greater or lesser extent in local instances, and may give rise to unusual mineralogies. Pneumatolysis is a process associated with a late stage in the cooling of an igneous mass and may therefore affect both the ◊ *country rock* and the main mass of igneous material. The development of boro-silicates, such as tourmaline (◊ *Cyclosilicates*) may occur in both granites and thermal ◊ *metamorphic aureoles*. Three main pneumatolytic processes may be recognised:

(1) TOURMALINISATION. When boron-rich volatiles react with pre-existing minerals, various different types of tourmaline are formed. Because tourmaline can have a wide range of cations in its structure, progressive tourmalinisation may take place, gradually affecting almost the entire rock. Two special types of tourmalinised granite have been named from localities in Cornwall: (a) luxullianite, in which the tourmaline occurs as radiating needles in quartz, associated with reddened feldspars; and (b) quartz-schorl rock or roche rock, which consists entirely of quartz and tourmaline, often rather fine-grained. Tourmalinised country rocks also occur, but in calcium-rich rocks tourmaline may be replaced by calcium boro-silicates, e.g. ◊ *axinite*.

(2) GREISENING. Greisens are mica-rich pneumatolysed rocks, sometimes containing topaz and/or tourmaline. They appear to result from the action of fluorine-rich vapours on granitic rocks, and are not uncommonly associated with kaolinisation (◊ *Hydrothermal processes*). The mica produced is often the yellowish fluorine-rich variety, gilbertite, or more rarely, a lithium-rich variety, zinnwaldite. Under the right conditions, ◊ *topaz* may develop progressively, yielding as an end product a quartz-topaz rock (with or without tourmaline), called topazfels. The effects of fluorine-rich vapours are more difficult to detect in the country rocks, although the development of topaz in argillaceous sandstones has been claimed.

(3) ORE MINERALISATION. There is some evidence that certain elements are concentrated and deposited as a result of pneumatolytic action; tin and wolfram are commonly quoted examples. It is suggested that they are transported as fluorides (SnF_4 and WF_6) which react with water to give the corresponding oxides (SnO_2 and WO_3) – the latter reacting with Fe and Mn to give ◊ *wolframite* – $(Fe,Mn)WO_4$ – or Ca to give ◊ *scheelite* ($CaWO_4$).

Many of the above minerals can also be found in environments which are commonly thought of as hydrothermal, e.g. ◊ *pegmatites*, mineral ◊ *veins*, etc. This association has been explained as due either to pneumatolysis succeeding hydrothermal activity (or vice-versa) or the simultaneous operation of the two processes – which seems a more likely possibility.

Podsol. A ◊ *soil* typical of considerable areas of cool temperate humid climate, with a strongly developed leached zone.

Poikilitic (Poecilitic). A textural term implying the occurrence of a number of orientated or unorientated crystals totally enclosed within a larger crystal, e.g. small rounded ◊ *olivine* crystals enclosed within large crystals of ◊ *hornblende*, giving rise to the phenomenon of lustre mottling. The use of the term is normally confined to igneous rocks. Poikiloblastic is a term used in a similar sense for a metamorphic texture.

Poikilo- (prefix). Spotted.

Poikiloblastic. ◊ *Poikilitic*.

Polarisation colours. When a ◊ *thin section* or grain of a mineral is viewed between crossed Nicols, or crossed polarisers, it usually appears coloured (unless it is isotropic or its own body colour is too strong, masking the polarisation colours). The polarisation colours are produced as follows. White light leaving the ◊ *polariser* and entering the mineral is resolved into light vibrating in two directions at right angles – the ordinary ray and the extraordinary ray (◊ *Vibration directions*). In passing through the mineral one ray will be retarded relative to the other. When these two rays reach the ◊ *analyser* they are recombined into a single plane. The recombined rays interfere with one another, with the suppression of certain frequencies and the reinforcement of others. This results in the production of coloured light. The colour produced is a function of the difference in refractive index (R I) of the two vibration directions of the mineral plate (the ◊ *birefringence*) and the thickness of the slice, so that the retardation, $\Delta, = (\mathrm{RI}_{max} - \mathrm{RI}_{min}) \times$ thickness in microns. Hence the importance for accurate optical work of controlling section thickness at 30 μ.

If a wedge-shaped slice of mineral (e.g. quartz) is examined under crossed polarisers a range of polarisation colours is seen, known as Newton's Scale. This scale is divided into orders, each red band marking the end of a particular order. Above the third order, colours are pale and indefinite, and above the fifth they are described as high-order whites. Each colour corresponds to a particular total retardation, and hence, by identifying the colour seen in a mineral and knowing the thickness of the slice, a value for the birefringence may be obtained. Alternatively, from a known mineral the thickness of the section may be calculated.

◊ *Isotropic* minerals yield no interference colours, i.e. they are black between crossed Nicols. This is because there is no difference in velocity between any two vibration directions in the section, and hence there is no retardation and all the frequencies are suppressed. Sections perpendicular to the optic ◊ *axes* in ◊ *tetragonal,* ◊ *trigonal* and ◊ *hexagonal* minerals also produce no relative retardation, and they, too, are isotropic. All other sections show ◊ *extinction* when rotated between crossed Nicols. Twinning (◊ *Twinned crystal*) may become apparent under these circumstances, since one 'unit' of a twin is in extinction when the other 'unit' is at maximum brightness.

Polarised light. Light in which the waves vibrate in only one plane, as distinct from ordinary light in which the vibrations take place in all directions perpendicular to the direction of propagation of the ray. For use in a petrological microscope, light is commonly polarised by passing it through 'Polaroid' or a Nicol prism. (◊ *Polariser*; *Analyser*; *Birefringence*.)

Polariser. The name given to the 'Polaroid' (or Nicol prism) in a petrological microscope, which polarises the light entering the microscope. The polariser is situated below the microscope stage and is arranged so as to make the resulting ◊ *polarised light* vibrate parallel to one of the microscope cross wires and at right angles to the vibration plane of the ◊ *analyser*. (◊ *Birefringence*.)

Polje. A large depression in a ◊ *karst* region with steep sides and a flat floor.

Pollen analysis. ◊ *Palynology.*

Poly- (prefix). Many.

Polymetamorphism. ◊ *Metamorphism.*

Polymict. A term describing detrital rock consisting of fragments of many different materials. The term is applied especially to ◊ *rudaceous rocks.* (Cf. ◊ *Oligomict.*)

Polymodal. A sediment which displays two or more maxima in its ◊ *particle size* distribution is said to be polymodal. Few examples of a more than bimodal distribution have been described.

Polymorphism. A substance which exists in two or more distinct forms having identical chemical compositions is termed polymorphic, e.g. there are three forms of titanium dioxide – rutile and anatase, both tetragonal, and brookite, which is orthorhombic. Calcite (hexagonal) and aragonite (orthorhombic) are polymorphs of calcium carbonate. The crystallographic differences are reflections of underlying differences in atomic structure.

(Dimorphism and trimorphism are related terms, implying two and three different forms respectively.)

Polyphyletic. As a result of convergent evolution, a group of organisms may consist of members which have evolved from different series of ancestral forms, e.g. the genus Monograptus (graptolites, ◊ *Chordata*). The group is then described as being polyphyletic.

Polyzoa (Bryozoa). (◊ Fig. 119.) Phylum consisting of aquatic

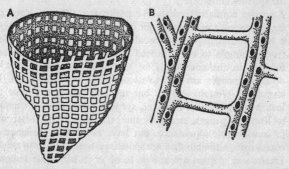

119. Polyzoa. (A) A complete colony (zoarium) of polyzoa (*Fenestrellina*); (B) An enlarged part of the colony showing apertures.

colonial organisms. The skeletal material may be of chitin or calcium carbonate, but only the calcareous forms are found as fossils. The colonies are usually small, ranging in size from 5 mm. to 15 cm., and are commonly fan-shaped, stick-like or lobate, and may be encrusting. They range from ◊ *Ordovician* to Recent, and are of no geological importance; however they occasionally occur in significant volume as reef-building organisms. Because of the apparent similarity between the soft parts of Bryozoa and ◊ *Brachiopoda*, it was formerly the practice of some workers to combine the two groups into a single phylum, the Molluscoidea. This grouping is now obsolete in view of the many differences between the two phyla.

Although 'Polyzoa' is the approved name the term Bryozoa is considered to be more descriptive, since some of the colonial ◊ *Coelenterata* could be considered as polyzoans.

Pontian. A stratigraphic stage name for the European Lower Pliocene (◊ *Tertiary System*).

Pore. (1) ◊ *Echinodermata* (Echinoidea) and Fig. 44.

(2) ◊ *Porosity*.

Porifera (Sponges). The most primitive of the many-celled (Meta-zoan) animals, assumed to have been derived from the ◊ *Protozoa* by their uniting in a colonial form. They are benthonic aquatic forms, having an internal skeleton consisting of spicules of calcite or silica, and fibres of chitin which are not generally preserved in fossils. Fossil sponges are divided into a number of groups, based on (a) the nature of the spicules, and (b) the way in which the spicules unite.

A sponge normally consists of a perforated structure, water pass-ing in through small pores all over the surface and leaving via a large terminal pore or osculum. The increase in complexity from a simple to an advanced sponge is generally seen as an increase in the complexity of the water-vascular system. Un-doubted sponges are said to occur in the ◊ *Precambrian* (Kat-angan) of Western Africa, but the only specimens have been lost. More generally, however, sponges range from ◊ *Cambrian* to Recent. Sponges, especially the calcareous forms, often occur in considerable abundance, but have never been designated as zone fossils, although they are sometimes used as horizon markers. Freshwater sponges are known to occur in lacustrine sediments. The following terms are used in the description of the sponges:

Asconid The simplest type of sponge, usually vase-shaped, in which the walls of the sponge are simple and perforated by small canals which connect directly into a central cavity.

Syconid Sponges in which the walls are folded to increase the area of the inner surface, the canals entering the individual folds.

Leuconid Sponges in which the folds of the syconid type become enclosed within the wall of the sponge, to give rise to separate chamberlets. Water entering the canals passes into these cham-berlets and then through a second series of canals to the cen-tral cavity.

Porosity. A rock is said to be porous if it possesses cavities between the mineral grains making up the rock which can contain liquid. The term 'porosity ratio' (or simply 'porosity') is given to the per-centage of void space that a rock contains. The porosity of a perfectly ◊ *graded* rock, consisting of spherical particles, perfectly close-packed, is 27%, and for open packing it rises to 47%. Be-cause ◊ *sedimentary rocks* are never perfectly graded or per-fectly packed, and may in addition be partly or wholly cemented, the porosity ratios range from less than 1% to greater than 50%. With sandstones (◊ *Arenaceous rocks*), which are perhaps the commonest porous rocks, it ranges from 5 to 15%, while in loose sand and gravel (◊ *Rudaceous rocks*) it may reach 45%. Clays are exceedingly porous rocks, with porosity sometimes reaching 50%. It should be noted that a porous rock is not necessarily permeable; e.g. sandstones, sands and gravels are commonly both porous and permeable, since they allow liquid to pass through, but clay is porous (since dry clay will absorb liquid), but impermeable, since it will not allow water to pass through. The opposite of porous is non-porous. (Cf. ◊ *Permeability*; ◊ *Pervious*).

Porphyritic. A texture term which describes igneous rocks containing relatively large crystals set in a finer-grained ◊ *groundmass*. Microporphyritic texture is a variety in which both the 'large' crystals and the groundmass (which may even be ◊ *glassy*) are only distinguishable using a microscope. Porphyritic crystals are usually ◊ *euhedral*. The name 'porphyritic' is derived from the Greek 'porphyreos' (purple, reddish), referring to the famous Imperial Porphyry (porfido rosso antico) much used in the ancient world as a decorative stone. This rock shows white feldspar ◊ *phenocrysts* in a reddish-purple groundmass, and the use of the term was gradually extended to other rocks showing a similar texture. (◊ *Porphyry*.)

Porphyroblastic (Porphyroblast). A term compounded from ◊ *porphyritic* and ◊ *-blastic*. It is applied to the larger more or less ◊ *euhedral* crystals found in metamorphic rocks, which have grown during the process of metamorphism.

Porphyry (Porphyrite). Rock names (porphyry being the more common) usually applied to ◊ *hypabyssal rocks* containing ◊ *phenocrysts* – in the strictest sense alkali ◊ *feldspar* phenocrysts. In practice the terms are used for any medium-grained rock containing phenocrysts of any mineral. 'Porphyrite' is now restricted to a porphyry containing andesine phenocrysts, but the term is practically obsolete.

The term 'porphyry' can usefully be employed in conjunction with mineral names, e.g. quartz porphyry, to distinguish the porphyritic mineral. It also appears in a contracted form, 'phyre', e.g. ◊ *lamprophyre*. There is a tendency at present to abandon porphyry as a rock name, and use porphyritic as a textural prefix to a micro-plutonic rock name, e.g. porphyritic microgranite is the equivalent of quartz porphyry. The combination of the term porphyry with a rock name, e.g. granite porphyry, diorite porphyry etc., is highly ambiguous, as the names have been used to mean both porphyritic granite, porphyritic diorite, etc. and porphyritic microgranite, porphyritic microdiorite, etc. For this reason it is recommended that this usage should be discontinued. Note that 'porphyry' should not be used as a synonym of 'phenocryst'.

Portlandian. A stratigraphic stage name for the top of the European Upper ◊ *Jurassic*.

Positive area. An area where deposition is not taking place. The opposite of a basin of deposition or ◊ *negative area*. (◊ *Shield*.)

Positive (optically). ◊ *Axes, optic*.

Post-kinematic, Post-orogenic, Post-tectonic. ◊ *Synkinematic*, etc.

Potsdamian. An obsolete stratigraphic stage name (◊ *Croixian* is the modern equivalent) for the North American Upper ◊ *Cambrian*.

Powell, John Wesley (1834–1902). American ex-Major who lost an arm in the Civil War and afterwards became a professor of geology. His expedition by small boat down the unknown rapids of the Colorado River made him something of a romantic hero, but it also produced the classic geological section across the Grand Canyon and was the basis for his introduction and description of the concepts of ◊ *base level*, and antecedent, consequent and superimposed ◊ *drainage*. Later he became Director of the U.S. Geological Survey, being largely responsible for its growth to a national body. He showed great foresight about the dangers of traditional farming in the arid lands of western U.S.A., although his advice on control and irrigation was largely ignored until after his death.

Praetiglian. A stratigraphic stage name for the base of the European Lower Pleistocene (◊ *Tertiary System*).

Prase. A green variety of chalcedonic silica. (◊ *Silica group of minerals*.)

Precambrian. That period of time from the consolidation of the Earth's crust to the base of the ◊ *Cambrian*. The terms Proterozoic, Azoic, and Archaean have been used either as synonyms or partial synonyms.
Because of the ◊ *unconformable* relationships and metamorphic state of much of the Precambrian, correlation on more than a local basis has proved exceptionally difficult. This has resulted in the erection of numerous regional stratigraphic sequences and local names for groups of strata. The advent of ◊ *radioactive dating* has done much to clear up numerous problems and the nucleus of a world-wide correlation scheme is beginning to appear. The duration of Precambrian time is probably not less than

4,000 million years, and during this time a number of orogenies are known to have occurred. Most Precambrian rocks, therefore, can be shown to have undergone one or more Precambrian orogenies, as well as post-Cambrian ones. However, relatively unmetamorphosed sediments of Precambrian age are known from a number of areas. In some of these (e.g. in S. Australia and Leicestershire) Precambrian fossils have been described, but their affinities are obscure.

Precambrian rocks are exposed at the surface in ◊ *shield* areas. The following is a list of stratigraphic names used for various parts of the Precambrian.

Stratigraphic name	Area	Position
Algoman	Canada	Pre-Algonkian
Algonkian	Canada	Youngest Precambrian
Bothnian	Baltic	Post-*Svionian*, Pre-*Karelian*
Briovenian	Brittany	
Charnian	Leicestershire, England	Volcanics. Precambrian
Dalradian	Scotland	Youngest Precambrian, upper part contains a Cambrian fauna
Grenville	Canada	Lower *Ontarian*
Huronian	Canada	Pre-*Algonkian*
Jatulian	Baltic	Youngest *Karelian*. Unmetamorphosed *Kalevian* + *Ladogian*
Jotnian	Baltic	Youngest Precambrian
Kalevian	Baltic	Younger *Karelian*
Karelian	Baltic	Post-*Sveccofennian*
Keewatin	Canada	Syn. *Ontarian* or Upper Ontarian
Keweenanwan	Canada	Syn. *Algonkian*
Ladogian	Baltic	Older *Karelian*
Laurentian	Canada	Pre-*Huronian* granite intrusive into *Ontarian*
Laxfordian	Scotland	Younger *Lewisian*. Cf. Scourian (below)
Lewisian	Scotland	Oldest British Precambrian, covered unconformably by *Torridonian*
Longmyndian	Shropshire, England	Sediments, Post-*Uriconian*
Malvernian	Herefordshire, England	Pre-*Uriconian* (?)

Stratigraphic name	Area	Position
Moinian	Scotland	Metamorphic equivalent of the *Torridonian*, passes up (? conformably) into Lower *Dalradian*
Ontarian	Canada	Oldest Precambrian
Saamian	Baltic	Pre-*Karelian*
Scourian	Scotland	Older *Lewisian*; division based upon the number of orogenies affecting the group (cf. *Laxfordian*)
Sparagmitian	Baltic	Post-*Jotnian* sandstones, which may be late Precambrian or Early Cambrian
Sveccofennian	Baltic	*Svionian + Bothnian*
Svionian	Baltic	Oldest Precambrian
Torridonian	Scotland	Unmetamorphosed equivalent of Moine, below *Dalradian*
Ukrainian	Russia	Pre-*Saamian*
Uriconian	Shropshire, England	Volcanic Precambrian underlying *Longmyndian*

(Italicised names are here used to indicate a cross-reference within the table.)

The multitude of African divisions of the Precambrian have been excluded from this list.

Stratigraphic names for the various parts of the Precambrian are given under those names, but a great deal of Canadian terminology has been abandoned in favour of a scheme based on ◊ *radioactive dating*. The table which follows gives some idea of the growing nomenclature.

Eon	Era	Approximate Duration (in millions of years)
Proterozoic	Hadrynian	570–880
	——————Grenville Orogeny——————	
	Helikian	880–1640
	——————Hudsonian Orogeny——————	
	Aphebian	1640–2390
	——————Kenoran Orogeny——————	
Archaean		2390 and above

Precipitates. A geochemical division of the ◊ *sedimentary rocks* including the chemically formed limestones.

362

Preferred orientation. Any rock in which the constituent units (whether minerals or rock fragments) are arranged in any but a random way is said to display a preferred orientation ◊ *fabric*. The units building up the rock may orient themselves according to (a) some shape characteristic, (b) some crystallographic characteristic, (c) their magnetic characteristics, or (d) some other physical characteristic.

In ◊ *clastic* sediments, preferred orientation is almost always a function of ◊ *particle shape*, disc-like and rod-like particles displaying it most obviously, commonly as a result of current action.

In igneous rocks, orientation of crystals generally arises as a result of the liquid parts of the magma flowing and causing the already formed crystals to align themselves along the flow direction (◊ *Flow structure*). In certain circumstances, iron-bearing minerals (in both igneous and sedimentary rock) may orientate themselves in a direction related to the local magnetic field existing at the time of formation (◊ *Palaeomagnetism*).

In metamorphic rocks, preferred orientation arises as a result of recrystallisation in a ◊ *stress* system. Many minerals in these conditions tend to grow with a particular crystallographic direction perpendicular to, or parallel to, the stress direction.

The detection of preferred orientations, where they are not obvious to a naked-eye examination in the field or of a hand specimen, is normally done by means of a study of ◊ *thin sections* (◊ *Fabric*). The study of preferred orientations in rocks can yield information about processes operating during and after their formation. In the case of metamorphic rocks, information about stress directions and other tectonic data may be obtained. It should be noted that study of preferred orientations is now largely on a statistical basis.

Pressure. Physicists define pressure as the force per unit area, i.e. ◊ *stress*. However the geologist commonly uses 'pressure' as a loose, all-embracing term to include concepts of stress, ◊ *strain*, and shear arising from external sources. He generally uses the phrase 'hydrostatic pressure' ('load pressure') as that force which acts upon a rock due to the mass of superincumbent material.

Priabonian. A stratigraphic stage name for the European Upper Eocene (◊ *Tertiary System*).

Primary. (1) An obsolete name for the ◊ *Precambrian*, subsequently the Precambrian + ◊ *Palaeozoic*, later the Palaeozoic only.

(2) In its ordinary sense of 'original', applied to those features

of a rock developed at the time of its formation. (Cf. ◊ *Secondary*.)

Prismatic body. ◊ *Sedimentary structures* (1).

Pro- (prefix). Before.

Prod marks. ◊ *Sedimentary structures* (3).

Proglacial lake. A lake formed directly adjacent to the 'snout' of a glacier. ◊ *Varves* are commonly found in such an environment.

Projection, crystallographic. A term used for any representation of a crystal. Purely 'pictorial' projections, such as the perspective or clinographic drawings which are commonly made, are of little use to a crystallographer, as he is unable to use them for measurements or calculations. The 'plan-and-elevation' projections are little better for these purposes. Since the crystallographer is mainly concerned with the angular relationships of the crystal faces, it is possible to dispense with a pictorial representation and use a purely geometrical projection. All such projections in common use involve the projection of normals to the crystal faces on to a sphere. This spherical projection can be reduced to a two-dimensional representation by the use of the same methods as are used in map projections for reducing the Earth's sphere to a flat map. The two main projections used are the stereographic and the gnomonic, which yield an 'angle-true' 'map' of the faces.

A. C. Bishop, *An Outline of Crystal Morphology*, 1967.

Propylitisation. The hydrothermal alteration of a fine-grained igneous rock (especially ◊ *andesite*) to a mass of ◊ *secondary* minerals such as ◊ *chlorite*, ◊ *epidote*, quartz, carbonates, and sub-◊ *micas* such as 'sericite'. (Cf. ◊ *Saussuritisation*; ◊ *Uralitisation*.)

Prospecting. ◊ *Geophysical prospecting*.

Proterozoic. ◊ *Precambrian*.

Proto- (prefix). First.

Protoclastic. A term used to describe the structure displayed by an igneous rock in which the early-formed crystals have been broken or deformed owing to movement of the still-liquid portion, either during intrusion or in a lava flow. The term has also been used to describe certain types of gneissose structure, in which fragmentation of crystals was presumed to be due to the flow of liquid material. It seems possible that some gneisses to which this adjective has been applied are in fact showing early mylonitisation, and not the effects of liquid injection. (For augen structure, ◊ *Gneiss*; ◊ *Mylonite*.)

Protoconch. ◊ *Mollusca* (Gastropoda) and Figs. 100 and 106.

Protore. Material containing ◊ *ore minerals* in too low a concentration for economic working, but which may be a workable ◊ *ore* where ◊ *secondary enrichment*, especially ◊ *supergene* oxidation, has occurred.

Protozoa. A unicellular animal, which may or may not secrete a test or skeleton, ranging in size from 0·1 mm. to 8 cm. Modern protozoa are divided into a number of groups, but only two of these are

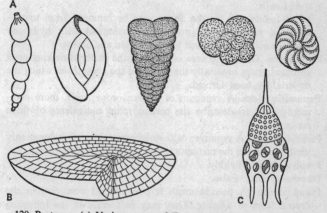

120. Protozoa. (A) Various types of Foraminfera; (B) A nummulitid (sectioned); (C) A radiolarian. (All greatly magnified.)

of geological importance – the Foraminifera and the Radiolaria (◊ Fig. 120):

FORAMINIFERA. The Foraminifera are mainly marine benthonic or planktonic forms, in which there is considerable morphological variation, the forms ranging from a single, non-chambered flask-shaped animal to the complex, chambered, form of the discoidal nummulites. The Foraminifera are subdivided into three groups, based on the character of the test, which may be of porcellaneous calcite, hyaline (clear) calcite, or arenaceous with a cement which is partly calcite. The test is usually perforated, allowing translocation of food particles from the external cytoplasm to the cytoplasm inside the test. The Foraminifera range from ◊ *Ordovician* to Recent, although some Cambrian forms may occur. They are important as zone fossils especially in the Tertiary, where they may be locally present in sufficient numbers to be major rock-building constituents.

RADIOLARIA. The Radiolaria are marine planktonic animals, in which a central capsule and surrounding spicules consist of silica, which may or may not be partially opaline. Again there is considerable variation within this group. The Radiolaria range from (? ◊ *Precambrian*) ◊ *Cambrian* to Recent. They are only rarely preserved in sediments, since their opaline tests are liable to be destroyed under alkaline conditions. They have no use as zone fossils.

Micro-fossils such as the above may be removed from unconsolidated sediments by preliminary treatment with hydrogen peroxide solution, followed by decantation and drying.

Provenance. The source area or areas of the material making up a sediment, more especially the nature of the rocks from which the material has been derived.

Psammitic rocks. A synonym of ◊ *arenaceous rocks*, more commonly used to describe the metamorphic equivalents of these rocks.

Psephitic rocks. A synonym of ◊ *rudaceous rocks*, more commonly used to describe the metamorphic equivalents of these rocks.

Psepho- (prefix). Pebbly.

Pseudo- (prefix). False.

Pseudomorph. A pseudomorph is one mineral occurring in the crystal form of another. They may be formed in the following ways:

(1) As a result of one ◊ *polymorph* of a compound converting to another polymorph, e.g. calcite pseudomorphs after aragonite, pyrite after marcasite. This type of pseudomorph is known as a paramorph.

(2) As a result of alteration of a mineral, e.g. gypsum after anhydrite (by hydration), malachite after azurite (hydroxyl replacing CO_2), limonite after pyrite (oxidation and hydration), kaolin after feldspar and serpentine after olivine (◊ *hydrothermal* action), and quartz after fluorite, cassiterite after feldspar (total replacement by unrelated substances; possibly the infilling of a negative pseudomorph – see below).

(3) As a result of the coating or encrustation of one mineral by another. Subsequently the first mineral is removed, leaving a mould or negative pseudomorph behind. This in turn may be filled by a later mineral totally unrelated to the original one. Negative pseudomorphs of quartz after fluorite, calcite, and barytes are known, and filling of these by calcite, pyrite, quartz, etc. can occur.

(4) As a result of the formation from an evaporating saline solution of crystals of soluble minerals, such as rock salt and gypsum (◊ *Evaporites*) which become partially embedded in soft mud. Subsequently, when the mud dries out and hardens into clay or silt, the soluble mineral may be dissolved, leaving behind a mould of the crystal, which is subsequently filled during deposition of the succeeding layers of sediment. It should be noted that this type of pseudomorph will be found on the under-surfaces of bedding planes.

Pseudonodule. ◊ *Balled-up structure.*

Pseudostratification. ◊ *Layered igneous structures.*

Pseudotachylite. Partially fused rock resulting from intense ◊ *cataclastic* activity. (◊◊ *Mylonite.*)

Psilomelane. A manganese mineral, $(Ba,H_2O)_2Mn_5O_{10}$, found as a secondary mineral in association with manganese deposits and in residual deposits due to weathering. (◊ Appendix.)

Pterocerian. A stratigraphic stage name for the European Lower ◊ *Kimmeridgian* (Jurassic System).

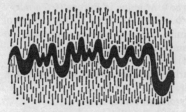

121. Ptygmatic folding.

Ptygmatite, Ptygmatic structures (Greek *ptygma*, folded matter). Originally applied to the kind of flow-folding common in ◊ *migmatites*, this term is now used essentially for highly contorted quartzo-feldspathic ◊ *veins* found in areas of intense ◊ *metamorphism* and ◊ *granitisation*. (◊ Fig. 121.) Their origin has been a matter of controversy, two different theories having been put forward: (1) The veins were originally intruded as flat sheets and the folding is due to deformation of the host rock. (2) Folding took place during the process of injection, which occurred under special circumstances.

It is possible to demonstrate experimentally that material being injected into a weaker (i.e. plastic, semi-fluid) host material will develop ptygmatic folding, if it encounters a more rigid mass. This is clearly what happens when ptygmatic veins are formed in granitised or migmatitic material. However, not all ptygmatites can be explained in this way, and Wilson (see reference below) has recognised three main divisions: (1) Contortion due to injection of the quartzo-feldspathic material into 'weak' host rock, as indicated above. (2) Some contortion due to injection-folding, but accentuated by deformation of the host rock. (3) Veins whose contortions conform to those of the host rock, of which three sub-divisions are recognised: (a) The veins follow a particular bed or plane which has previously been tightly folded. (b) The contortions are due to post-intrusion deformation. (c) The veins were intruded at the same time as folding was taking place. Thus ptygmatic structures may be pre-, syn-, or post-tectonic in their formation.

G. Wilson, 'Ptygmatic Structures and Their Formation', *Geological Magazine*, Vol. 89, Pt 1, 1952.

Pudding stone. Popular name for a ◊ *conglomerate.*

Puercan. A stratigraphic stage name, the North American equivalent of the Lower ◊ *Montian* in Europe (◊ *Tertiary System*).

Pulaskite. A ◊ *syenite* consisting almost entirely of alkali feldspar, but containing some ◊ *aegirine* and other ◊ *ferromagnesian minerals,* and often ◊ *nepheline.*

Pull-apart. (1) (tectonic) ◊ *Boudinage.*
(2) An early stage during the formation of a sedimentary ◊ *collapse* structure.

Pumice. ◊ *Pyroclastic rocks.*

Pumpelly's Rule. ◊ *Fold.*

Purbeckian. A stratigraphic stage name for the top of the European Upper ◊ *Jurassic.*

Puy. A volcanic ◊ *plug,* especially as found in the Auvergne district of the Central Massif of France, where the volcanic necks stand up as steep-sided pillar-like hills.

Pygidium. ◊ *Arthropoda* (Trilobita) and Fig. 3.

Pyralspite. ◊ *Garnets.*

Pyriboles. A 'portmanteau' word for ◊ *pyroxenes* plus ◊ *amphiboles.*

Pyrite. The most widespread sulphide mineral, FeS_2, found as an ◊ *accessory mineral* in igneous rocks; in hydrothermal ore veins, ◊ *contact metamorphic* deposits, and ◊ *anaerobic* sediments; and perhaps as a result of ◊ *magmatic differentiation.* (◊ Appendix.)

Pyroclastic rocks. Pyroclastic rocks consist of fragmental volcanic material which has been blown into the atmosphere by explosive activity. They are generally produced from volcanoes whose ◊ *lava* is of a more viscous type. (Less viscous lava produces the characteristic lava flows.) They are of a number of different types, but two major groups may be distinguished:

(1) Material which, having been thrown out of the volcano as liquid globules, has solidified in the air and been deposited as solid particles.

(2) Material which has been thrown out of the volcano as solid fragments, being solid material which has been fractured by the explosive activity.

The first group may be subdivided into four classes:

PELÉES HAIR. A fine mass of hair-like glass which is formed as a result of ◊ *lava* being exuded through a small orifice and blown about by the wind.

BOMBS. Larger masses of liquid lava which have been flung through the air so that they rotate and take on characteristic internal structures and shapes. Spindle-bombs and breadcrust bombs are well-known examples. They commonly have a well-developed crust, and range in size from small droplets to objects which are several cubic metres in volume. They are generally ◊ *vesicular*.

PUMICE. A highly ◊ *vesicular* material derived from acidic lavas and produced in very large quantities. It may accumulate as lumps or as abraded fragments of pumice, called shards. Fragments of pumice can be recognised in pyroclastic rocks by the fact that they are commonly bounded by concave surfaces, and in thin section this gives them a trifid appearance.

SCORIAE. Basic lavas tend to produce scoriaceous material, and scoriae consist of highly ◊ *vesicular* basic material having a much higher density than pumice (pumice often floats on water but scoriaceous material is too dense). Large flat masses of scoriae are very often found and may be anything up to 3 m. across; these are simply masses of vesicular lava which have almost completely solidified in the atmosphere but are still soft when they hit the ground, so that they flatten out like pancakes. Scoriaceous material and pumiceous material both represent the solidified 'froth' of a volcanic rock.

The second group, solid fragments derived from the fracture of pre-existing rocks, come either from the ◊ *country rock* of the volcano or from earlier-formed volcanic rocks broken up by subsequent explosive activity of the volcano. The rocks are generally classified according to particle size, composition, and mode of deposition.

TUFF. The general name for the unconsolidated material is ash, which on consolidation (◊ *lithifaction*, ◊ *diagenesis*) is called a tuff. The material forming such a tuff or ash may consist of (a) crystals ejected from the volcano – hence 'crystal tuffs'; (b) small fragments (less than 4 mm.) of lava, sedimentary rock or other country rock of the volcano – hence 'lithic tuffs'; (c) fragments (4–32 mm.) of rocks, known as lapilli – hence 'lapilli tuffs'; (d) fragments of a glassy nature – hence 'vitric tuffs'. In general, the term 'tuff' is limited to rocks containing a predominance of fragments not greater than 2 cm. in diameter. For rocks consisting mainly of fragments larger than this, the term agglomerate is employed (although objections have been raised to this usage). Water-lain tuffs generally show excellent bedding, and many sedimentary features such as ◊ *grading* and current-bedding (◊ *Crossbedding*) may be detected. A useful word for the mixed clastic sedimentary-pyroclastic material which not infrequently occurs is tuffite.

IGNIMBRITES (WELDED TUFFS). These are a special group formed as a result of deposition by ◊ *nuées ardentes* at high temperatures consisting of layers of tuff material (including pumice, lapilli, crystals, etc.) which were so hot when deposited that the edges of the fragments tended to weld together, giving rise to rocks having an appearance quite distinct from normal tuffs. In many cases well-marked banding occurs, and glass shards and other fragments may become flattened and drawn out, giving the appearance of flow-banding. It is almost certain that many so-called ◊ *rhyolites* and some related rocks are in fact ignimbrites and not true lavas. This is particularly true of some of the extensive sheets of rhyolitic lava which have been described. In a typical sheet of welded tuff the upper portion, where preserved, can be seen to consist of typical pyroclastic material; below this, in the layers where heat was not lost as rapidly as in the upper layers, compaction and welding of the tuff fragments can be observed, until, in the lowest portions, a pseudo-lava may apparently develop. Walker has described from Iceland a welded tuff sheet which passes downwards into a glassy pitchstone-like mass produced by almost complete fusion of the tuff material. In this case the temperature of the ejected matter must have been very high and the heat was retained for a considerable period. Ignimbrites can form only as a result of sub-aerial volcanic activity.

PALAGONITE TUFF. This is another type of tuff which develops when eruptions take place under water or ice-sheets. It consists mainly of glassy material formed by the rapid break-up of chilled magma during eruption into a cold environment. They are not strictly tuffs, since they have not been produced by explosive activity, and the descriptive term hyaloclastic has been suggested instead. Palagonite was originally thought to be a distinct mineral. Pyroclastic material which is ejected explosively may suffer two separate phases of particle size separation – once when it is flung through the air, when the larger particles tend to travel shorter distances than the finer ones (fine ash may travel thousands of kilometres before being finally deposited), and for a second time if it falls into water, when there will be separation as the particles of different sizes sink through the water layer. This can give rise to graded bedding and many graded beds of tuffs and ash are known. The coarsest types of pyroclastic material, the agglomerates or volcanic breccias, are generally found in very close proximity to, or actually within, the volcanic ◊ vent. (◊ Fig. 122.) (◊ Volcanoes; Lava.)

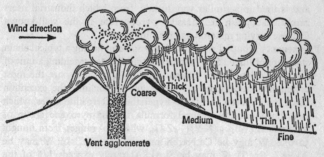

Wind direction

Coarse
Thick
Medium
Thin
Fine
Vent agglomerate

122. Pyroclastic 'fallout'.

Pyroelectric. If when one end of a crystal is heated, the other end develops an electric charge, the crystal is said to be pyroelectric. If one then heats the end on which the electric charge has developed, an electric charge of opposite sign develops at the other end. Tourmaline (◊ Cyclosilicates) is the best-known example. Only crystals belonging to classes lacking a centre of symmetry can be pyroelectric. (Cf. ◊ Piezoelectric.)

Pyrolusite. A manganese mineral, MnO_2, found in bog and lake deposits associated with other manganese and iron oxides. (◊ Appendix.)

Pyrometamorphism. An intense form of thermal ◊ *metamorphism* developed at or near a contact with a high-temperature igneous intrusion.

Pyrometasomatic. A term used to describe ◊ *metasomatic* changes occurring in rocks under conditions of high temperature and pressure at or near contacts with high-temperature igneous intrusions. The effects are most easily observable in limestones, and high-temperature ◊ *skarns* form under these conditions. The formation of some ore bodies, e.g. ◊ *magnetite*, ◊ *pyrrhotite*, has been ascribed to this process.

Pyromorphite. A lead mineral, $(PbCl)Pb_4(PO_4)_3$, found in the oxidised zones of lead deposits. (◊ Appendix.)

Pyrope. ◊ *Garnets*.

Pyrophyllite. A ◊ *layer-lattice mineral*, $Al_2Si_4O_{10}(OH)_2$. Both fibrous and lamellar forms are known. As with ◊ *talc*, the cleavage lamellae are flexible but not elastic. It is very difficult to distinguish pyrophyllite from talc except by a chemical test. It occurs in low- and medium-grade metamorphosed aluminium-rich rocks, and is used in a similar way to talc, from which industrial users rarely distinguish it. Massive pyrophyllite is the well-known Chinese carving material agalmatolite.

Pyroxenes. A group of rock-forming ◊ *silicates* having a typical chain structure of linked SiO_4 tetrahedra (inosilicates), yielding a unit of the form $(SiO_3) \infty$; for convenience, Si_2O_6 or Si_4O_{12} are the most commonly used. Most pyroxenes are monoclinic, the exception being the important enstatite–hypersthene–ferrosilite series, which is orthorhombic. A general formula for the pyroxene group of minerals is $(W)_{1-p}(X,Y)_{1+p}Z_2O_6$, where 'p' ranges from nought to one; W may be Ca,Na; X may be Mg, Fe'', Li; Y may be Al,Fe''', Cr, Ti; Z may be Si,Al; Al rarely exceeds $1/5$ of the total Z-ions.

The pyroxenes show many features in common with the ◊ *amphibole* group, the main difference being a 90° cleavage as opposed to the 124° of the amphiboles. Fibrous pyroxenes are very rare. Pyroxenes are found most abundantly in ◊ *basic* and ◊ *ultrabasic* igneous rocks, but certain types occur fairly commonly in metamorphic rocks; they are rather rare as detrital grains. Composition of the pyroxene minerals is as follows:

Monoclinic Pyroxenes (Clino-pyroxenes)		Orthorhombic Pyroxenes (Orthopyroxenes)
*Clino-enstatite	$Mg_2Si_2O_6$	Enstatite
*Clino-hypersthene	$(Mg,Fe'')_2Si_2O_6$	Hypersthene / Bronzite
*Clino-ferrosilite	$Fe''_2Si_2O_6$	Ferrosilite
Diopside	$CaMgSi_2O_6$	
Sahlite	$Ca(Mg,Fe'')Si_2O_6$	
Hedenbergite	$CaFe''Si_2O_6$	
Augite	$Ca(Mg,Fe'',Al,Fe'''),(Si,Al)_2O_6$	
Omphacite	$(Ca,Na)(Mg,Fe'',Fe''',Al)Si_2O_6$	
*Pigeonite	$(Mg,Fe'',Ca)(Mg,Fe'')Si_2O_6$	
Aegirine	$NaFe'''Si_2O_6$	
Jadeite	$NaAlSi_2O_6$	
Spodumene	$LiAlSi_2O_6$	

* These forms develop at higher temperatures, and normally occur only in rocks which have been rapidly cooled, e.g. lavas. Under conditions of slower cooling, pigeonite inverts to an orthopyroxene and augite, while clino-enstatite and clino-hypersthene, etc. invert to their orthorhombic equivalents. The following prefixes are used to describe certain pyroxene varieties:

Magnesio-	Mg-rich (magnesio-diopside = endiopside)
Ferro-	Fe-rich
Chrome-	Cr-bearing
Titan-	Ti-bearing
Subcalcic-	Low Ca content

Intermediate types between augite and aegirine are referred to as aegirine-augite. Some pyroxenes in the diopside-augite group develop a parting parallel to the front pinacoid (100) of the crystal, which produces a characteristic bronzy ◊ schiller. The apparent parting may be exsolution lamellae. In old textbooks the term diallage is sometimes used as a synonym for augite, especially when occurring in gabbro.

Pyroxenite. A ◊ perknite consisting of ◊ pyroxenes. (◊ Ultrabasic rocks.)

Pyroxenoids. A group not related structurally to the ◊ pyroxenes, although chemically they have identical formulae. They have a single chain of linked SiO_4 tetrahedra which is not the simple chain of the pyroxenes. A more restricted range of cations enters the structure and replacement of Si by Al is very limited. The following are the commonest pyroxenoids; all are triclinic:

		Occurrence
Wollastonite	$CaSiO_3$	Metamorphosed siliceous
Pectolite	$Ca_2NaH(SiO_3)_3$	◊ *Zeolites*
Rhodonite	$MnSiO_3$	⎱ Both formed by hydrothermal
Bustamite	$MnCa(SiO_3)_2$	⎰ metasomatic processes

Pyrrhotite. A mineral, $Fe_{1-x}S$, found in basic igneous rocks, pegmatites, and contact metamorphic deposits. (◊ Appendix.)

Quaquaversal. Radiating (dipping) in all directions away from a centre.

Quarter-wave plate (Mica plate). ◊ *Accessory plate.*

Quartz. ◊ *Silica group of minerals* and Appendix.

Quartzite. For orthoquartzite, ◊ *Arenaceous rocks.* Metaquartzite is a metamorphosed arenaceous rock. The constituent grains recrystallise and develop an interlocked mosaic texture, with little or no trace of cementation. Impurities in the original rock may give rise to metamorphic minerals, e.g. argillaceous impurity will yield ◊ *chlorite* or ◊ *biotite*; ◊ *calcite* may give ◊ *wollastonite.* Quartzites are usually thought of as thermally metamorphosed rocks, but regional metamorphism also produces them, usually with the development of an orientated ◊ *fabric* which distinguishes them from the thermally metamorphosed kinds (◊ *Metamorphism*). Under intense regional metamorphism, a quartz ◊ *schist* may develop.

Quartz porphyry. ◊ *Microgranite.*

Quartz wedge. ◊ *Accessory plate.*

Quaternary. The latest period of time in the stratigraphic column, 0–2 m.y., represented by local accumulations of glacial (Pleistocene) and post-glacial (Holocene) deposits which continue, without change of fauna, from the top of the Pliocene (Tertiary). The Quaternary appears to be an artificial division of time to separate pre-human from post-human sedimentation. As thus defined, the Quaternary is increasing in duration as man's ancestry becomes longer. The Quaternary is here regarded as a division of the ◊ *Tertiary.*

Radioactive dating (Radiometric dating). The most reliable method of obtaining a 'date' for a rock depends upon the observation that the rate of decay of a radioactive element is a constant (◊ *Age of the Earth*). The earliest methods, utilising uranium and thorium minerals as the starting material, yielded evidence that the extent of geological time was at least 2,000 m.y. The development of knowledge concerning radioactive processes since 1939 has made available a number of refined techniques for radioactive dating which are nowadays routine processes.

The basic principle underlying all methods is that a radioactive 'parent' element decays into a stable 'daughter' element at a constant rate. The rate of decay (λ) is the fraction of the initial number of 'parent' element atoms which decay in unit time (for geological purposes the unit time is one year). An alternative way of expressing this is the half-life period (T), which is the period of time necessary for one half of the number of 'parent' element atoms to decay. The relationship between T and λ is $T = 0.693/\lambda$.

In order to use a radioactive element for dating purposes, it is essential that T or λ should be accurately known. If, knowing this constant, a chemical analysis for the 'parent' and 'daughter' elements is made, the elapsed time can be computed. In order to minimise sources of error it is often necessary to use minerals which contain only minute proportions of 'parent' and 'daughter' elements, and hence analytical techniques of maximum sensitivity and accuracy have had to be devised.

It should be noted that the decay of 'parent' to 'daughter' may take place in one step or in a series of steps. Serious errors will arise if, at any stage, any 'parent', intermediate or 'daughter' element is introduced into, or removed from, the system. The initial presence of 'daughter' elements of non-radiogenic origin may also give rise to unreliable results.

In general, the most reliable dates are derived from igneous rocks, principally because the moment of crystallisation of the mineral yields a sharp starting point. However, it is often difficult to relate dates derived from intrusive igneous rocks to a stratigraphic scale. These methods can also be applied to metamorphic rocks, which in general yield a date for the last metamorphic event. Attempts to date ◊ *authigenic* minerals in sediments (e.g. glauconite) have yielded somewhat inconsistent results except under rather favourable conditions. The following are the more important radioactive systems now used for dating purposes.

URANIUM-LEAD METHODS: Two isotopes of uranium can be used:

$$U^{238} \text{ yielding } Pb^{206} + 8He^4$$
$$\text{and } U^{235} \text{ yielding } Pb^{207} + 7He^4$$

The half-lives of these two processes are 4,498 m.y. and 713 m.y. respectively. Early attempts to use the helium accumulation as a measure of time were unsuccessful because of the ease with which helium leaks from a rock. Nowadays ◊ *zircon*, which usually contains a minute trace of uranium, is the favoured mineral for this determination, since the amount of primary lead (i.e. the amount of non-radiogenic lead) is very low and the mineral itself is highly resistant, preventing the leaching of uranium.

THORIUM-LEAD: Th^{232} yields $Pb^{208} + 6He^4$, half-life 13,900 m.y. For some reason or other, this method yields erratic results which are not regarded as reliable by themselves.

LEAD-LEAD: Because Pb^{207} accumulates approximately six times as fast as Pb^{206} an age can be obtained if one can rely on the material used having been free, initially, from these two isotopes of lead. Non-radiogenic lead (Pb^{204}) is always found associated in lead ores with Pb^{206} and Pb^{207}. On the assumption that all the Pb^{206} and Pb^{207} is radiogenic, it is possible to obtain an estimate of the age of the Earth since the crust crystallised by comparing the isotope ratios of lead ores of differing geological ages.

POTASSIUM-ARGON: 11% of K^{40} yields A^{40}, the remaining 89% yielding Ca^{40}: the half lives of these two processes are 11,850 m.y. and 1,470 m.y. respectively. Unfortunately the $K^{40} \rightarrow Ca^{40}$ system is difficult to use because of the abundance, under normal circumstances, of non-radiogenic Ca. Surprisingly, the $K^{40} \rightarrow A^{40}$ system yields good results, providing materials are carefully chosen, even though argon is a gas which might be expected to escape from the system. It is possible, by employing refined techniques, to use minerals which contain only small quantities of potassium. This is the method which has been tried for glauconite dating.

RUBIDIUM-STRONTIUM: Rb^{87} yields Sr^{87}: half-life about

50,000 m.y. Because of the long half-life this method is mainly used for Precambrian rocks, although Palaeozoic ones have yielded some information. It is a particularly valuable method for metamorphic rocks. Although neither Rb nor Sr is an abundant element, contamination with primary Sr or loss of Rb or Sr from the system are both rather unlikely occurrences, and this makes the method fairly reliable.

CARBON 14: N^{14} (in the atmosphere) become C^{14} as a result of cosmic ray reactions. The C^{14} is radioactive, with a half-life of 5,570 years, the decay yielding N^{14} again. This short half-life makes the system especially useful for dating relatively recent materials, up to about 70,000 years. All organisms take in C^{14} and it can be shown that a constant level of this isotope is maintained by all living organisms. At death the C^{14} intake ceases, and its proportion decreases at a constant rate. Knowing the initial concentration, the present proportion and the decay constant, an age correct to about $\pm 5\%$ can be obtained. The combustion of fossil fuels containing no C^{14} has diluted the present-day C^{14} content of the atmosphere; on the other hand, nuclear explosions have slightly increased the concentration. On balance, modern plants contain slightly less C^{14} than those of a century or so ago.

It should be noted that dates (absolute ages) obtained by different methods from the same material commonly show some discrepancies. It is anticipated that improvements in techniques will remove these.

As the Committee on the Measurement of Geological Time said in 1950, 'These figures (i.e. dates) are, as railway timetables say, subject to change without notice.'

Radiolaria. ◊ *Protozoa.*

Radiolarian cherts. Another name for ◊ *radiolarites.*

Radiolarian ooze. ◊ *Abyssal deposits* (Siliceous oozes).

Radiolarites. A name sometimes given to certain ◊ *cherts* from geosynclinal deposits which contain abundant radiolaria (◊ *Protozoa*).

Rain prints. Raindrops falling on soft sediment (usually mud) often form a characteristic 'micro-crater', which, under certain circumstances, may be preserved in the rocks. (◊ *Sedimentary structures*; *Way-up criteria.*)

Raised beach. A wave-cut platform, with or without a covering of beach deposits, which is now raised above the present sea-level. (◊ *Rejuvenation*; *Marine erosion.*)

Ramsay, Andrew Crombie (1814–91). One of Britain's outstanding field geologists, who was responsible for establishing the evidence for the former glaciation of the British Isles. He was prominent during his life for his theory of the glacial erosion of rock basins, and as an advocate of the fluvial theory of denudation. He also produced a geological map of England and Wales, and a textbook of physical geography and geology.

Rang. ◊ *CIPW Classification.*

Rank. The percentage of carbon in dry mineral-free ◊ *coal.*

Rapakivi texture. A texture originally described from certain Scandinavian ◊ *granites* in which relatively large oval ◊ *orthoclase* crystals occur, mantled with a soda-rich ◊ *plagioclase.*

Rauracian. A synonym of ◊ *Argovian* (Jurassic System).

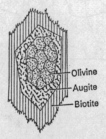

123. An example of a reaction rim.

Reaction rim. A crystal which is surrounded by a zone of secondary mineral is said to possess a reaction rim. (◊ Fig. 123.) Two main kinds of reaction rim are developed: firstly by late-stage or metasomatic fluids attacking the crystal, e.g. a rim of ◊ *serpentine* developed around an ◊ *olivine* crystal; secondly, the reaction of the crystal with its environment, either the liquid from which it crystallised or the solid crystals with which it is in contact, e.g. early-formed olivine crystals may develop a rim of ◊ *hypersthene* by reaction with a later, more siliceous, molten phase; ◊ *garnets* may develop a fringe of ◊ *biotite* where they are in contact with ◊ *orthoclase* in a metamorphic rock. (Cf. ◊ *Corona structure;* ◊ *Epitaxy.*)

Realgar. A mineral, AsS, found in association with ◊ *orpiment* in hydrothermal sulphide veins and as a deposit from hot springs and vulcanicity. (◊ Appendix.)

Recent. A synonym of Holocene. (◊ *Quaternary.*)

Recrystallisation. The process whereby a mass of crystals passes through a solution phase in developing a new set of crystals of the same kind. The change requires the presence of a suitable solvent at a suitable energy level. Probably small amounts of solvent circulating in pore spaces can bring about recrystallisation. It is not envisaged that more than a minute proportion of the material is in solution at any one time. Recrystallisation commonly results in an increase in crystal size. Under pressure it may give rise to an orientated ◊ *fabric*. The process sometimes

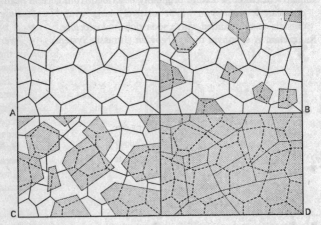

124. Recrystallisation. (A) Original material; (B) Growth of new crystals (dotted); (c) Enlargement of new crystals; (D) Complete recrystallisation.

operates in response to localised stress and this may give rise to localised development of coarser and/or orientated material. Recrystallisation may take place as a result of ◊ *supergene* or ◊ *metamorphic* processes. (◊ *Crystalloblastic*; Fig. 124.)

Recumbent fold. ◊ *Fold*, Fig. 62.

Red beds. A term applied to an assemblage of sedimentary rocks formed in a highly oxidising environment, so that the iron present is in the form of red ferric hydroxide. The term has acquired a meaning almost synonymous with arid continental sediments, since most red beds were probably formed in such an environment. The meaning has also been extended not only to the typical red sandstones, red shales and red marls, but also to include associated ◊ *evaporites*, ◊ *breccias*, and ◊ *cornstones*. Secondarily-reddened sediments should not be described as red beds.

Red clay. ◊ *Abyssal deposits.*

Reduzates. A geochemical division of the ◊ *sedimentary rocks*, including the sedimentary sulphides, coal, and petroleum.

Reef. (1) A gold-bearing quartz vein (especially in Australia).

(2) The Precambrian gold-bearing conglomerates of Africa (◊ *Banket*).

(3) A ridge of rocks, the top of which lies at just about sea level, so that the rocks are submerged at high tide and nearly submerged at low tide.

(4) A mass of organic skeletal material consisting in part of organisms which have grown *in situ*, in part of material derived from organic debris which has been transported to the site of the reef, and in part of chemically precipitated material. This latter, in general, forms a very small proportion of the whole. At the present time, the commonest reef-building (hermatypic) organisms are the corals, although the calcareous algae constitute a larger proportion of most reefs than is generally appreciated. ◊ *Polyzoa* are locally significant. In the geological past, stromatoporoids, tabulate corals (◊ *Coelenterata*), rudistid lamellibranchs (◊ *Mollusca*), crinoids (◊ *Echinodermata*) and ◊ *Brachiopoda* have been important as reef builders. Many other organisms live on, in, or around reefs, and their skeletons may contribute to the total mass of reef-material. The following types of reef occur (◊ Fig. 125):

REEF KNOLLS (REEF MOUNDS, BIOHERMS). Dome-like reefs of varying sizes, commonly surrounded by a mass of eroded reef-debris. Where a reef knoll has persisted at a particular site, growing upwards so as to keep pace with the deposition of the surrounding sediments, large columnar reef-masses may develop. ◊ *Quaquaversal* dips may develop in the debris surrounding the reef knoll, giving an impression of folding. A bioherm is a large reef knoll.

FRINGING REEFS (BARRIER REEFS). A reef which develops offshore so that a lagoon is formed between the reef and the coastline. The seaward extension of the reef is limited by the maximum depth of water in which the reef-building organisms flourish. Erosion of the seaward face of the reef produces a mass of debris upon which living reef can develop in the optimum depth of water. As the growth of most reef-building organisms is inhibited by muddy water, gaps in barrier reefs commonly develop opposite the mouths of rivers.

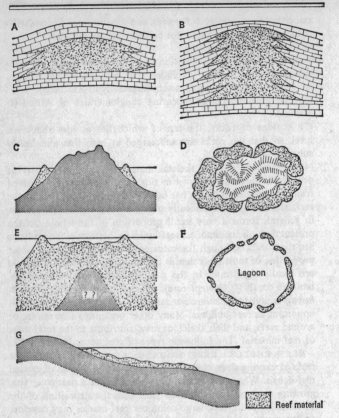

125. Types of reef. (A) A reef knoll; (B) A bioherm; (C) A fringing reef (section); (D) A fringing reef (plan); (E) An atoll (section); (F) An atoll (plan); (G) An apron reef.

ATOLLS. More or less circular reef-masses, with one or more breaches joining the central lagoon to the open sea. Many atolls can be shown to have formed around a central island which has slowly sunk beneath the sea, the upward growth of the coral keeping pace with the sinking. In a number of atolls, however, very great thicknesses of reef material are encountered, which makes this explanation more difficult to accept.

APRON REEFS. Tabular sheet-like masses of indigenous reef

material, as distinct from the eroded reef-debris, but sometimes interbedded with it. They are thus one type of ⋄ *biostrome*.

At the present time reefs are found only in clear marine waters and their much wider distribution in rocks of all geological ages suggests strongly that the average temperatures of the present-day oceans is much less than the geological norm.

Reflection, seismic. Energy which is propagated down into the Earth and which returns after having undergone a single reflection from a subsurface velocity (density) contrast. (⋄ *Geophysics*.)

Regolith. The loose, incoherent mantle of rock fragments, soil, blown sand, alluvium, etc. which rests upon solid rock, i.e. the ⋄ *bedrock*. Some authors also include volcanic ash, glacial drift, and peat under this heading. It is probably not accurate to use regolith as a synonym for ⋄ *soil*.

Regression, marine. The withdrawal of the sea from a large area of land in a relatively short space of time (geologically speaking). It is the precise antithesis of marine ⋄ *transgression*. (⋩ *Unconformity*.)

Rejuvenation. Rejuvenation takes place with the relative uplift with respect to sea level of a region which has developed a mature drainage system (⋄ *Cycle of erosion*). Streams whose ⋄ *base level* has fallen cut down very rapidly and become incised (entrenched) into the land in an attempt to re-create the previous ⋄ *long-profile* of equilibrium. An uplift of the land (or a fall in sea level) causes rivers to reach the sea by means of waterfalls. In time these regress upstream as the streams adjust themselves to the new profile. Breaks, such as waterfalls, in the long-profile of a river are known as knick (nick) points, and represent the junction of the new, adjusted, profile and the old, now maladjusted, one.

On the coast rejuvenation produces elevated beaches known as raised beaches some metres or tens of metres above sea level. Inland, rivers cut into their flood plains, subsequently leaving only remnants of the old valley floor preserved above the new one. Such elevated areas are called ⋄ *river terraces*. Successive uplifts produce multiple terraces, raised beaches and knick points. The oldest terrace, or raised beach, is that which is highest above the river, or sea. Terraces should grade upstream into knick points and pass downstream into raised beaches, but these relationships are not always clearly preserved. (⋄ Fig. 126.)

Relict structure. A near-synonym of ⋄ *palimpsest structure*.

Relief, optical. In a ⋄ *thin section* or grain mount, colourless min-

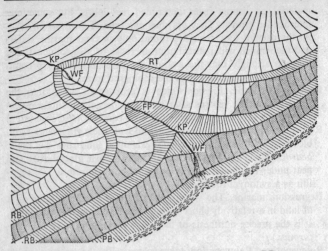

126. Rejuvenation. FP, flood plain; KP, knick point; PB, present beach; RB, raised beach; RT, river terrace; WF, waterfall.

erals are visible when there is a difference in refractive index between the mineral and the mounting medium or adjacent minerals. The greater the difference in refractive index the stronger will be the demarcation of grain boundaries; this phenomenon is termed the relief of the mineral and is produced by optical effects at the mineral/mountant or mineral/mineral interface. Minerals with a very high or very low refractive index will respectively show a positive or negative relief. For minerals of low relief, the ◊ *Becke* and ◊ *shadow tests* can be used to determine relative refractive index.

Remanié fossils. Organic skeletal material, accumulating over a considerable period of time, which has been rolled and abraded, more or less *in situ*, before its eventual burial. The high concentration of fossils is brought about either by a paucity of sediment or by current action preventing the deposition of sediment. Many ◊ *bone beds* are produced by these methods.

Reniform. Kidney-shaped. (◊ *Botryoidal.*)

Replacement. A term widely used in geology to describe the process whereby one constituent in a system is progressively substituted by another. The following examples will make the scope of the term clear (◊ Fig. 127 and ◊ *Ghost stratigraphy*).

Chalcocite (white) replacing
sphalerite (black)
Reflected light

Crystal growth by replacement

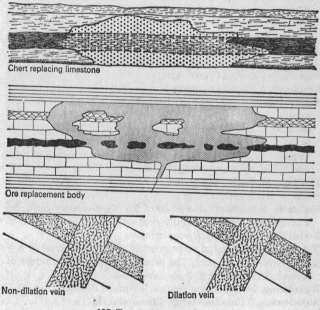

Chert replacing limestone

Ore replacement body

Non-dilation vein   Dilation vein

127. Types of replacement.

(1) The replacement of one ion by another in an atomic lattice,
e.g. Al''' can replace Si'''' in the ◊ *silicates*. (◊ *Geochemistry*.)
(2) The replacement of one crystal by another. (◊ *Pseudo-morph*.)

(3) The replacement of the skeletal material of fossils by another material. (◊ *Fossils.*)

(4) The replacement of a rock body, e.g. chert replacing limestone.

(5) ◊ *Hydrothermal* replacement, the important process by which deposits of sulphide minerals (and sometimes others) replace pre-existing rocks, usually limestones. In Fig. 127 note the continuity and retention of the orientation of the blocks of ◊ *country rock* isolated by the replacing ore body.

(6) Veins and dykes are sometimes regarded as replacing *country rock* (non-dilation vein) rather than mechanically invading it (dilation vein).

(7) The structures and textures of one metamorphism replace those of an earlier event (overprinting).

(8) It is considered by some authorities that many ◊ *batholiths* are emplaced by replacement of the country rock. (◊ *Granitisation.*)

(9) In palaeontology one fauna may be said to replace another either horizontally in space or vertically in time.

Reptilia. ◊ *Chordata* (Vertebrata).

Resequent. ◊ *Drainage pattern.*

Residual deposit. Material left behind when part of a rock is removed by chemical ◊ *weathering* processes, usually solution or leaching. ◊ *Laterite and bauxite* are types of residual deposits, as is the ◊ *terra rossa* of limestone regions, which consists of the insoluble residue after the carbonate has been removed in solution. (Cf. ◊ *Clay-with-flints;* ◊◊ *Soil.*)

Resilium (Mollusca). ◊ Fig. 101.

Resin. ◊ *Hydrocarbon minerals.*

Resistates. A geochemical division of the ◊ *sedimentary rocks,* including the ◊ *arenaceous* and ◊ *rudaceous rocks.*

Resistivity. ◊ *Geophysical prospecting.*

Resorption. Partial refusion or solution of a ◊ *phenocryst* in a ◊ *porphyritic* igneous rock. The process may result in a border of secondary minerals. (Cf. ◊ *Reaction rim.*)

Retardation. ◊ *Polarisation colours; Vibration directions.*

Rhabdosome. ◊ *Chordata* (Graptolithina) and Fig. 17.

Rhaetian. The stratigraphical stage name for the European ◊ *Rhaetic.*

Rhaetic Series. Named from the Rhaetic Alps, the Rhaetic forms a transitional series between the ◊ *Trias* and the Lower ◊ *Jurassic,* being included with the Trias by European geologists, but in the Lower Jurassic by most British stratigraphers. It marks that period of time after the first marine transgression in the Mesozoic and before the later deepening of the seas which led to the deposition of the Lower Jurassic with its typical fauna of ammonites. The Rhaetic deposits are extremely variable, consisting of black shales, sandstones, and thin limestones. ◊ *Bone beds* are common throughout the succession and these include both aquatic and terrestrial bone debris. The rest of the fauna is largely one of small lamellibranchs and gastropods, and is characterised by the presence of the form Rhaetavicula contorta. Remains of the first undoubted mammals occur in the Rhaetic.

Rheo- (prefix). Streaming or flowing.

Rheomorphism. A term more or less synonymous with viscous flow, as applied to rocks which have been deformed when partial melting takes place. The use of the term has been extended to many types of plastic or plastico-viscous ◊ *deformation.*

Rhodocrosite. A hexagonal ◊ *carbonate* mineral, $MnCO_3$, found in lead and silver–lead ore veins and in ◊ *metasomatic* deposits. (◊ Appendix.)

Rhodonite. A manganese mineral, a ◊ *pyroxenoid*, $MnSiO_3$, sometimes found in ◊ *hydrothermal* veins or ◊ *metasomatic* deposits. (◊ Appendix.)

Rhombohedral system. A synonym of ◊ *trigonal system.*

Rhomb porphyry. A name given to a series of lavas and minor intrusions of ◊ *syenite* character, originally described from the Oslo region of Norway. They are characterised by the presence of large pale ◊ *phenocrysts* of feldspars which commonly display rhomb-shaped sections on surfaces of the rock.

Rhyodacite (Toscanite). ◊ *Rhyolite.*

Rhyolite. Fine-grained to glassy acid volcanic rocks. Mineralogically they are similar to ◊ *granites* and ◊ *microgranites*, although chemically they appear somewhat richer in SiO_2. Occasionally quartz in rhyolites is replaced by the high-temperature beta-form, and very rarely by tridymite or cristobalite (◊ *Silica group of minerals*). ◊ *Ferromagnesian minerals* are less obvious than in the corresponding plutonic rocks. Rhyolites in the strict sense are divided into sodic and potassic forms, according to the type of feldspar present.

Quartz keratophyre is a type of sodic rhyolite or quartz-◊ *trachyte* with practically no ferromagnesian mineral present, and is related to the ◊ *spilites.*

Pantellerite contains a soda-potash feldspar, quartz, aegirine, and sometimes sodic amphibole. Some pantellerites are low enough in quartz to be considered varieties of trachyte.

Volcanic rocks which are equivalents of adamellite are termed toscanite or rhyodacite; the volcanic equivalent of granodiorite is dacite. It is probable that both these types are more common than is usually thought, owing to the difficulty of distinguishing them from rhyolites.

Glassy rocks of the acid group are termed obsidian or ◊ *pitchstone*. Obsidian is a black, wholly glassy rock and displays a conchoidal fracture; pitchstone is a glassy rock which contains a much higher proportion of ◊ *crystallites* and displays a flatter fracture. It may be black, dark grey, red or brown. ◊ *Phenocrysts* are not uncommonly present.

Pitchstones are unusual in occurring commonly as dykes as well as lava flows. Both obsidian and pitchstone may show the beginnings of ◊ *devitrification* by the formation of radiating masses of crystals called ◊ *spherulites* (hence 'spherulitic pitchstone'). Progressive devitrification of these glassy rocks produces ◊ *felsites*. If the initial process of devitrification gives rise to spherulites, a spherulitic rhyolite or ◊ *variolite* may result, which should be carefully distinguished from an amygdaloidal rhyolite which, if the ◊ *amygdales* are filled with quartz, has a superficially similar appearance. Many rhyolites show well-marked ◊ *flow* textures and structures.

With decreasing quartz, rhyolites pass into ◊ *trachytes*, and dacites into ◊ *andesites*. Rhyolite lava is notably very viscous, and at the present time never forms extensive flows around volcanic vents. In the geological past extensive areas of rhyolitic rocks appear to have been produced, and it seems likely that they formed in one of two ways: (1) By the unroofing of a mass of acid magma by a process of ◊ *cauldron subsidence*; (2) By the development of an ignimbrite or welded tuff (◊ *Pyroclastic rocks*).

Rhythmic sedimentation. The term applied to a sequence of sediments which change their character progressively from one extreme type to another, a change which is followed directly by a return to the original type. A typical example would be the Coal Measure rhythm:

$$\left.\begin{array}{l}\text{Marine} \\ \text{Swamp} \\ \text{Deltaic} \\ \text{Lagoonal} \\ \text{Marine} \\ \text{Swamp}\end{array}\right\}\ \text{Rhythmic unit}$$

The essential features of the rhythmic unit is a steady sequence of changes followed by an abrupt return to the starting point. This may be symbolically expressed:

$$
\begin{array}{c}
d \\
c \\
b \\
a \\
\hline
d \\
c \\
b \\
a
\end{array}
$$

Rhythmic units are sometimes as large and as extensive as cyclic units, but many examples of very small-scale local rhythms are known. (Cf. ◊ *Cyclothem*; ◊ *Cyclic sedimentation*.)

Rib. (Brachiopoda) ◊ Fig. 11; (Mollusca) ◊ Figs. 100, 105.

Ribbon diagram. ◊ *Fence diagram*.

Richmondian. A stratigraphic stage name for the North American mid Upper ◊ *Ordovician*.

Ridge, ocular (Arthropoda). ◊ Fig. 3.

Riebeckite. ◊ *Amphiboles*; *Asbestos*; Appendix.

Riedel fracture. A branching ◊ *fault* developed in the cover by movement in the basement.

Rift valley. An elongated trough bounded by ◊ *faults*; in all cases the tectonic depression and a geomorphological valley coincide; the term 'graben' appears to be synonymous. Taphrogeosyncline is also a synonym.

Various theories concerning rift valleys and the volcanic activity characteristically associated with them have been stated: Cloos suggested that they resulted from the slow up-doming of ◊ *kratonic* areas, while others have explained them as due to tension from the breaking-up of continental masses (e.g. ◊ *Gondwanaland*), deep-seated lateral compression, or the 'displacement of the ◊ *sima* in the substratum'. A full explanation of their origin and the nature of their connection with vulcanism has yet to be produced. (◊ Fig. 128.) Outstanding examples are the East African System, which can be traced over a distance of 4,000 km. (3,000 miles) from Syria to the Zambesi, although the heights of the floor and opposing walls vary considerably (Kilimanjaro and Mt Kenya are notable examples of the associated vulcanism); and the Rhine Graben, with a main area 280 km. (180 miles) long and

30–40 km. (20–25 miles) wide, part of a much larger system. The Midland Valley of Scotland is the best example in Great Britain, although it differs from the previous systems in that it is associated with ◊ *orogenic* belts rather than being wholly within a ◊ *shield* area, and its boundary faults coincide with the margins of a major depositional basin.

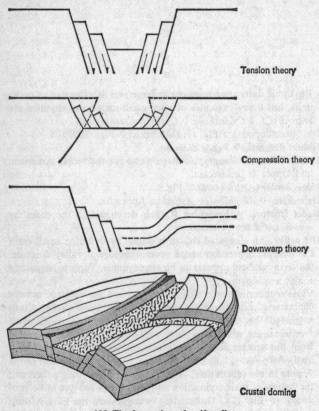

Tension theory

Compression theory

Downwarp theory

Crustal doming

128. The formation of a rift valley.

Ring structure. A composite igneous intrusion in which the individual members appear as circular or part-circular outcrops. The two main constituents of such structures are ring ◊ *dykes* and ◊

cone-sheets. (⟡ *Cauldron subsidence.*) Ring structures are commonly associated with volcanic episodes. Examples of granitic ring complexes may be seen in Glencoe and Ben Nevis, Scotland, and in Nigeria and adjoining territories. In the Oslo district of Norway, granite–alkali syenite complexes occur, while over a large portion of eastern and southern Africa alkali syenites with or without ⟡ *carbonatites* occur in a similar way. Basic ring complexes are best known from the Inner Hebrides and adjacent mainland of West Scotland.

Ripple-drift bedding (Ripple-drift cross-bedding). ⟡ *Cross-bedding.*

Ripple marks. Undulations produced by fluid movement over sediments. The nomenclature of the parts of a ripple is shown in Fig. 129. Oscillatory currents produce symmetric ripples whereas a well-defined current direction produces asymmetrical ripples. The crest lines of ripples may be straight or sinuous. The characteristic features of ripples depend upon current velocity, particle size, persistence of current direction, and whether the fluid is air or water. Sand ⟡ *dunes* may be regarded as a special kind of 'super'-ripple. The term 'mega-ripple' is used for ripples having a wavelength greater than 50 cm. Lobate ripple marks are termed linguoid or lunate ripples according to whether they point down or up current respectively. During sedimentation ripples migrate 'up each other's backs', producing the type of ⟡ *cross-bedding* known as ripple-drift bedding.

River capture. The abstraction of the head-waters of one river by a more actively eroding stream – usually by the breaching of a local watershed. Certain cases of capture arise as a result of ⟡ *glacial breaching.* ⟡ *Rejuvenation* may also produce this phenomenon and a number of repeated captures of successive streams by a single master stream have resulted from the favourable location of the master stream on easily eroded rocks. The point at which capture occurs is known as the elbow of capture, and associated with this point is usually found a dry valley or ⟡ *misfit stream.*

River terrace. A river terrace marks the position of the former ⟡ *flood plain* which developed when the river flowed at a higher level. It usually consists of a bench on the side of the valley, covered with the usual flood plain deposits of clay, sand, and gravel. Terraces may be developed: (1) By ⟡ *rejuvenation* of the river causing downcutting through the existing flood plain and the development of a new one at a lower level. This type is symmetrical, in as much as the terraces occur at the same level

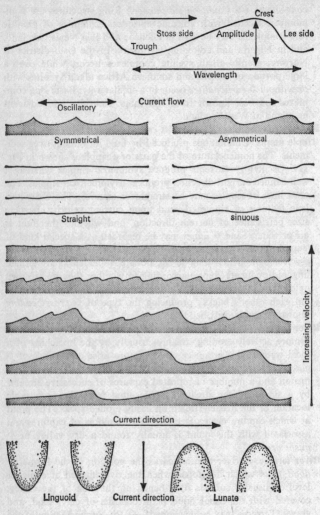

129. Ripple marks.

on either side of the stream ($\diamond$ Fig. 126). (2) As a result of meandering during the downcutting stage of a river, which yields terraces at different levels on either side of the river ('asymmetrical'). (3) Where a river has developed a flood plain at a temporary level owing to the development of a lake along its course – or where it has encountered a layer of resistant rock over a considerable stretch of its length which inhibits downcutting. When the local base level is destroyed by drainage of the lake or the eventual penetration of the resistant layer, terraces will remain as evidence of their former existence.

In most cases where terraces are due to rejuvenation they can be traced upstream to a $\diamond$ *knick point* and downstream to a $\diamond$ *raised beach*. It will be realised that terraces do not remain at a constant height above the river as they are traced upstream, and hence terms like '25-ft' terrace and '50-ft' terrace are rather misleading.

Roche rock. $\diamond$ *Pneumatolysis* (Tourmalinisation).

Roches moutonnées. (French, 'sheep rocks', from the similarity to recumbent sheep.) Glacial erosion features, consisting of asymmetrical mounds of rock of varying size, with a gradual smooth abraded slope on one side and a steeper rougher slope on the other. They are produced by the action of advancing ice, the smooth slope being on the side of ice advance, and the steep face on the lee side, where shattering by ice-plucking has occurred. Roches moutonnées commonly occur in swarms, often with a common alignment. ($\diamond$ *Glaciers and glaciation*.)

Rock. (1) To the geologist any mass of mineral matter, whether consolidated or not, which forms part of the Earth's crust is a rock. Rocks may consist of only one mineral species, in which case they are called monomineralic, but they more usually consist of an aggregate of mineral species.

(2) The civil engineer regards rock as something hard, consolidated, and/or load-bearing, which, where necessary, has to be removed by blasting. This concept also accords with the popular idea of the meaning of the word.

Rockallite. A soda-rich $\diamond$ granite containing $\diamond$ *aegirine*.

Rock crystal. A clear variety of quartz ($\diamond$ *Silica group of minerals*).

Rock flour. Rock material which has been ground to a uniformly fine particle size of clay or fine silt grade. It generally results from either $\diamond$ *cataclasis* or glacial erosion where rocks are ground together. It should be noted that, unless the rocks producing the rock flour are themselves of an argillaceous character, rock flour does not behave as a normal clay, despite its particle size. Rock flour forming during the production of a ◊ *fault breccia*, or ◊ *crush breccia* is known as gouge or fault gouge. (Cf. ◊ *Mylonite*; ◊ *Argillaceous rocks*.)

Rock-forming minerals. Those minerals which occur in sufficient abundance and frequency as to constitute the major bulk of a rock. The main rock-forming minerals are the silicates (including silica), carbonates, and oxides, with sulphates, chlorides and phosphates in lesser proportions. (◊ *Accessory minerals*; *Essential minerals*.)

Rock mechanics. The study of the mechanical properties of a rock, especially those properties which are of significance to the civil engineer. It includes the determination of physical properties such as crushing strength, bending strength, shear strength, moduli of elasticity, internal angle of friction, porosity and permeability, density, etc., and their interrelationships. The correlation of these properties with similar properties of the constituent minerals of the rock is also significant. Care must be taken to distinguish between properties derived from the study of a rock specimen and those obtained by the study of a rock mass *in situ*. The influence of bedding planes, schistosity, cleavage, joints, etc., may vitally change the bulk properties of a rock compared with those of a single, selected, hand specimen.

Rock phosphate. ◊ *Phosphatic deposits*.

Rod, Rodding. ◊ *Mullion structure*.

Roof pendant. A large mass of the ◊ *country rock* forming the roof of a major igneous ◊ *intrusion*. They are most commonly found around the margins of ◊ *batholiths*. They are in effect giant ◊ *xenoliths*.

Ropey lava. A synonym of ◊ *pahoehoe*.

Rosenbusch, Karl Heinrich (1836–1914). German professor who published the first comprehensive account of petrological methods, including the use of the microscope. Later he dropped mineralogical classification in favour of one based upon rock-structure (cf. ◊ *Zirkel*), mainly as a result of his findings on the origin of eruptive rocks.

Rossi-Forell Scale. An arbitrary scale of ◊ *earthquake* intensity, now supplanted by the Modified Mercalli Scale.

Rostrum. (Echinodermata) ◊ Fig. 45; (Mollusca) ◊ Fig. 105.

Rotliegende. A stratigraphic stage name for the European Lower and Middle ◊ *Permian*.

Rotten stone. A siliceous residue from the weathering or partial weathering of certain limestones which, when suitably crushed and graded, can be used as a mild abrasive.

Roundness. A property of the individual fragments of a ◊ *clastic* sediment which is independent of the shape of the fragment, and is a function of the radius of curvature of the various edges and corners of the particle. It may be quantitatively determined as the ratio of the mean radius of curvature of the edges and corners of the fragment to the radius of the maximum possible inscribed sphere. It should be noted that roundness is distinct from ◊ *sphericity*. For most practical purposes particles may be clas-

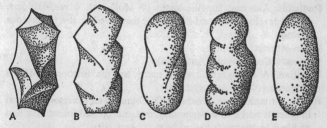

130. Degrees of roundness. (A) Angular; (B) Sub-angular; (C) Sub-rounded; (D) Rounded; (E) Well-rounded.

sified qualitatively on the scale: Angular; Sub-angular; Sub-rounded; Rounded; Well-rounded. (◊ Fig. 130.) In general, increasing roundness may be correlated with increasing duration of transport from the source. (◊ *Particle shape*.)

Rubellite. A pink lithium-rich variety of ◊ *tourmaline*.

Ruby. A red, transparent variety of ◊ *corundum*, Al_2O_3. (◊ Gem.)

Rudaceous rocks. A group of detrital sedimentary rocks in which the particles range in size from 2 mm. upwards. The term psephitic is precisely equivalent to rudaceous but is now hardly ever used. Rudaceous rocks can be deposited in either water or sub-aerially, and have generally not been transported far from their point of origin. Two main classes of rudaceous rocks may be recognised: ◊ *breccias* and ◊ *conglomerates*.

Boulder bed is a term reserved for rocks containing fragments larger than 256 mm. (10 ins.) in diameter. The term 'basal conglomerate' has the stratigraphic implication of a conglomerate which lies above an unconformity, and is chiefly formed of fragments derived from below this plane.

The essential feature of a breccia is that its constituent fragments are all angular to sub-angular. In a conglomerate they are rounded to sub-rounded. The term breccio-conglomerate may be used for rocks containing both angular and rounded fragments.

For conglomerates and breccias composed essentially of calcareous material (calci-rudite), ◊ *Limestone* (Clastic). (It should be noted that the presence of limestone fragments or a calcareous cement in a rudaceous rock is not by itself sufficient to warrant the use of the name.) For cementation, sorting, grading and maturity, ◊ *Arenaceous rocks*. (◊ *Banket*; *Particle shape*; *Sphericity*; *Roundness*.)

Rudistids. Aberrant lamellibranchs (◊ *Mollusca*); ◊ *reef*-builders, which developed a coral-like form; they lived in the Cretaceous period.

Rudus- (prefix). Pebbly.

Rugosa. ◊ *Coelenterata* (Anthozoa).

Rupelian. A stratigraphic stage name for the European Middle Oligocene (◊ *Tertiary system*).

Rutile. A titanium ore mineral, TiO_2, found as an accessory mineral in igneous rocks, in ◊ *pegmatites*, metamorphosed limestones, and as fibres in quartz (◊ *Venus' hair*). (◊ Appendix.)

Ryazanian. A stratigraphic stage name for the base of the Lower ◊ *Cretaceous* in Eastern Europe.

Saalian. A stratigraphic stage name for the base of the European Upper Pleistocene (◊ *Tertiary System*).

Saamian. A stratigraphic division of the Baltic ◊ *Precambrian* – Pre-◊*Karelian*.

Sabkha (Sabka, Sabkhah, Sabkhat). Strictly speaking, the Arabic term sabkha refers to the broad, salt-encrusted, supra-tidal surfaces or coastal flats bordering lagoonal or inner shelf regions. The 'type area' is on the Trucial Coast of Arabia (Abu Dhabi). An essential feature of the sabkha is that it is only flooded occasionally. There are two types, coastal and continental, which are both equilibrium surfaces in the geomorphological sense.

Coastal sabkhas are in part the product of offshore deposition and in part accumulation from the landward side. In arid regions the area becomes a salt flat. The typical coastal sabkhas consist of ◊ *carbonate* sediments, mainly aragonite. Salt water is drawn into the pores of the sediment and evaporation from the sabkha surface causes concentration of the seawater solution. Seawater also sinks into the sediment during the infrequent flooding of the area. ◊ *Gypsum* is extensively deposited, and ultimately redissolves and migrates to be re-deposited as anhydrite, together with some primary anhydrite. The concentrated brine also ◊ *dolomitises* the aragonite, and occasionally other minerals are formed such as ◊ *celestite* and ◊ *magnesite*. ◊ *Halite* (rock salt) forms as a superficial crust, most of which is removed by the periodical flooding. However, some is carried down into the sabkha sediments and may ultimately crystallise.

Continental sabkhas are formed on the landward side of coastal sabkhas and seem to consist of a mixture of dune sand and wind-blown carbonate material from the coast. Surface crusts formed by ◊ *evaporite* deposition occur in deflation hollows which have reached the water table. The sabkha sediments rest on a basement of hard rock which seems to be responsible for the water table being near the surface. Deposition of evaporite minerals occurs in the same way as on the coast. The source of the brine is not clear. Some salts may be blown from the coastal region inland and some may be due to leaching by the ground water. Continental sabkha deposits differ in detail from the coastal ones. It will thus be seen that deposition in the sabkha can yield the same sequence of evaporite deposition as does the evaporation of seawater in an enclosed basin. Extensive replacement of one evaporite mineral by another is recorded in the sabkha sediments and it seems possible that some at least of the fossil evaporites may have been formed in the sabkha environment.

D. Kinsman, 'Modes of Formation, Sedimentary Association, and Diagnostic Features of Shallow-Water and Supratidal Evaporites', and C. Kendal and P. A. d'E. Skipworth, 'Holocene Shallow-Water Carbonate and Evaporite Sediments of . . . the South-West Persian Gulf', in *Bulletin of the American Association of Petroleum Geologists*, Vol. 53/4, April 1969.

Saccharoidal. A texture term meaning 'having a sugary appearance'; i.e. a fine or medium grain size with closely interlocking crystals. It is commonly applied to ◊ *marbles* and ◊ *quartzites*.

Saddle reef. An ore deposit in the form of a ◊ *phacolith* (e.g. at Bendigo, Australia).

Sahlite. ◊ *Pyroxenes*.

Sakmarian. A stratigraphic stage name for the base of the East European and Russian ◊ *Permian*.

Salic. A mnemonic word made up from *si*lica and *al*uminium for a particular group of arbitrarily calculated 'minerals' in the ◊ *CIPW Classification*, essentially feldspar, feldspathoids, and quartz. (Cf. ◊ *Femic*.)

Salopian. Obsolete stratigraphic stage name for the ◊ *Ludlovian* plus. ◊ *Wenlockian* in Britain (Silurian System).

Salt domes. ◊ *Diapiric* masses of salt. Under high pressure, salt deforms plastically and behaves like an intrusive ◊ *magma*, deforming and piercing the overlying sediments. (◊ Fig. 131.) Occasionally salt domes reach the surface, giving rise to salt glaciers. Salt domes are of considerable economic importance as they may develop ◊ *oil* traps, while in Texas and Louisiana sulphur occurs in association with them.

Saltation (Latin *saltare*, 'to jump'). (1) A term used to describe the movement of a particle being transported by wind or water which is too heavy to remain in suspension. The particle is rolled forward by the current, generates lift and rises, loses the forward momentum supplying the lift and settles to the floor, where the process is repeated. The size of particles which can be saltated depends upon the velocity of the current and its density, e.g. water will saltate larger particles than air at the same velocity. (◊ Fig. 132.)

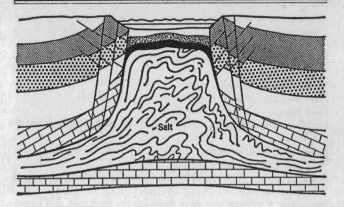

131. A salt dome.

(2) A very slow form of saltation occurs on slopes which are subjected to intermittent freezing and thawing. When water below a particle freezes, it expands and carries the particle upwards at right angles to the surface; when the ice film subsequently melts, the particle falls vertically downwards under the influence of gravity.

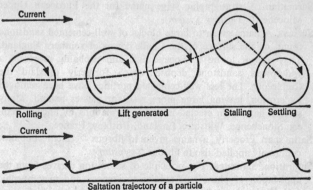

132. Saltation.

399

(3) Evolution which progresses by jumps rather than by a steady progression.

(4) ◊ *Longshore drift* is a form of saltation.

Sand. ◊ *Arenaceous rocks*; *Silica group of minerals* (Quartz); *Particle size.*

Sandstone. ◊ *Arenaceous rocks.*

Sandstone dyke. ◊ *Injection structure*; *Neptunean dyke.*

Sand volcano. ◊ *Injection structures.*

Sanidine. High-temperature potassium or potassium-sodium ◊ *feldspar.*

Sannoisian. A stratigraphic stage name for the European Lower Oligocene (◊ *Tertiary System*).

Santa Catherina. A stratigraphic name for the Brazilian equivalent of the ◊ *Karroo.*

Santonian. A stratigraphic stage name for the European mid Upper ◊ *Cretaceous.*

Sapelite. ◊ *Phlogopite-*◊*hornblende* peridotite (◊ *Ultrabasic rocks*).

Sapphire. Blue transparent or translucent ◊ *corundum* (Al_2O_3). (◊ *Gem.*)

Sapropelic. ◊ *Coal.*

Sard. A reddish or brownish variety of chalcedonic silica. (◊ *Silica group of minerals.*)

Sardonyx. A banded chalcedonic silica. (◊ *Silica group of minerals.*)

Sarmatian. A stratigraphic stage name for the European Upper Miocene (◊ *Tertiary System*).

Sarsens. A name given to large blocks of well-cemented sandstone found on the surface of the Chalk downs of southern England. They appear to be of Eocene age, and probably represent a widespread sandstone deposit which formerly overlaid the Cretaceous. The less well-cemented portions have been removed by erosion, leaving large more resistant masses behind. Many sarsens have been erected into ring formations by ancient man, e.g. Stonehenge, Wiltshire, England; Brittany, France.

Satin spar. Properly, a name given to fibrous ◊ *calcite*, but now commonly applied also to fibrous ◊ *gypsum.*

Saturation, Saturated. The principle of saturation derives from the application of phase rule studies to igneous and metamorphic rocks. It has been shown that if there is an excess or a deficiency of a certain component in the rock, particular minerals can or cannot develop. An alternative statement of this is that certain minerals are stable in the presence of an excess of a component, while other minerals are stable only if this component is not in excess.

In rocks, the ubiquitous nature of the SiO_2 component makes it inevitable that saturation is almost always described in relation to the silica concentration of a rock. However, occasionally it may be convenient to refer to 'saturation with respect to' Al_2O_3, $Fe_2O_3 + FeO$, K_2O, or Na_2O. The presence of an unsaturated (with respect to SiO_2) mineral in association with free SiO_2 in a rock may be regarded as good evidence of a lack of chemical equilibrium in the rock. A most convenient division of igneous rocks is into ◊ *oversaturated* (containing free SiO_2 minerals) and ◊ *undersaturated* (containing unsaturated minerals, e.g. olivine and/or feldspathoids). It is theoretically possible for a rock to be saturated, i.e. neither oversaturated nor undersaturated, but this state is rare and local.

Saussuritisation, Saussurite. The alteration of a basic ◊ *plagioclase* to a fine-grained aggregate of sodic plagioclase and ◊ *epidote*, together with ◊ *calcite*, ◊ *micas*, or other ◊ *layer-lattice minerals*, and occasionally a colourless ◊ *amphibole* which may replace the epidote.

This mixture was originally thought to be a single mineral, saussurite. It commonly results from the low-grade metamorphism of basic igneous rocks. (Cf. ◊ *Uralitisation*; *Propylitisation*.)

Saxonian. A stratigraphic stage name for the West European Middle ◊ *Permian*.

Saxonite. An ◊ *ultrabasic rock* consisting of ◊ *olivine* and ◊ *enstatite*.

Scaphopoda. ◊ *Mollusca*.

Scheelite. A tungsten ore mineral, $CaWO_4$, found in ◊ *pneumatolytic* and hydrothermal veins and in contact-metamorphic deposits. (◊ *Fluorescence of minerals* and Appendix.)

Scheuchzer, Johann (1672–1733). Swiss geologist who was one of the early advocates of the theory that fossils were the organic remains of the Biblical Flood. He is chiefly remembered for his identification of a Tertiary giant salamander fossil as a skeleton of pre-Deluge 'sinful' man – 'Homo diluvii testis' – but he was also responsible for some of the first sectional drawings of mountain structure.

Schiller. A peculiar play of light which appears when certain minerals are examined so that their crystal faces or cleavage faces are at a particular angle to the incident illumination. It may be related to the colour of the mineral and sometimes has a metallic quality, or may have a colour unrelated to that of the mineral. Minerals which commonly display schiller are the ◊ *pyroxenes* (bronzites and the diallage types) and ◊ *feldspars* (labradorite, microcline, and anorthoclase). The phenomenon may be due to the regular inclusion of minute (sometimes sub-microscopic), platy crystals in a regular orientation which all reflect simultaneously at a particular angle as the crystal is rotated. However, other causes are possible, e.g. regular discontinuities in the crystal structure, closely spaced incipient cleavage planes, exsolution lamellae. It is probable that some kind of an interference effect plays a part in producing the coloured types.

Schist. A regionally metamorphosed rock characterised by a parallel arrangement of the bulk of the constituent minerals. Schists are generally distinguished from ◊ *phyllites* by having a coarser grain size and a tendency towards an undulose cleavage. The common minerals which give rise to schistosity are the ◊ *micas* in the case of the platy (◊ *lepidoblastic*) schists, while ◊ *amphiboles* give rise to the linear (◊ *nematoblastic*) schists. Schists are named according to their most prominent minerals, e.g. garnet-mica-schist, staurolite-andalusite-schist. (◊ *Hornblende schist*; *Gneiss*; *Foliation*.)

Schistosity. ◊ *Cleavage, rock* (Flow cleavage).

Schlieren (German, 'streaks'). The term is generally applied to elongated, 'streaked-out', patches in an igneous rock. The patches differ in composition from the host rock and generally have rather diffuse boundaries. They may represent partly mobilised ◊ *xenoliths* or fragments derived from an earlier phase of crystallisation of the igneous rock.

Schorl. Black tourmaline, commonly in the form of radiating clusters of needles. (◊ *Pneumatolysis*, Tourmalinisation; *Cyclosilicates*.)

Schorlomite. ◊ *Garnet*.

Schuppen structure. ◊ *Fault* (Thrust fault).

Scleractinia. ◊ *Coelenterata* (Anthozoa).

Scolecodonts. ◊ *Annelida*.

Scolecoid. ◊ *Coelenterata*.

Scoriaceous. A term used to describe a ◊ *lava* or ◊ *pyroclastic rock* containing empty cavities, or ◊ *amygdales*.

Scourian. A stratigraphic division of the Scottish ◊ *Precambrian* older ◊ *Lewisian*. The division between it and the ◊ *Laxfordian* is based on the number of orogenies affecting these groups.

Scree (Talus). The accumulation formed by the fragments resulting from the mechanical ◊ *weathering* of rocks. The mass of scree is formed more or less *in situ* or as a result of transport by gravity over a short distance. Scree usually forms heaps of coarse debris at the foot of cliffs and steep slopes, the free face (◊ *Slope*) adopting the angle of repose for the material, usually 25–35°. The finest particles are usually removed from a scree by percolating water. A cemented scree is a ◊ *breccia.*

Scrope, George Poulett (1797–1876). English geologist and M.P. whose ◊ *uniformitarian* account of the origin and composition of volcanoes is basic to modern attitudes. He was one of the first to make a scientific study of primary magma, and produced a classic account of the vulcanicity of the Auvergne district of France.

Scyelite. A ◊ *hornblende-◊biotite* peridotite (◊ *Ultrabasic rocks*) in which the biotite ◊ *poikilitically* encloses the olivine.

Scyphozoa. ◊ *Coelenterata.*

Scythian (Skythian). A synonym of ◊ *bunter.*

Seat earth. A type of fossil ◊ *soil.*

Sebkha. Alternative spelling of ◊ *sabkha.*

Secondary. (1) An obsolete name for the ◊ *Mesozoic* Era.
(2) A term approximately equivalent to 'late' or 'subsequent', e.g. as in the phrase 'secondary alteration products' especially referring to ◊ *weathering* or ◊ *hydrothermal* action. It may also imply a second generation of a mineral, e.g. secondary calcite. (◊◊ *Secondary enrichment.*)

Secondary enrichment. Secondary (supergene) enrichment is a term used especially of ore deposits and is applied to parts of the ore body in which the metal content of the ore (or, in the case of non-metallic material, the valuable mineral content) has been increased as a result of the downward percolation of waters carrying material in solution. The additional material may (a) be deposited as a separate new mineral, (b) be deposited as a further deposition of the existing ore mineral, or (c) replace the original ore mineral by one richer in the valuable constituent. As an example, a vein containing chalcopyrite – $CuFeS_2$ (34·5%Cu) – may be enriched either by deposition of more chalcopyrite, by deposition of bornite – Cu_5FeS_4 (63·3% Cu) – or by conversion of the chalcopyrite to chalcocite – Cu_2S (79·8% Cu). (◊ Fig. 133.)

Sedgwick, Adam (1785–1873). Cambridge professor of geology who is remembered chiefly for his work on what he termed the Cambrian series. He disentangled the structure of the Lake District and North Wales, but could find no palaeontological proof of the independence of these strata from ◊ *Murchison*'s Lower Silurian. He was unable to convince his colleagues, particularly Murchison, of the validity of his theories, and the problem was only resolved after his death by ◊ *Lapworth*.

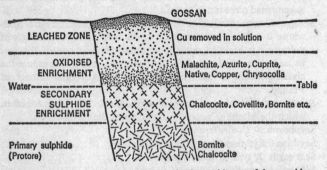

133. Secondary enrichment. Gossan is the residuum of iron oxides (limonite) from which mobile elements have been removed, e.g. copper, sulphur, as sulphates, etc. In the leached zone the sulphides are oxidised to sulphates and transported in solution. In the zone of oxidised enrichment, the reaction of the sulphate solution with the original ore in an oxygen-rich environment (carbon dioxide is also usually present) results in the formation of carbonates and oxides, native metals, and (rarely) silicates. Below the water table, secondary sulphide enrichment occurs in an oxygen-free environment.

Sedimentary iron ores. Sedimentary iron ores are of considerable economic importance and supply a major proportion of iron ore. They may be classified as:

(1) Detrital (black sands – ◊ *Arenaceous rocks*)
(2) Residual (◊ *Laterite*)
(3) ◊ *Replacement* and late ◊ *diagenetic*
(4) Marine
(5) Freshwater

Groups 4 and 5 include both ◊ *primary* and early ◊ *diagenetic* types. Sedimentary iron pyrites is not regarded as iron ore. The role of bacteria and possibly other organisms in the deposition of iron minerals in sediments is not fully understood. The main minerals found in sedimentary iron ores are ◊ *limonite*, ◊ *siderite* and chamosite, an oxy-◊*chlorite* having a characteristic

greenish colour. ◊ *Haematite* and ◊ *magnetite* also occur in undoubted sediments but in many cases their presence marks a
degree of metamorphism.

The main requirement for the formation of a primary iron sediment is a supply of iron-rich water. Diagenetic and replacement
ores arise by reaction with iron solutions, either contained in the
pores of a sediment or percolating from above. Very many sedimentary iron ores are oolitic; ◊ *ooliths* have been reported consisting of most of the minerals mentioned above and sometimes
with two or more of them in alternating layers.

In Britain there is a concentration of sedimentary iron ores in the
◊ *Jurassic System*, with some in the Lower ◊ *Cretaceous*, but
sedimentary iron ores of a wide variety of ages are known.

Freshwater iron ores are rarely oolitic and include bog limonites,
◊ *clay ironstones*, and ◊ *black-band ironstones*.

Sedimentary rocks. Rocks formed from material derived from pre-
existing rocks by processes of ◊ *denudation*, together with
material of organic origin. The term includes both consolidated
and unconsolidated material; the latter is also referred to as
'sediment'. They are classified on the basis of the character of the
material and the process which leads to its deposition:

Clastic	*Organic*	*Chemical*
◊ *Arenaceous*	◊ *Limestones* (in part)	Limestones (in part)
◊ *Argillaceous*	◊ *Abyssal deposits* (in part)	◊ *Evaporites*
◊ *Rudaceous*	◊ *Bone beds*	◊ *Sedimentary iron ores*
	◊ *Chert* (in part)	Chert (in part)
	◊ *Coal*	
	◊ *Phosphatic deposits*	

The process of conversion of unconsolidated sediments to
coherent sedimentary rocks is called ◊ *diagenesis*. A high proportion of sedimentary rocks contain contributions from all three
sources listed above, usually with one kind dominant. Sedimentary
material may also be classified on a geochemical basis:

RESISTATES (Si) correspond to the arenaceous and rudaceous
rocks.

HYDROLYSATES (Al, Si, Fe″) are mainly the clay minerals
and correspond to the argillaceous rocks.

OXIDATES (Fe‴, Mn‴) are the sedimentary iron and manganese ores.

REDUZATES (Fe″, S, C) are the sedimentary sulphides, coal and petroleum.

PRECIPITATES (Ca, Mg) are the chemically formed limestones.

EVAPORATES (Na, K, Ca,Mg) are the evaporites.

The elements indicated in each class are the most important ones present; oxygen is present in all classes except the reduzates. There is a considerable element of environmental control of sedimentation, and in broad terms a characteristic sedimentary assemblage exists for each environment. Some of the more important are:

Environment	Sediments
Geosynclines	Greywackes (◊ *Arenaceous rocks*), dark shales (◊ *Argillaceous rocks*), ◊ *polymict* ◊ *conglomerates*. (Often ◊ *graded* beds)
Shelf seas	Orthoquartzites (◊ *Arenaceous rocks*), ◊ *limestones*, shales, ◊ *oligomict* conglomerates. (Often ◊ *cross-bedded*)
Basins with a restricted connection to the ocean	Black shales, often pyritic
Basins not receiving clastic sediments where evaporation is greater than the inflow of water	◊ *Evaporites*
Basins on continental margins subject to intermittent marine invasion	Rhythmic deposition of sandstones and shales etc. (Coal measures)
Piedmont areas, intermontane basins	Arkoses (◊ *Arenaceous rocks*), ◊ *breccias*, red sandstones
Deltas	Sandstones and shales showing cross-bedding and lensing
Deserts	◊ *Loess*, dune-bedded sandstones
Ice-sheet margins	◊ *Till*, ◊ *varved* clays, sands and gravels (◊ *Loess*)
Lakes	Evaporites (some), clays and sandstones
Oceanic depths	Oozes (◊ *Abyssal deposits*)

Estimates of the relative proportions of sediments in the Earth's crust based on measurements of stratigraphic sections are as follows: shale 47%; sandstone 30%; limestone 22%; others

1%. Those based on geochemical calculations, however, are: shale 77%; sandstone 13%; limestone 10%. The discrepancies between these two sets of figures (which are based on averages of the most recent estimates) arise partly because rock names tend to be over-simplifications (e.g. many rocks recorded as limestone contain appreciable quantities of sand and/or clay), and partly because clay material is extensively lost to the deeper parts of the oceans. Re-working of metamorphosed sediments may also introduce errors into the calculations of the geochemical data. (For sedimentary textures, ◊ *Grade*; *Particle shape*; *Particle size*; *Roundness*; *Sphericity*.)

Sedimentary structures. In recent years a voluminous vocabulary of terms relating to the structural features of sedimentary rocks has grown up. In general, the terms used relate to processes taking place during, or immediately after, deposition of a sediment, prior to ◊ *diagenesis*. The following outline classification of sedimentary structures has been adapted from the work quoted at the end of the entry.

(1) EXTERNAL FORM OF THE BEDDING, i.e. the shape, thickness and continuity of sedimentary units. ◊ *Bedding* is described in terms of the thickness of layers, the variation of thickness, both in an absolute sense and in relation to variation in thickness of other layers. The shape of the body of sediment, e.g. sandstones, may be thin and extensive in one horizontal dimension – the so-called shoe-string sands – or may be thin and horizontally extensive in two dimensions – a blanket sand. Bodies in which the ratio of thickness to areal extent lies between these two extremes are called prismatic bodies. Continuity of deposition is also important.

(2) INTERNAL STRUCTURE OF BEDS, i.e. the various types of ◊ *bedding*, including ◊ *cross-bedding*, ◊ *graded bedding*, and sedimentary fabrics such as ◊ *imbricated* pebble beds.

(3) BEDDING PLANE AND SOLE MARKINGS, covering a wide variety of markings and surface irregularities to be seen on the top and bottom of a bed. (◊ *Ripple marks*.) Amongst the commoner markings of the upper surfaces are organic marks (tracks, trails, footprints, etc.), raindrop pits (◊ *Rain prints*), desiccation cracks or sun cracks (although desiccation is not necessarily associated with heat). The term sole structure is used for a structure preserved on the base of a bed which is sharply differentiated lithologically from the bed below, e.g. a sandstone overlying a shale. Some of the features referred to previously may occur as

sole structures. (◊ Fig. 134.) Two structures especially associated with ◊ *turbidites* are (a) flute casts – grooves eroded by turbulent flow and subsequently filled with coarser sediment; they are usually seen as bulbous or ramifying casts; and (b) tool marks – which include grooves, bounce marks, prod marks, etc. – pro-

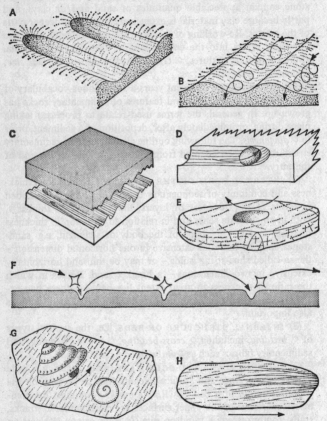

134. Sole markings. (A) Flute casts; (B) Flute cast formation; (C) Groove casts and marks; (D) Groove formation; (E) The bounce mark of a pebble; (F) Prod marks produced by ◊ *saltation* transport of a vertebra; (G) A roll mark of a gastropod and a prod mark of the spire of a gastropod; (H) A brush mark produced by trailing seaweed.

duced by an object ('tool') such as a pebble, shell, bone, even masses of seaweed, being dragged or bounced along the sea floor. Many complex patterns are formed in this way and are found commonly preserved as casts. Many of the above features may be subsequently deformed by loading, which may itself produce characteristic features (◊ *Load cast*).

(4) PENECONTEMPORANEOUS DEFORMATION OF BEDDING, covering all kinds of deformation, ranging from a slight distortion of the upper surface of the bed, due to the weight of the overlying sediment, to complete destruction of the original organisation to give contorted bedding. Four main types may be recognised: (a) ◊ *slumping*, (b) ◊ *collapse* structures, (c) ◊ *injection structures*, (d) ◊ *convolute bedding*.
P. E. Potter and F. J. Pettijohn, *Atlas and Glossary of Primary Sedimentary Structures*, 1964.

Segregation. A process which tends to produce a local concentration of one mineral or group of minerals in a rock of contrasting mineralogy. Examples are (a) in sedimentary rocks, lenses and streaks of ◊ *heavy minerals*, such as magnetite, in a sandstone; calcareous ◊ *nodules* or ◊ *doggers*; (b) in igneous rocks, an early formed mineral, e.g. magnetite or olivine, concentrated in one part of an igneous rock mass; (c) in metamorphic rocks, diffusion of ions resulting in the development of a concentration of a mineral or minerals in a limited number of sites, e.g. certain types of ◊ *gneiss* develop clots or lenses of quartzo-feldspathic material in the normal banded structure. (◊ *Differentiation*.)

Seif. A longitudinal ◊ *dune*.

Selenizone (Mollusca). ◊ Fig. 100.

Senecian. A stratigraphic stage name for the base of the North American Upper ◊ *Devonian*.

Senonian. A stratigraphic stage name for the European Upper ◊ *Cretaceous*.

Sensitive tint (Gypsum plate). ◊ *Accessory plate*.

Septum. ◊ Fig. 13 (Brachiopoda); *Coelenterata* (Anthozoa) and Figs. 20–23; Fig. 106 (Mollusca).

Sequanian. A stratigraphic stage name for the European Upper ◊ *Corallian* (Jurassic System).

Sericite. ◊ *Micas*.

Series. ◊ *Stratigraphic nomenclature*.

Serpentine. (1) A ◊ *layer-lattice mineral*, $Mg_6Si_4O_{10}(OH)_8$, of which two forms occur, a fibrous one known as chrysotile, and a lamellar one, antigorite. Serpentine is the main alteration product of ◊

olivines and ◊ *pyroxenes*. Cleavage flakes and fibres of serpentine are flexible but not elastic. Nickel-bearing serpentine is garnierite, while iron-rich serpentine (rare) is greenalite – both are ore minerals. Chrysotile is perhaps the most important of the asbestiform minerals of commerce (◊ *Asbestos*). (◊ Appendix.) (2) A rock (more correctly serpentinite). Serpentinites are named according to the parent rock, e.g. dunite serpentine (from olivine rock), bastite serpentine (from hyperstone-rich rocks). Serpentine rocks are cut, turned, and polished as ornamental material. The process of serpentinisation is essentially one of auto◊ *metamorphism*, by late stage hydrothermal action on ◊ *ultrabasic rocks*, especially ◊ *olivine* and orthorhombic ◊ *pyroxene*-rich types.

$$4Mg_2SiO_4 + 4H_2O + 2CO_2 = Mg_6Si_4O_{10}(OH)_8 + 2MgCO_3$$

Olivine   Serpentine   Magnesite

Common associates of serpentine in such rocks are various carbonates, and iron oxides; veins of ◊ *chrysolite* and ◊ *asbestos* often occur. ◊ *Marbles* containing serpentine after forsterite (olivine) are known as ophicalcites.

Serpentinite. ◊ *Serpentine* (2).

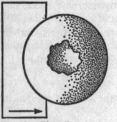 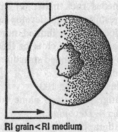

Direction in which card is introduced below stage

RI grain > RI medium   RI grain < RI medium

135. The shadow test.

Sessile. A term used to describe a non-mobile organism, which may or may not be fixed. (◊ *Benthos*.)

Shadow test. A useful technique for determining the relative refractive index of a grain and the medium in which it is mounted. It is sometimes called the oblique illumination test. The effect is as follows (◊ Fig. 135): If a piece of card is inserted immediately below the stage of a microscope (the condenser lens is best removed), one half of the field of view will appear darkened and the mineral grains will acquire a shadowed edge. For a mineral with a higher refractive index than the mounting medium, the shadow edge is on the same side as the card, whereas a mineral with a lower refractive index will have the shadow on the opposite side to the card. (Cf. ◊ *Becke test*.)

Shale. ◊ *Argillaceous rocks*.

Shape. ◊ *Particle shape*.

Shard. Abraded fragments of pumice. (◊ *Pyroclastic rocks*.)

Shear. ◊ *Stress*.

Shear cleavage. A fracture cleavage. (◊ *Cleavage, rock*.)

Shelf facies. A term applied to those sediments and their associated fauna which are found in the shallower, marginal parts of a ◊ *basin* or geosyncline.

Shield. A major structural unit of the Earth's crust, consisting of a large mass of ◊ *Precambrian* rocks, both metamorphic and igneous, which have remained unaffected by later ◊ *orogenies*. (◊ *Positive area*.) Well-known examples are the Canadian and Baltic Shields. The term is practically synonymous with ◊ *kraton*.

Shingle. Material of gravel or pebble grade (◊ *Particle size*) accumulated on beaches or off-shore bars.

Shoe-string sand. A body of sediment which is thin and extensive in one horizontal dimension. (◊ *Sedimentary structures*.)

Shonkinite. A ◊ *melanocratic* feldspar-rich ◊ *syenite*, consisting largely of ◊ *pyroxenes*, a little ◊ *olivine*, and potassic feldspars (traces of ◊ *nepheline* may also occur).

Shoulder (Mollusca). ◊ Fig. 105.

Shutter valley. A valley which has been affected by a tear ◊ *fault* at right angles to the axis of the valley so that the two halves no longer coincide.

Sial. The upper portion of the ◊ *crust of the Earth*, composed predominantly of *si*lica and *al*uminium.

Sicula. ◊ *Chordata* (Graptolithina) and Fig. 18.

Siderite. (1) A ◊ *meteorite* composed entirely of metal.
(2) (Mineral) $FeCO_3$. (◊ *Carbonates*; ◊ Appendix.)

Sidero- (prefix). Iron.

Siderolite. A *meteorite* consisting of both metal and silicate.

Siderophile. Elements with weak affinities for oxygen and sulphur, but soluble in molten iron. They are presumed to be concentrated in the ◊ *core of the Earth*, and are found in metal phases of ◊ *meteorites*. (◊ *Geochemistry*.)

Siegenian. A stratigraphic stage name for the European mid Lower ◊ *Devonian*.

Silesian. A stratigraphic stage name for the European Upper ◊ *Carboniferous*.

Silica group of minerals. Although silica is chemically an oxide – SiO_2 – the general structures and properties of its various forms are more closely allied to those of the silicates. The silica group of minerals have structures consisting of three-dimensional lattices of SiO_4 tetrahedra, in which all four oxygens of each tetrahedron are shared by adjoining tetrahedra. This of course does not apply to the natural silica glass, lechatelierite. There are three main crystalline ◊ *polymorphs* of silica (quartz, tridymite, and cristobalite), several high-pressure varieties (notably coesite), one crypto-crystalline variety (chalcedony), and an amorphous variety (opal):

QUARTZ. Common quartz ($α$ or low quartz) is stable up to 573°C., and is a common mineral in all kinds of rocks and mineral veins. Crystallographically, quartz is trigonal trapezohedral. When quartz crystals occur showing trapezohedral faces, a division can be made into left-handed and right-handed according to the position of the 'indicator' faces. Many coloured varieties have been named:

Violet/purple	Amethyst
Brown	Cairngorm
Yellow	Citrine
Pink	Rose quartz
White	Milky quartz
Clear/Transparent	Rock crystal

A characteristic feature of quartz is the absence of ◊ *cleavage* and the presence of a well-developed conchoidal fracture; it is the standard for ◊ *hardness* 7 on Mohs' scale.

At 573°C. low quartz gives place to high quartz ($β$ quartz), which is hexagonal trapezohedral. When high quartz occurs in crystals, the prism faces (common in low quartz) are absent. Many of the reported occurrences of $β$ quartz crystals have been found to consist of pseudomorphs of low quartz after high quartz. High quartz has been reliably reported in acid volcanic rocks, where it is presumably metastable.

Quartz in the form of sand has a vast number of well-known uses, e.g. in glass-making, as an abrasive, and as a constituent of concrete. Optically clear quartz is used in the manufacture of lenses, prisms, etc., while some of the coloured varieties may be cut and used as semi-precious stones; specially cut thin plates of quartz are used as oscillators in electronics.

TRIDYMITE. Above 870°C. high quartz inverts to tridymite,

which is very probably hexagonal holohedral. Tridymite is only found in acid volcanic rocks, where it is metastable. In other cases, quartz ◊ *pseudomorphs* after tridymite are known.

CRISTOBALITE. At 1,470°C. tridymite inverts to cristobalite, which is cubic. It is probably even rarer than tridymite, and occurs in somewhat similar environments.

COESITE. A monoclinic polymorph of silica which develops at very high pressures (20 kilobars). Although originating as a laboratory substance, it was subsequently found in quartzose rocks in craters formed by the impact of large meteorites.

CHALCEDONY. A ◊ *cryptocrystalline* variety of silica, consisting essentially of fibrous or ultrafine quartz, some opal, together with water, which is either enclosed in the lattice or in the macrostructure of the mineral. It is possible that some of the quartz has had oxygen ions replaced by hydroxyl ions. A very large number of varieties of chalcedony occur. Banded varieties include agate, onyx and sardonyx; reddish or brownish varieties are called sard or carnelian, while green varieties are called prase or chrysoprase. Jasper is a red ◊ *chert*-like variety. Chalcedony is usually regarded as a low-temperature material, occurring mainly in sediments, low-temperature ◊ *hydrothermal* veins, and as an ◊ *amygdale* filling. Several varieties of chalcedony are used as semi-precious stones.

OPAL. A hydrated amorphous variety of silica, probably derived from a silica gel. It contains more water than chalcedony and is considerably softer than quartz. Opal occurs mainly as a secondary deposit formed by the action of percolating ground waters; shells of various types are known in which replacement of the shell material by opal has occurred. Silica deposited by hot springs (siliceous sinter and geyserite) is opaline in character. Sponges, radiolaria, and diatoms secrete opaline skeletons. Precious opal is a well-known gemstone showing a characteristic play of colours. ◊ *Diatomite* (kieselguhr), a rock made up almost entirely of diatom skeletons, is important as an insulator, an abrasive, and a filtering agent. (◊ Appendix.)

Silicates. Silicates are the most important group of compounds occurring in the crust of the Earth, and probably make up 95% of the crust (if one counts the ◊ *silica group*, SiO_2, as silicate). They are classified according to their atomic structure; as listed in the table overleaf, the various structures may be regarded as being derived from a tetrahedral unit, SiO_4, by linking them together with the elimination of an oxygen at each linkage.

413

Name	Structural Group	Unit	Example	Typical Formula
Nesosilicates	Independent tetrahedra	SiO_4	◊ Olivines	Mg_2SiO_4
Sorosilicates	Two tetrahedra sharing one oxygen	Si_2O_7	Melilite	$Ca_2MgSi_2O_7$
Cyclosilicates	Closed rings of tetrahedra each sharing two oxygens	$(SiO_3)_n$ $n = 3,$ $4,6$	◊ Beryl (6-fold) ◊ Axinite (4-fold) ◊ Benitoite (3-fold)	$Be_3Al_2(SiO_3)_6$ $Ca_2(Mn,Fe'')Al_2BO_3(SiO_3)_4(OH)$ $BaTi (SiO_3)_3$
Inosilicates	(a) Continuous single chains of tetrahedra, each sharing two oxygens (b) Continuous double chains (ribbons) of tetrahedra alternately sharing two and three oxygens	$(SiO_3)\infty$ Si_4O_{11}	◊ Pyroxenes ◊ Pyroxenoids ◊ Amphiboles	$MgSiO_3$ $CaSiO_3$ $Mg_7Si_8O_{22}(OH)_2$
Phyllosilicates	Continuous sheets of tetrahedra sharing three oxygens	Si_4O_{10}	◊ Micas ◊ Talc	$KAl_2(Si_3Al)O_{10}(OH,F)_2$ $Mg_3Si_4O_{10}(OH)_2$
Tektosilicates	Three-dimensional framework of tetrahedra with all four oxygen atoms shared.	SiO_2	Quartz ◊ Feldspars	SiO_2 $KAlSi_3O_8$

The ability of Al to replace Si in the tetrahedral unit greatly increases the number of possible types of silicates; normally an SiO_4 tetrahedra possesses four excess negative charges, which have to be balanced by four units of positive charge provided by cations. If Al substitutes for Si, an additional unit of negative charge is 'available', thus allowing additional or different cations to enter the structure. More complicated structures arise by the inclusion of two or more silicate 'units', e.g. idocrase and epidote contain both SiO_4 and Si_2O_7 units. ($\Diamond$ Fig. 136.)

($\Diamond$ *Aluminium silicates*; *Amphiboles*; *Chlorite*; *Clay minerals*; *Cyclosilicates*; *Epidotes*; *Feldspars*; *Feldspathoids*; *Garnets*; *Layer-lattice minerals*; *Micas*; *Olivines*; *Pyroxenes*; *Pyroxenoids*; *Silica group of minerals*; *Zeolites*.)

Siliceous sinter. Silica deposited by hot springs. ($\Diamond$ *Silica group of minerals*; *Volatiles*.)

Silicification. The process of introducing silica (commonly $\Diamond$ *crypto-crystalline*) into a non-siliceous rock, either by filling pore spaces or as a $\Diamond$ *replacement*, e.g. of calcite in limestones. The silica may be introduced either by ground-water solutions or from igneous sources. ($\Diamond$ *Fossils*; *Silica group of minerals*.)

Sill. A sheet-like body of igneous rock which conforms ($\Diamond$ *Conformable*) to bedding or other structural planes. (Cf. $\Diamond$ *Dyke*.) The 'original' sill is the Great Whin Sill of Northern England, which has a very wide extent, underlying much of Northumberland and Durham. Sills may be $\Diamond$ *composite* or $\Diamond$ *multiple*. Sills which have a small area may be confined to a single plane, but those which have a large geographical extent can commonly be found occupying different planes in different places. Sills of this kind are said to be transgressive. Sills my be $\Diamond$ *differentiated*, and it has been suggested that $\Diamond$ *laccoliths* may grade into sills by increasing lateral extent and decreasing convexity upwards. This has also been claimed for $\Diamond$ *lopoliths* and $\Diamond$ *phacoliths*. Sills are normally medium-grained but large ones may be coarse enough to be described as plutonic. Basic sills ($\Diamond$ *Dolerite*) are by far the commonest. ($\Diamond$ Fig. 137; $\Diamond$ *Lava*.)

Sillimanite. $\Diamond$ *Aluminium silicates* and Appendix.

Silt. $\Diamond$ *Argillaceous rocks*.

Silurian System. Named from the Silures, an ancient Celtic tribe of the Welsh Borderland. The period extended from 435 to 395 m.y., having a duration of 40 m.y. The lowest beds of the Silurian are the Lower Llandovery Series or Valentian, defined by the graptolite zone Glyptograptus persculptus. The upper limit is the

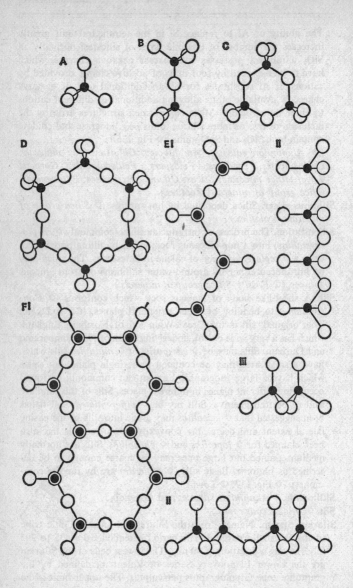

A   B   C

D   EI   II

FI   III

II

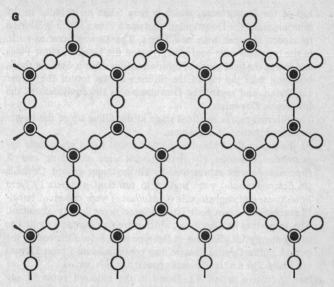

136. Silicates. (A) Neosilicate; (B) Sorosilicate; (C) A three-membered-ring cyclosilicate; (D) A six-membered-ring cyclosilicate; (E) A single-chain inosilicate. i, viewed along the a-axis; ii, viewed along the b-axis; iii, viewed along the c-axis; (F) A ribbon ino-silicate. i, viewed along the a-axis; ii, viewed along the c-axis; (G) A phyllosilicate.

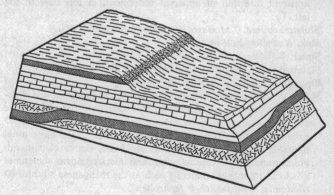

137. A sill.

top of the Downtonian, which is here taken as included in the Silurian, since the Downtonian contains a fauna which is Silurian in aspect, together with new forms. The Downtonian, as such, in the type area rests conformably upon the Upper Ludlow Flags and marks the final phase of Silurian sedimentation. Some authors, however, take the top of the Silurian as the top of the Upper Ludlovian, and regard the Downtonian as the equivalent of the lowermost Devonian.

The Silurian marks the final stage in the filling up of the Lower Palaeozoic basins of deposition.

At the top of the Valentian the first jawed fishes made their appearance. Trilobites (◊ *Arthropoda*) were abundant and ◊ *Brachiopoda* were represented by all the major groups. Crinoids (◊ *Echinodermata*) were present in sufficient numbers to form limestones, and cephalopods (◊ *Mollusca*) were common; lamellibranchs and gastropods (◊ *Mollusca*) were locally abundant. The graptolites (◊ *Chordata*), which were represented by varieties of Monograptids, died out in the type area before the end of the period, but in Central Europe they continued into Lower Devonian times. The first land plants appeared in this period.

Silver. A native metal, Ag, found in the oxidised zones of ore deposits and in ◊ *hydrothermal* veins. (◊ Appendix.)

Sima. The lower portion of the ◊ *crust of the Earth*, composed predominantly of *si*lica and *ma*gnesium.

Sinemurian. A stratigraphic stage name for the base of the European Lower ◊ *Jurassic*.

Sinistral. A term applied to tear (strike) ◊ *faults* to describe the apparent direction of apparent movement, in this case to the left.

Sinistral coiling. ◊ *Mollusca* (Cephalopoda) and Fig. 100.

Sink. A swallow-hole or ◊ *doline*.

Sinter. ◊ *Siliceous sinter*.

Siphonal canal, anterior and posterior (Mollusca). ◊ Fig. 100.

Siphuncle (Mollusca). ◊ Figs. 104–106.

Skarn. A thermally ◊ *metamorphosed* impure limestone (or dolomite) in which ◊ *metasomatism* has also occurred. Skarns are generally characterised by the presence of minerals such as ◊ *axinite* or datolite (CaBSiO$_4$(OH)) in addition to the more usual minerals of a ◊ *calc-silicate hornfels*. Many skarns contain sulphide minerals of various sorts, and metasomatism sometimes gives rise to unusual minerals such as the manganese silicates (◊ *rhodonite*, ◊ *bustamite*, ◊ *tephroite*).

Tactite is nearly synonymous with skarn, the term being used mainly for metasomatic calc-silicate rocks formed directly at an igneous contact. The term skarn is no longer reserved exclusively for iron-rich calc-silicate rocks (i.e. those containing heden-bergite and iron-garnet in place of diopside and magnesium-garnet).

Skiddavian. A local stratigraphic stage name for the ◊ *Arenigian* (◊ *Ordovician*)+part of the ◊ *Cambrian* in Great Britain.

Skythian (Scythian). A synonym of ◊ *Bunter* (Triassic System).

s.l. ('sensu latu'). In the broad sense, as opposed to s.s., 'sensu strictu' (in the strict sense).

Slates. Low-grade regionally metamorphosed ◊ *argillaceous rocks* which have developed a well-marked ◊ *cleavage* but have suf-fered little recrystallisation, so that the rocks are still very fine-grained. Where slates are subjected to thermal ◊ *metamorphism* they may develop spots or clots of incipient new minerals (spotted or knotted slates), or recognisable crystals of new minerals such as ◊ *pyrite* or the chiastolite variety of ◊ *andalusite*. The term slate appears therefore as part of the name of certain thermally metamorphosed rocks, but it should be noted that this is purely because of their cleavage. Thermal metamorphism does not pro-duce slaty cleavage; in fact it tends to destroy it, with the pro-duction of ◊ *hornfels*. Slate is an important roofing material.

Slaty cleavage. ◊ *Cleavage, rock.*

Slickensides. When one surface of a rock moves over another sur-face in close contact under pressure, the two surfaces develop a kind of polish, with linear grooves and ridges parallel to the direction of movement. This is termed slickensiding and is com-monly seen on fault surfaces. It should be noted that slickensides

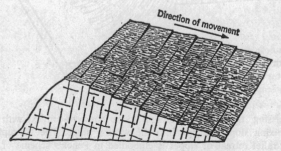

138. A slickensided surface.

may develop with surprisingly little movement if the rock type is suitable. In many cases the slickensided surfaces will be found to be coated with a ◊ *layer-lattice mineral* produced by the breakdown of the original rock. The grooves and ridges tend to stop abruptly in the form of minute steps facing the direction of movement. Hence, if one runs one's fingers over the surface in the direction of movement it will feel smooth, whereas against the direction of movement the slickensided surface will feel rough and irregular. (◊ Fig. 138.)

Slide. Low-angle (nearly horizontal) ◊ *fault.*

Slope. There are four recognised elements of hillside slopes: (1) A convex or waxing slope; (2) A free face or outcrop of bare rock; (3) A uniform slope of talus or detrital material; (4) A concave or waning slope – the pediment. (◊ Fig. 139.) It is possible for one or more of these elements to be missing, most commonly the free face, since it depends for its existence on a hard, resistant band of rock. If both the free face and the constant slope are

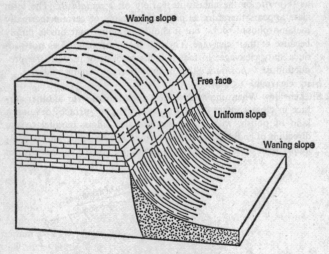

139. Slope terminology.

missing, the landscape is said to be concavo-convex, with the waxing slope merging into the waning slope. The concept of parallel retreat of slopes is embodied in Penck's scheme of the ◊ *cycle of erosion.*

Slump. (1) ◊ *Gravity transport.*

(2) As a sedimentary structure, the contorted structures produced by the movement of a mass of incoherent sediment down a slope. The term slump-bedding is often loosely applied to any sort of contorted bedding whether produced by slumping in the strictest sense or not.

Smectite group of minerals. The montmorillonite group of ◊ *clay minerals.*

Smith, William (1769–1839). One of the most outstanding practical British geologists, he first recognised the importance of fossils in identifying the chronology of rock-strata on his country-wide travels as a surveyor for the construction of canals and bridges. All the information he collected he eventually combined in his classic 1815 map of the British Isles, and in the books published afterwards which related drawings of fossils to the strata in which they were found. His work is all the more remarkable for his being self-taught, and receiving little professional or financial support from others in carrying it out. (◊ *Law of Strata Identified by Fossils*; *Superposition, Principle of.*)

Smithsonite. ◊ *Carbonates* (Hexagonal) and Appendix.

Soapstone. The popular name for any very soft rock having a greasy 'feel'. Most soapstones are ◊ *talc* rocks.

Sodalite. ◊ *Feldspathoids.* (◊ Appendix.)

Soil. Soil is regarded by the geologist as the accumulation of loose ◊ *weathered* material which covers much of the land-surface of the Earth to a depth ranging from a fraction of an inch to many feet. Between the soil proper and the ◊ *bedrock* is a layer of shattered and/or partly weathered rock – the subsoil. The term ◊ *regolith* is a convenient one to cover both soil and subsoil.

A vertical section through the soil–subsoil–bedrock sequence is termed a soil profile. In the nomenclature of soil scientists (pedologists) the D horizon corresponds to bedrock and the C horizon to the subsoil. The soil proper is divided into an upper A horizon and a lower B horizon, which are further subdivided as required. The upper part of the A horizon contains much organic matter and is strongly leached; the lower portion of the A horizon has much less organic matter and has suffered maximum leaching – the leaching (eluviation) mainly removes Ca, Fe, etc. The B horizon is largely a zone of deposition (illuviation) of leached material and fine clay and silt particles.

Soil is essentially a mixture, in varying proportions, of organic matter (largely vegetable) called humus, and inorganic (mineral) particles derived by weathering from rocks. The inorganic part of a soil may be derived *in situ* or as a result of the transport of debris from elsewhere.

Because weathering and vegetation cover are so closely controlled by climatic factors, the distribution of soil types is markedly parallel to the distribution of climatic types. However, even under uniform climatic conditions, soils will vary considerably as a result of differences in the underlying rocks.

Soils may be divided into: (a) ZONAL. Soils which are climate-controlled and fully developed so as to be in equilibrium with the weathering regime. They may be further subdivided into pedalfers, which are leached, formed in regions of high rainfall, and pedocals, which are unleached, containing $CaCO_3$ and essentially formed in regions of low rainfall. (b) INTRAZONAL. Modified zonal soils, containing unusual quantities (in relation to the general circumstances) of such things as water, soluble salts, iron oxides, etc. (c) AZONAL. Immature soils, not yet developed to a stage of equilibrium with the local weathering regime. Zonal soils tend to occur in large tracts, whereas intrazonal soils occur in rather localised areas.

Fossil soils are rare; the best examples are the seat-earths (◊ *gannisters* and ◊ *fire clays*) of ◊ *coal* seams. 'Dirt beds' not uncommonly occur in some fresh-water series, e.g. the Purbeck and Wealden of southern England.

The civil engineer uses the term soil (engineering soil) for any soft, unconsolidated, deformable material. Thus Tertiary sands and clays are 'soils' by this definition.

(For types of soil, ◊ *Brickearth*; *Brown soil*; *Chernozem*; *Glei*; *Laterite and bauxite*; *Loam*; *Loess*; *Peat*; *Podsol*; *Terra rossa*; *Tundra*.)

Sole markings. ◊ *Sedimentary structures.*

Solfatara. ◊ *Volatiles.*

Solid. A term used to describe ◊ *geological survey* or other maps from which the superficial deposits (drift, river gravel, etc.) are omitted, thus leaving the outcrops of the pre-superficial formations unobscured. (Cf. ◊ *Drift* (4).)

Solifluxion. The slow downhill movement of soil or ◊ *scree* cover as a result of the alternate freezing and thawing of the contained water. (◊ *Gravity transport.*)

Soma- (prefix). Body.

Sorby, Henry Clifton (1826–1909). British pioneer of the microscopic study of rocks. He made his first thin sections of rock in

1849, but despite publication in 1851 and 1858 his work received little recognition until he came into contact with ◊ *Zirkel*.

Sorosilicates. ◊ *Silicates*.

Spar. A miner's term for any white or light-coloured mineral which displays a good cleavage and a more or less vitreous ◊ *lustre*. The word is preserved in such names fluorspar, heavy-spar (barytes), Iceland or calc-spar (calcite), and feldspar.

Sparagmite. A Scandinavian term for a ◊ *Precambrian* arkosic sandstone (◊ *Arenaceous rocks*). Used also as a name for a late Precambrian formation in Scandinavia, which although largely 'sparagmatic' in the sense defined above contains conglomerates, breccias, quartzites and argillaceous limestones.

Sparite. ◊ *Limestone*.

Sparnacian. A stratigraphical stage name for the European Upper Palaeocene (◊ *Tertiary System*).

Specular iron (Specularite). The popular name for ◊ *haematite* (Fe_2O_3), crystallised in the form of brilliant, black, metallic crystals.

Spessartite. (1) ◊ *Lamprophyre*.
(2) A manganese ◊ *garnet*.

Spiralium (Brachiopoda). ◊ Fig. 13.

Sphalerite (Zinc blende). The main zinc ore, ZnS, found in ◊ *metasomatic* deposits with ◊ *galena*, in ◊ *hydrothermal* vein deposits, and in ◊ *replacement* deposits. (◊ Appendix.)

Sphene (Titanite). $CaTiSiO_5$, found as an ◊ *accessory* mineral in acid igneous rocks and in metamorphosed limestones. (◊ Appendix.)

Sphericity. A measure of the shape of a particle and also of the deviation of its shape from an equivalent sphere. (Cf. ◊ *Roundness*; ◊ *Particle shape*.) Strictly, it is measured as the ratio of the surface area of a sphere having the same volume as the fragment to the actual surface area of the fragment. For all solids other than a sphere this ratio is less than 1. However, since it is easier to measure lengths than surface areas, it is common practice to define sphericity as the ratio of the diameter of a sphere having the same volume as the particle to the diameter of a sphere circumscribing the particle. This also produces a value of less than 1 for non-spherical fragments. The nearer to 1 that the ratio approaches, the higher is the sphericity.

Spherulite, Spherulitic texture. A more or less globular mass of crystals, generally of an ◊ *acicular* habit, having a radial arrangement. Spherulites form as a result of the ◊ *devitrification* of volcanic ◊ *glass* and may occur in sufficient numbers to constitute a large part of some rocks. ◊ *Rhyolite*, ◊ *pitchstone*, and obsidian (◊ *Rhyolite*) not uncommonly show spherulitic texture. Spherulites are distinct from mineral aggregates filling ◊ *vesicles* or ◊ *amygdales*. Occasionally hollow spherulites may be found: to these the name 'lithophysae' has been applied. (◊ *Variolitic structure*.)

Spilite. A ◊ *basaltic* rock type, containing ◊ *chlorite* in lieu of ◊ *augite* and ◊ *olivine*, and ◊ *albite* as the ◊ *plagioclase*. Trachytic and rhyolitic types containing sodi-potassic feldspars, and sometimes quartz, have been termed keratophyres. Spilites and their associates are generally found as pillow ◊ *lavas* interbedded with ◊ *geosynclinal* sediments. Much controversy has raged concerning their origin; some authors regard them as primary types, some suggest that reaction has taken place between the lava and the sodium ions in sea water, and others consider that later ◊ *metasomatism* has occurred. The problem remains unsolved.

Spindle-bombs. ◊ *Pyroclastic rocks*.

Spine, genal, pleural and pygidial (Arthropoda). ◊ Fig. 3.

Spinel group of minerals. A group of cubic minerals with the general composition $R''R'''_2O_4$, where R'' may be Mg, Fe'', Mn'', and Zn, R''' may be Al, Fe''', Cr. Ti also occurs in some varieties. The commonest spinel mineral is ◊ *magnetite* ($Fe''Fe'''_2O_4$); spinel (s.s.) ($MgAl_2O_4$ – ◊ Appendix) and hercynite ($Fe''Al_2O_4$) are the end members of a continuous series which includes the varieties ceylonite and picotite. Other important spinels are ◊ *chromite* ($Fe''Cr_2O_4$), gahnite ($ZnAl_2O_4$), and franklinite (Zn,Fe'')Fe'''_2O_4. Spinels commonly occur as octahedral crystals. Magnetite is ubiquitous and chromite is a common constituent of ◊ *ultrabasic rocks*; these minerals are important ores of iron and chromium respectively. Transparent red spinel crystals are known as balas ruby. Spinels are most often found in metamorphic rocks or detrital sediments derived from them, apart from the occurrences mentioned above for magnetite and chromite.

Ulvöspinel (Fe''_2TiO_4) is a rare form, of rather different character, occurring mainly as exsolution patches or lamellae in magnetite. Hausmannite ($Mn''Mn'''_2O_4$) appears to be the Mn analogue of magnetite; in fact it has a distorted spinel structure and is tetragonal. Numerous synthetic spinels have been prepared containing ionic combinations which are unknown in nature.

The spinel lattice is capable of existing even though there is a deficiency in the number of cations. The best example of this is maghemite, which has a composition approximating to Fe_2O_3 (this may be written $Fe_{2.66}O_4$ – cf. magnetite Fe_3O_4 – an iron deficiency of 11%). The spinel structure of maghemite inverts to the haematite structure on heating. Maghemite occurs mainly as an alteration product of magnetite.

Spiracle (Echinodermata). ◊ Fig. 42

Spiralium (Brachiopoda). ◊ Fig. 13.

Spire, Spire angle (Mollusca, Gastropoda). ◊ Fig. 100.

Spit. Spits are formed along coasts where ◊ *longshore drift* is in progress. Where the coastline is indented by bays or where the coast changes direction, sediment being transported by longshore drift is carried on in more or less a straight line and deposited across the bay mouth, building up into a ridge above the surface

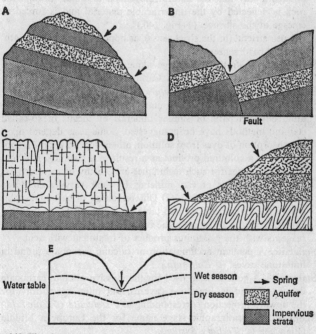

140. The formation of springs. (A) At the junction of strata; (B) At a fault; (C) As an emergence of underground drainage; (D) At an unconformity (below) and at an igneous rock layer (above); (E) A seasonal spring (bourne).

of the water, terminating in open water. A bar is a spit connecting two sides of a bay. Occasionally a cross-current will cause the spit to change direction, giving a recurved spit.

Spodumene. ◊ *Pyroxenes.*

Spondylium (Brachiopoda). ◊ Fig. 13.

Springs. A spring is water from an underground source issuing at the surface. Very small quantities of water may not actually flow but may simply create a local marsh or bog, and these may be called seepages. In contrast, cases are known where water fountains out of the ground in the form of an ◊ *artesian* spring. Intermittent or seasonal springs are known as bournes. The essential condition for the formation of a spring is the juxtaposition of a ◊ *permeable* rock mass and an impermeable rock mass in such a way that the natural movement of water through the rock is deflected by the impermeable material and is forced to emerge at the surface. (◊ Fig. 140.)

s.s. 'Sensu strictu' (in the strict sense), as opposed to s.l., 'sensu latu' (in the broad sense).

Staffordian. A stratigraphic stage name, the British equivalent of the European ◊ *Morgannian* (Carboniferous System).

Stage. ◊ *Stratigraphic nomenclature.*

Staining tests. At various times numerous techniques for differentiating between pairs of similar minerals by means of so-called staining methods have been suggested. Some tests depend upon the absorption of dyes from solution, others depend upon the production of a coloured product as a result of a chemical reaction. Three main uses for such techniques are worthy of mention: (1) The differentiation of clay minerals by their varying ability to absorb dyes from solution. (2) Differentiation of carbonate minerals by various reactions (see table opposite). (3) Distinction between various ◊ *silicates* on the basis of the reaction of dyes and reagents with the gelatinous product of treatment with acid.

Stalactite. A pendant accumulation of calcium carbonate found in limestone caves. (◊ *Calc tufa.*)

Stalagmite. A general term used for the massive calcium carbonate deposits found in limestone caverns, but more specifically for the upward projection corresponding to ◊ *stalactite.* (◊ *Calc tufa.*)

Stampian. A stratigraphic stage name for the European Middle Oligocene (◊ *Tertiary System*).

Standing water level. ◊ *Water table.*

Staurolite. ◊ *Aluminium silicates* and Appendix.

Steatite. ◊ *Talc.*

STAINING TESTS FOR CARBONATE MINERALS

	Alizarin Red-S	Potassium ferricyanide	Together
Aragonite	pale pink	—	—
Witherite	red	—	—
Calcite	pink to pale pink	—	pink to pale pink
Ferroan calcite	pink	turquoise	mauve-purple
Dolomite	—	—	—
Ferroan dolomite	—	turquoise	turquoise
Siderite	—	—	—

(Ferroan = iron-bearing)

The reagents are applied in dilute hydrochloric acid solution to thin sections or polished surfaces of the material.

Steinmann trinity. The supposedly 'inevitable' association of ◊ *spilites*, ◊ *serpentines*, and radiolarian ◊ *cherts* found in ◊ *geosynclinal* regions. Numerous examples are known however where one or two of the members are missing.

Steno, Nicolaus (1631–87). A Danish-born physician at the court of the Grand Duke of Tuscany, much famed as a scientist during his life although his views were cramped by his religious beliefs. Convinced of the organic origin of fossils, he attempted to draw up stratigraphic laws to explain this, and also produced some of the earliest attempts to explain the origin of unconformities and the influence of folding and faulting in orogenesis.

Stephanian. A stratigraphic stage name for the top of the European Upper ◊ *Carboniferous*.

Steptoe. An isolated hill surrounded by a lava flow, i.e. an upstanding part of the original relief projecting above the lava flow which has otherwise submerged it. The original example is Steptoe Butte, in the Columbia River lavas, Oregon, U.S.A. (Cf. ◊ *Nunatak*.)

Stereozone (Coelenterata). ◊ Figs. 21 and 23.

Stibnite. An antimony ore mineral, Sb_2S_3, found in low-temperature ◊ *hydrothermal* veins. (◊ Appendix.)

Stipe. ◊ *Chordata* (Graptolithina) and Figs. 17 and 18.

Stock. An intrusive mass of plutonic igneous rock smaller in size than a ◊ *batholith* and usually possessing a more or less circular or elliptical cross-section. The contacts with the ◊ *country rock* may vary from steep to low angle. (Cf. ◊ *Plug*.) Some stocks are probably ◊ *cupolas* of hidden batholiths.

Stockwork. A large-scale ramifying and dichotomising series of fissures filled with mineral matter. The term is not generally applied to a small-scale system such as might occur in a brecciated zone. If the veinlets contain an ore mineral it is generally necessary to work the mass as a whole, since the veins are too thin and too closely spaced to be worked individually. (◊ Fig. 141.)

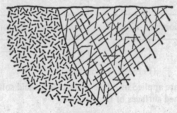

141. Stockwork.

Stolotheca. ◊ *Chordata* (Graptolithina) and Fig. 17.

Stomach stone (Gastrolith). Certain groups of reptiles, e.g. plesiosaurs, seem to have behaved in much the same way as chickens, but instead of swallowing small pieces of grit and retaining them in the gizzard, they seem to have swallowed quite sizeable stones, presumably as an aid to digestion. Examples have been found with skeletal remains, and isolated ones are occasionally found.

Stone. In geology the word 'stone' is admissible only in combinations such as limestone, sandstone, etc. or where it is used as the name for extracted material – building stone, road stone. It should not be used as a synonym for rock or pebble. (For 'stones' and 'stony', ◊ *Meteorites*.)

Stony-irons. ◊ *Meteorites* consisting of both metal and silicate.

Stoping. (1) In mining, a method of extracting ore from a vertical or steeply dipping vein by driving tunnels along the ◊ *strike* of a vein and extracting the ore from above and below the tunnel – overhand and underhand stoping respectively.

(2) A suggested mechanism for the emplacement of large masses of igneous rocks. The process is thought to operate by upwelling

◊ *magma* forcing its way along joints and other fissures in the ◊ *country rock*, causing blocks to become detached and sink into the magma, thus allowing igneous material to advance upwards. Evidence that such a process operates can be seen in some igneous contacts (◊ Fig. 142), but whether the process could emplace large

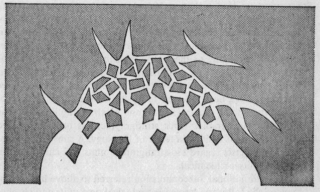

142. Stoping.

volumes of igneous rock by such piecemeal stoping is disputed. Objections which have been raised include doubts whether country rocks would have a high enough specific gravity to sink into the fairly mobile magma necessary for the process. It also seems unlikely that the stoped material would not, in fact, react with and be assimilated by the magma.

Stormberg. A stratigraphical term for the Upper ◊ *Karroo*.

Stoss. The less steep side of a ripple. (◊ *Cross-bedding*; *Ripple marks*.)

Strain. The deformation of rocks caused by ◊ *stress*. It may be dilation, a change in volume; distortion, a change of shape, or both. (◊ *Pressure*.)

Homogeneous strain (affine deformation) is deformation in which the particles move uniformly with respect to each other. This results in undistorted straight lines and planes, but circles become ellipses.

Heterogeneous strain (non-affine deformation) is deformation in which particles do not move uniformly with respect to each other, and thus straight lines and planes are contorted and folded. (◊ *Fold*, Shear folding; Fig. 73.)

Strain-slip cleavage. ◊ *Cleavage, rock.*

Stratification. ◊ *Stratum.*

Stratigraphic nomenclature. The stratigrapher has a peculiarly complex problem in attempting to set up a series of divisions and subdivisions of his material. These complexities are fully expressed in the various reports of the American Commission on Stratigraphic Nomenclature (1949–61) and the International Sub commission on Stratigraphic Terminology (1952 to date). From the earliest times geologists have used separate sets of terms for the division of geological time into convenient portions and for the rocks belonging to these time divisions. A complication has been introduced by the establishment of a separate set of terms based upon the fossil assemblages of the rocks. These three categories have been referred to as chrono-stratigraphic or geological time units, time-rock units and biostratigraphic units. The term litho-stratigraphic (rock stratigraphic) has been introduced to describe units defined in terms of lithology. A substantial number of geologists regard bio-stratigraphic units and chrono-stratigraphic units as identical.

The International Subcommission referred to above has recommended the following terms for geological time units and chrono-stratigraphic units:

Rank	Geological Time Units	Chrono-stratigraphical Units
1st Order	Era	(Erathem)
2nd Order	Period	System
3rd Order	Epoch	Series
4th Order	Age	Stage
5th Order	(Time)	Substage

(Terms in brackets are hardly ever used)

There appear to be no properly recognised biostratigraphical terms precisely equivalent to the high-order terms above; 'bioseries' has been suggested. However, it appears that the term 'zone', defined as rock strata characterised by a closely defined fossil content, corresponds approximately to the 5th-order terms. The term zone (or biozone) is, however, used informally to mean any

rock stratum which can be precisely defined in terms of palaeontological characters or properties, and even palaeontologically defined zones may be further subdivided into:

(1) Assemblage zones (coenozones, faunizones). Strata characterised by a faunal assemblage.

(2) Range zones (acrozones). Strata defined by the time-range
of a species.

(3) Peak zones (epiboles). Strata in which a species reaches its
acme. The term hemera, as originally defined, is the time unit
equivalent to an epibole.

Since many lithostratigraphic units are diachronous, the terms
used must not be equated with any of the preceding ones. The
recommended terms, arranged in order of decreasing size, are:

Group
Formation
Member
Bed

The prefixes 'sub-' or 'super-' have been applied to most of the
preceding terms, with the usual meanings.

C. H. Holland, *Science Progress*, July 1964.

Stratigraphy. The study of the stratified rocks (sediments and vol-
canics) especially their sequence in time, the character of the
rocks and the correlation of beds in different localities. (◊ *Historical geology*.)

Stratum (plural, strata). Synonymous with bed. Rocks which display layering or bedding may be described as displaying stratification. (◊ *Sedimentary structures*.)

Stratum contour. A synonym of ◊ *Strike line*.

Streak. The streak of a mineral is its colour in a finely divided state.
The most convenient way of determining the streak is to scratch
the material across the surface of a piece of hard unglazed por-
celain – a streak plate. Care must be taken to distinguish between
a mineral yielding a white streak and a mineral yielding no streak
(i.e. a mineral that is harder than the streak plate). A file is an
alternative means of obtaining a streak. Minerals may be divided
into five groups according to their streak characteristics:

(1) No streak.

(2) White or pale-coloured minerals yielding a white streak, e.g.
◊ *gypsum*, ◊ *calcite*, ◊ *feldspar*.

(3) Black or strongly coloured minerals yielding a white streak,
e.g. ◊ *augite*, ◊ *hornblende*.

(4) Black or strongly coloured minerals yielding a streak of

431

the same colour, e.g. ◊ *malachite*, ◊ *pyrolusite*, ◊ *graphite*. (5) Black or strongly coloured minerals yielding a streak of different colour, e.g. black or brown ◊ *haematite* yields a red streak. Black or brown ◊ *limonite* gives a yellow streak. Brass-yellow ◊ *chalcopyrite* gives a greenish-black streak.

Streamer. A non-piercing ◊ *injection structure*.

Stream tin. ◊ *Cassiterite* (SnO_2) occurring as detrital grains in ◊ *alluvial deposits*.

Stress. Stress is the system of internal forces within a body which is established as a reaction to an external force tending to change the shape or volume of the body. Any stress system can be resolved into two forces acting in opposite directions on either side of a plane – the plane of maximum stress. Geologists tend to use the word stress to describe effects arising from external forces.

Stress may be compressive, tensile, or shearing. Shearing stress (tangential stress), often abbreviated to 'shear' by the geologist, is a force tending to deform by the translation of one part of a body relative to another part.

The various forces acting upon a point which is subjected to pure stress without shear may be resolved into the three principal axes of stress, which are mutually at right angles to one another. If these compressive stresses act upon a sphere, the tendency would be to reduce its size but not to alter its shape. If, however, the first principal axis of stress is less than or greater than the others, the sphere will change shape and become an ellipsoid, and shear will arise as a result of this. (◊ *Ellipsoid*; *Strain*; *Pressure*.)

Stress mineral. A term applied to minerals formed under conditions of metamorphism when ◊ *stress* was present to a greater or lesser degree. It was supposed that 'stress' inhibited the growth of some minerals (anti-stress minerals) and promoted the growth of others. Doubt has been cast on the validity of the concept, as many examples of 'stress' minerals occurring in 'unstressed' environments have been recorded.

Striae, striations. Small grooves. The term is commonly applied to grooves formed by glacial action.

Strike. The direction in which a horizontal line can be drawn on a plane. In geological usage, the strike is important in determining the direction in which to measure the true ◊ *dip*. The term is also used in the sense of 'the general trend or run of the beds'; e.g. one might say that the strike of the beds in a particular region is east–west, ignoring the fact that there are minor variations in the strike (cf. regional ◊ *dip*). Another use of the term is as an adjective to describe the direction of a structure, e.g. a strike fault is parallel to the strike of the beds. (◊ *Fault*, Strike, Dip, and Strike Slip Movement.)

Strike line. A strike line (stratum contour and structure contour are synonyms) is a line joining points of equal height above or below a datum, on a planar structure. For a flat, evenly dipping plane, the strike lines will be straight, parallel, and evenly spaced. They are much used as a means of illustrating structural features without the complicating effect of topography. (◊◊ *Three-point problem.*)

Strombolian eruption. ◊ *Volcano*.

Strontianite. A strontium mineral, $SrCO_3$, found in veins, as nodules in limestones and as concretionary masses. (◊ *Carbonates* and Appendix.)

Structural bench. A flattening of dip without any change in direction. (◊ *Fold*, Fig. 62.)

Structure. (1) (petrology) The relationship between different parts of a rock (cf. ◊ *Texture*; ◊ *Fabric*). Examples of structures are ◊ *flow*, ◊ *amygdaloidal*, ◊ *bedding*, ◊ *jointing*, ◊ *cleavage*.

(2) A term used to describe the overall relationship of rock masses, e.g. ◊ *folding*, ◊ *faulting*, ◊ *unconformities*.

(3) (noun) Implying structural feature; e.g. a petroleum geologist talks about 'drilling a structure', meaning to drill a structural feature which may be an oil trap.

(◊◊ *Atomic structure*; *Strike line*.)

Structure contour. A synonym of ◊ *strike line*.

Strunian. A stratigraphical stage name for the transition beds of the European ◊ *Devonian*/◊*Carboniferous Systems*.

Stylolite. An irregular suture-like boundary developed in some limestones (also in some ◊ *evaporites*) which is generally independent of the bedding planes. The rock masses on either side of the boundary appears to fit into one another by means of a series of irregular 'teeth and sockets'. It is generally accepted that stylolites are formed by some kind of pressure-controlled solution followed by immediate local redeposition. Since the structure can sometimes be seen to cut across fossils, primary sedimentary features and late diagenetic cements, it is presumably developed in fully consolidated material.

Sub- (prefix). Under.

Subhedral. Showing some traces of crystal form. (◊ *Texture*.)

Subjacent. Bottomless. A mass of igneous rock is said to be subjacent if no evidence of a base can be actually seen, or inferred, at depth.

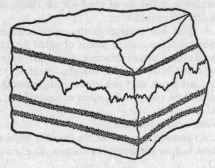

143. A stylolite.

Sub-littoral. That area between the lowest limit of the ◊ *littoral* zone and the ◊ *continental shelf*. A synonym is neritic.

Subsequent. ◊ *Drainage pattern.*

Subsoil. A layer of shattered and partly weathered rock between the ◊ *soil* proper and the ◊ *bedrock*.

Subsolvus granites. ◊ *Granites* containing two feldspars. (Cf. ◊ *Hypersolvus granites.*)

Substage. ◊ *Stratigraphic nomenclature.*

Sudetan. A stratigraphic stage name for the top of the ◊ *Viséan* (Carboniferous System).

Suess, Edward (1831–1914). Viennese professor who is remembered for his work on igneous intrusions (for one form of which he used the word 'batholite') and on earthquake origins and other crustal movements. His book *The Face of the Earth* provided a comprehensive survey of the extant knowledge of surface formations, while the later volume, *The Origin of the Alps*, put forward a theory of mountain structure which had immense influence. He was also responsible for the first sub-division of the ammonites (◊ *Mollusca*).

Sulcus (Brachiopoda). ◊ Fig. 11.

Sulphur. A non-metallic native element, S, found in areas of recent volcanic activity and around hot springs, and in sedimentary rocks with ◊ *gypsum* and limestone. (◊ Appendix.)

Sun cracks. Desiccation cracks. (◊ *Sedimentary structures.*)

Super- (prefix). Above.

Supergene. A word suggesting an origin literally 'from above'. It is used almost exclusively for processes involving water, with or without dissolved material, percolating down from the surface. Typical supergene processes are solution, hydration, oxidation, deposition from solution, reactions of ions in solution with ions in existing minerals (i.e. ◊ *replacement*).

Supergene enrichment. ◊ *Secondary enrichment.*

Superimposed drainage. ◊ *Inconsequent drainage.*

Superposition, Principle of. One of the two principles of ◊ *stratigraphy* enunciated by William ◊ *Smith*, stating that if one series of rocks lies above another then the upper series was formed after the lower series, unless it can be shown that the beds have been inverted as a result of tectonic action. (Cf. ◊ *Law of Strata Identified by Fossils*; ◊◊ *Way-up criteria*.)

Superstructure. A series of folded sedimentary or metamorphosed sedimentary rocks overlying a ◊ *migmatitic* zone (cf. ◊ *Infrastructure*). The two zones are folded disharmonically with the migmatite developed in the antiformal ◊ *folds*.

Suspension. The process by which material which is light enough is carried in the zone of ◊ *turbulent flow* and does not settle out of this zone. (Cf. ◊ *Saltation*; *Traction*.)

Suture, facial. ◊ *Arthropoda* (Trilobita) and Fig. 3.

Suture line. ◊ *Mollusca* (Cephalopoda) and Fig. 105.

Sveccofennian. A stratigraphic stage name for the ◊ *Svionian* + ◊ *Bothnian* (Precambrian).

Svionian. The oldest stratigraphic division of the Baltic ◊ *Precambrian*.

Swallow-hole. A synonym of ◊ *doline*.

Syenite. Coarse-grained ◊ *intermediate* igneous rocks characterised by the presence of alkali ◊ *feldspars*, and/or ◊ *feldspathoids*. Syenites fall naturally into two major groups, the ◊ *saturated* and ◊ *oversaturated*, and the ◊ *undersaturated*. This latter group is here regarded as equivalent to the ◊ *alkali syenites*.

The essential minerals are alkali feldspars; the presence of different types gives rise to the three main groups of saturated and oversaturated syenites – the sodic, sodi-potassic, and potassic groups. Oversaturated syenites contain quartz in accessory amounts, e.g. akerite, nordmarkite, umptekite, etc. Typical ferromagnesian minerals are ◊ *hornblende*, ◊ *biotite*, ◊ *augite*, ◊ *aegirine*. ◊ *Zircon*, ◊ *sphene*, and ◊ *apatite* are common accessory minerals.

A few types consist almost entirely of alkali feldspar – albitite, perthosite, and pulaskite; the last two, however, contain some aegirine and other ferromagnesian minerals, and often ◊ *nepheline*. Nepheline is commonly recorded in very small quantities in a number of syenitic rocks which are nevertheless described as saturated. In these rocks there is much local variation between just oversaturated and just undersaturated limits, and the types may be regarded as transitional to the alkali syenite group.

Syenites characteristically have a rather high sodium plus potassium content, and the more leucocratic varieties, especially, are low in FeO, MgO, and CaO. In certain varieties Al_2O_3 is exceptionally high, and titanium is an element which is often unusually abundant.

In addition to the sub-division into sodic, sodi-potassic, and potassic, it is convenient to distinguish ◊ *leucocratic*, ◊ *mesotype*, and ◊ *melanocratic* types based upon the ◊ *colour index*. The feldspar-rich types previously mentioned are leucocratic examples, while shonkinite is a typical melanocratic syenite, consisting largely of ◊ *pyroxenes*, a little ◊ *olivine*, and potassic feldspars (traces of nepheline may also occur).

With increasing quartz, the rocks grade into ◊ *granite*, and with the occurrence of labradorite and sometimes olivine the rock becomes an ◊ *alkali gabbro*. The presence of ◊ *oligoclase* or ◊ *andesine* produces a ◊ *monzonite*. Syenites are the plutonic equivalents of the ◊ *trachytes*. Separate bodies of syenite sometimes occur, and usually involve undersaturated and saturated/oversaturated types. ◊ *Lopolithic* forms and ◊ *ring structures* have been described, the latter often being associated with ◊ *carbonatites*. Small amounts of syenite commonly occur as marginal features around granite, monzonites, and gabbros.

◊ *Dykes* and ◊ *sills* of microsyenites occur, the ◊ *rhomb porphyries* of the Oslo District and elsewhere being a well-known type characterised by large alkali feldspar ◊ *phenocrysts*. Certain microsyenites consisting almost entirely of alkali feldspars are known as bostonites, and often display trachytic features. Syenite ◊ *aplites* (which include some bostonites) and syenite ◊ *pegmatites* are of rather uncommon occurrence. The syenite pegmatites often contain rare minerals, especially ones containing the rare earths. (For melanocratic microsyenites, ◊ *Lamprophyres*.)

Few syenites have any direct economic importance, with the exception of the well-known Norwegian rock larvikite (or laurvigite), which is used extensively as an ornamental stone. It is very coarse-grained, and consists largely of anorthoclase feldspars which show a characteristic blue ◊ *schiller* when polished. Larvikite contains clots of ferromagnesian minerals, which include titanaugite, ferro-olivine, euhedral apatite, biotite and magnetite. Lardalite is a nepheline-bearing larvikite.

Syenodiorite. A synonym of ◊ *monzonite.*

Syenodolerite. A close synonym of alkali dolerite. (◊ *Alkali gabbro and alkali dolerite.*)

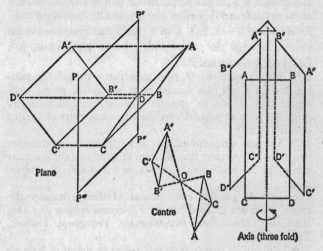

144. Types of symmetry.

Syenogabbro. A close synonym of ◊ *alkali* gabbro and olivine ◊ *monzonite.*

Syenoids. ◊ *Alkali syenites* containing no feldspar.

Symmetry. In crystallography, symmetry depends upon the distribution of angular elements and does not relate to the shape or size of faces. Four elements of symmetry are recognised in crystallography (◊ Fig. 144):

(1) Plane of symmetry. A plane such that a line perpendicular to it will pass through identical points on either side of the plane and at equal distance from it. Thus the crystal is divided into two equal halves, one being the mirror image of the other.

(2) Axis of rotation symmetry. A line about which a crystal may

be rotated so as to bring it into identical orientation 2, 3, 4, or 6 times in a single rotation of 360°. (A five-fold axis of symmetry is impossible.)

(3) Axis of rotary inversion symmetry (inversion axis). This involves the compound operation of rotating a face clockwise about an axis through a certain fraction of 360°, followed by translation through the ◊ *origin* of the crystal to a point equidistant on the opposite side. The process is repeated until the starting point is reached. Translation implies the carrying of a face along a straight line perpendicular to it to a similar position at the opposite end. Inversion axes are usually symbolised thus: $\overline{X}$-fold where X = 1, 2, 3, 4, or 6. The $\overline{1}$-fold operation implies rotation through 360°; $\overline{2}$-fold, 180°; $\overline{3}$-fold, 120°; $\overline{4}$-fold, 90°; $\overline{6}$-fold, 60°, before translation.

(4) Centre of symmetry. A point such that a straight line passing through it will pass through pairs of identical points, equidistant from it, and on opposite sides. A centre of symmetry generates pairs of parallel opposite edges and pairs of parallel opposite faces.

If in a crystal an even-fold axis of symmetry is perpendicular to a plane of symmetry then a centre of symmetry automatically arises. The converse is true, except in the ◊ *triclinic system*, where a centre exists alone.

There are 32 possible combinations of these symmetry elements, which are grouped into the 7 ◊ *crystal systems* (◊ *Cubic, Hexagonal, Monoclinic, Orthorhombic, Tetragonal, Triclinic, Trigonal*).

Symmetry can be conveniently referred to by means of a simple international notation as follows:

m = plane

X, where X = 2, 3, 4, or 6 = rotation axis

$\overline{X}$, where X = 1, 2, 3, 4, or 6 = inversion rotary axis

In combination:

X/m = a rotation axis perpendicular to a plane of symmetry.

Xm   a rotation axis co-planar with a plane of symmetry.

X/mm   a rotation axis with both kinds of planes of symmetry.

X2   a rotation axis with a two-fold axis perpendicular to it.

(Similar symbols can be constructed using $\overline{X}$.)

It should be noted that certain combinations of these symbols duplicate others which are preferred: sometimes either form may be used, e.g. mm in the orthorhombic system (= 2 planes of symmetry mutually at right angles) may be written 2m, which is also equivalent to $\bar{2}$m.

All properties of a crystalline substance, including physical and optical ones, and also atomic structure, conform to the symmetry of the crystal.

Symplectic texture. The texture produced by the intergrowth of two different minerals, e.g. ◊ *graphic texture*, ◊ *poikilitic* texture, ◊ *ophitic texture*.

Symplektite. Any secondary intergrowth of two minerals in which one is contained as worm-like inclusions within the other. ◊ *Myrmekite* is a special case. The texture is described as symplektitic.

Syn- (prefix). With, at the same time.

Synantectic. A synonym of ◊ *deuteric*.

Syncline, Synclinorium, Synform. A basin-shaped ◊ *fold* (or fold system).

Synkinematic, Synorogenic, Syntectonic. Terms often used as though they were synonymous, although strictly they are not. Synorogenic implies an event taking place simultaneously with a major period of mountain building, e.g. granite emplacement and regional metamorphism. Syntectonic and synkinematic imply an event taking place coextensively with a definite period of deformation, which may be a single phase of an orogeny or a closely associated set of phases, e.g. the formation of ◊ *mantled gneiss domes*, ◊ *flysch* sediments. The opposite of these three terms are postkinematic, post-orogenic, post-tectonic, implying events occuring after the movement and deformation periods. (◊ *Orogeny*; *Tectonic*.)

Synonym. If two distinct names are given to the same organism or group of organisms the names are said to be synonymous. In general the more recent synonym is judged to be invalid. (Cf. ◊ *Homonym*.)

Synrhabdosome. ◊ *Chordata* (Graptolithina) and Fig. 18.

Syntexis. A term which is used to include both pure melting and assimilation of ◊ *country rock*. (Cf. ◊ *Palingenesis*; ◊ *Anatexis*.)

Syntype. A series of type specimens of a genus selected to show the range of variation in the organism. (◊ *Type*.)

System. (1) ◊ *Stratigraphic nomenclature*.

(2) ◊ *Crystal system*.

Tabula (Coelenterata). ◊ Figs. 20 and 21.

Tabulata (Coelenterata). ◊ Fig. 20.

Tachylite. ◊ *Basalt* glass.

Tactite. A near-synonym of ◊ *skarn*.

Taghanican. A stratigraphic stage name for the top of the North American Middle Devonian.

Talc. A ◊ *layer-lattice mineral*, $Mg_3Si_4O_{10}(OH)_2$, the softest common mineral, 1 on the Mohs' scale (◊ *Hardness*). Its cleavage flakes are flexible but not elastic. It is found in low or medium-grade metamorphosed ◊ *basic* or ◊ *ultrabasic* rocks, and in certain cases talc constitutes the greater part of a rock, producing the material known as steatite or soapstone. The rare iron analogue of talc is called minnesotaite. Apart from the use of ground talc as talcum powder and other cosmetic preparations, it is of considerable importance as an insulator in the electrical industry, and as a filler. (◊ Appendix.)

Talus. A synonym of ◊ *scree*.

Taphrogenesis. Essentially vertical movements in the Earth's crust which result in major high-angle faulting such as is commonly associated with ◊ *rift valleys*.

Taphrogeosyncline. A synonym of ◊ *rift valley*.

Tarn. A small lake formed in a ◊ *cirque*.

Tar pits. Relatively small areas where soft asphaltic (or bituminous) 'tar' wells up to the surface and fills a hollow. The material is almost certainly derived from some underground source of hydro-carbons. Some tar pits contain a mass of bones of animals which were trapped in the tar. The most famous of these are at Rancho El Brea, California, and contain a rich fauna of Pliocene and Pleistocene animals. Very large tar pits are known as asphalt lakes, e.g. Trinidad Asphalt (or Pitch) Lake. (Cf. ◊ *Asphalt*.)

Tartarian (Tatarian). A stratigraphic stage name for the top of the East European Upper ◊ *Permian*.

-taxis (suffix). Arrangement.

Tecto- (prefix). Building.

Tectogene. A hypothetical deep, tightly-folded downwarp below an orogenic belt, suggested as an explanation of the greater thickness of sialic material beneath mountain ranges.

Tectonic. An adjective used to relate a particular phenomenon to a structural or orogenic concept, e.g. 'tectonic control of sedimentation' implies that the process of sedimentation was controlled by orogenic activity; a 'tectonic map' is a map designed to demonstrate structural features rather than stratigraphical or lithological ones. (◊ *Axis, tectonic*; *Transport, tectonic*.)

Tectonics. The study of the major structural features of the Earth's crust or the broad structure of a region.

Tectonite. A rock in which crystallisation has taken place under stress, resulting in the mineral constituents taking up a preferred orientation or ◊ *fabric*. (◊ *Axis, tectonic*.)

Tectosilicates. ◊ *Silicates*.

Teeth. (Brachiopoda) ◊ Fig. 11; (Mollusca) ◊ *Mollusca* (Lamellibranchiata) and Figs. 101 and 103.

Tegmen (Echinodermata). ◊ Fig. 43.

Tektites. Glassy ◊ *meteorites*.

Tektosilicates. ◊ *Silicates*.

Telemagmatic. This term has been used to describe ore bodies which, while displaying typical features suggesting an igneous origin, are nevertheless not obviously associated with any observable igneous body. It is suggested that they are deposits formed by ore fluids travelling unexpectedly long distances from a parent igneous mass. The mechanism of formation of such deposits is largely unexplained and some workers have denied their igneous origin.

Telson (Arthropoda). ◊ Fig. 3.

Tenor. The proportion of ore mineral in an ◊ *ore*.

Tension gash. A ◊ *joint* which opens as a result of tensional forces developed during deformation. It normally becomes filled with quartz, more rarely with other minerals. Tension gashes occur (a) at the noses of folds, and (b) at an oblique angle between shear planes (cf. ◊ *Boudinage*).

Tephra. A term for all fragmental ◊ *volcanic products* which are ejected through the vent, e.g. ash, cinders, lapilli, scoriae, pumice, bombs, etc. (◊ *Pyroclastic rocks*.)

Tephrite. ◊ *Alkali basalt*.

Tephroite. A mineral of the ◊ *olivine* group, Mn_2SiO_4.

Terminal curvature. It is commonly found that on a hill slope the 'ends' of the layers of rock are bent over in a downhill sense. This is due to the downhill creep of the mantle of soil and rubble and in favourable circumstances this can produce a convincing approximation to tectonic folding. The optimum arrangement is for the layers to lie as nearly at right angles to the slope surface as is possible, and the effect is inoperative where the angle between the layers and the slope surface is less than 45°.

Terra rossa. An insoluble red clay-like soil left as a ◊ *residual deposit* after the carbonate has been dissolved out of limestone by the ground water. It is found particularly in ◊ *karst scenery*. (◊◊ *Clay-with-flints*.)

Terrigenous sediments. Literally 'land-formed' sediments. The term is applied (a) to sediments formed and deposited on land (e.g. soils, sand-dunes) and (b) to material derived from the land when mixed in with purely marine material (e.g. sand or clay in a shelly limestone).

Tertiary System. The period of time which elapsed between the end of the ◊ *Cretaceous* and the present time, having a duration of 65 m.y., from 65 m.y. to 0 m.y. – although the precise limits are defined variously by different authors. A division of the Tertiary may be made as follows:

Holocene	(Youngest)
Pleistocene	
Pliocene	
Miocene	
Oligocene	
Eocene	
Palaeocene	(Oldest)

Because the majority of the sediments are shallow-water in origin and are often ◊ *diachronous*, many of the minor as well as the major boundaries are difficult to define. Usually the Montian is taken to be the lowest stage of the Palaeocene, but some authors consider it to be the equivalent of the Upper Cretaceous Danian stage. The Montian together with the Thanetian constitutes the Palaeocene, and is generally characterised by a fauna of Cretaceous survival forms, prior to the arrival of the true Tertiary fauna. The Eocene commences with the Sparnacian, and heralds the arrival of the true Tertiary faunas in most parts of the world. During the Eocene period there was a general increase in temperature over the Earth's surface, culminating in the Bartonian stage, which was characterised mainly by tropical and subtropical forms.

The succeeding Oligocene, which commences with the Sannoisian, showed a reversal of the conditions pertaining during the Eocene, with a general reduction in average temperature. Only the lower part of the Oligocene is represented in Great Britain; elsewhere, the upper part of the Oligocene shows a regression of the seas, which left a number of isolated basins each with their characteristic sediments and fauna.

During Miocene and Pliocene times, the withdrawal of the seas continued, resulting in the absence of the Miocene in Great Britain and a meagre development of the Pliocene. Elsewhere the Miocene and Pliocene are represented either by freshwater sediments in basins, or by the marine sediments deposited near to the existing coast lines. The gradual reduction in average temperature was continued throughout this time.

During Pliocene times the continuing drop in average temperature caused the extinction of many groups of mammals and the migration of other forms to warmer regions. Pliocene deposits are represented in Great Britain by accumulations of shallow-water shell gravels in East Anglia, together with some high-level gravels in the southern part of Great Britain and occasional small basins of deposition, as at St Erth, Cornwall. Between 80% and 90% of the fossil forms occurring in the Pliocene in this country still exist at the present day.

A gradual deterioration of climate throughout the Pliocene led eventually to the Ice Ages of the Pleistocene.

In Great Britain it is virtually impossible to draw a boundary between the Pliocene and Pleistocene, although this has been done in Italy and elsewhere.

The glacial period can be divided into two parts separated by the Great Interglacial, the length of which has been estimated at 200,000 years. It has been shown that two periods of glaciation occurred before the Great Interglacial and two afterwards. In the interglacial periods the climate was, on occasions, appreciably warmer than the climate at the present day. The last ice sheet to cover Great Britain receded about 11,000 years ago, and the average annual temperature gradually increased until it reached an optimum about 5,000 years ago, when the bulk of Great Britain was covered with extensive deciduous forest. Since that time the climate has deteriorated again.

The simple picture of four ice ages cannot be readily recognised everywhere, since each ice age consists of numerous major and minor fluctuations in temperature. The ice ages most probably were represented in warmer regions by pluvial or higher rainfall periods.

443

PLEISTOCENE – GLACIAL AND INTERGLACIAL STAGES

	British Isles	Alps	Northern Europe	Central North America
Last glacial	NEWER DRIFT	WÜRM	WEICHSEL	WISCONSIN
Interglacial		Riss/Würm	Eemian	Sangamon
Glacial	GIPP-ING TILL	RISS	SAALE	ILLINOIAN
Interglacial	Hoxnian	Mindel/Riss	Needian	Yarmouth
Glacial	LOWES-TOFT TILL	MINDEL	ELSTER	KANSAN
Interglacial	Cromerian	Günz/Mindel		Aftonian
Glacial	WEY-BOURNE CRAG	GÜNZ		NEBRASKAN
Interglacial	Norwich Crag	Donau/Günz		
Glacial	PRE-CRAG	DONAU		PRE-NEBRASKAN?

(The "GIPPING TILL ... LOWESTOFT TILL" group is bracketed as **OLDER DRIFT**.)

It was under these somewhat hostile conditions that man evolved to his present form.

During this time, spreads of glacial debris accumulated north of a line from the Thames estuary to the Bristol Channel. South of this line, Pleistocene sediments are represented largely by spreads of gravel and river terraces, which represent the re-sorting of debris from the ice sheets. It was during this period that the major physiographic features of Great Britain approached their present form. Fossils include remains of horses, elephants, and pigs.

The term Holocene has been used for post-glacial deposits and the term Quaternary for Pleistocene plus Holocene. However, attempts to establish the term Quaternary as a 'new era' appear unjustified. They were based upon the concept of defining a new era at a major faunal change, in this case the appearance of hominid fossils. The discovery of undoubted Pliocene hominids appears to invalidate this criterion.

Teschenite. An ◊ *alkali gabbro* similar to ◊ *essexite*, but containing analcite to the exclusion of nepheline. Barkevicite is common, and olivine-teschenites are of frequent occurrence; the presence of alkali feldspar gives rise to the variety glenmuirite.

Bekinkinite is a nepheline-bearing teschenite often containing barkevicite. Crinanite is a term which has been used for rocks so similar to teschenite that it appears to be superfluous. Microteschenites have been described. (◊ *Theralite*.)

Test. Skeleton. ◊ *Echinodermata* and *Protozoa*.

Tetartohedral. The term applied to the class of a crystal system which has a reduced symmetry, so that the general form has one-quarter the number of faces of the corresponding form in the ◊ *holohedral* class, e.g. in the cubic system the tetartohedral general form has 12 faces as opposed to 48 in the holohedral class. (Cf. ◊ *Hemihedral*.)

Tethys. A Mesozoic ◊ *geosyncline* which developed between ◊ *Laurasia* and ◊ *Gondwanaland*, and covered southern Europe, the Mediterranean, North Africa, Iran, and the Himalayan region, possibly extending into Burma and south-east Asia. It was the site of the Alpine orogeny. (Adjective: Tethyan.)

Tetracoralla. ◊ *Coelenterata* (Anthozoa).

Tetragonal system. A ◊ *crystal system* divided into seven symmetry classes, as in the following table:

	Symbol	Centre	Planes of symmetry	Axes of rotation symmetry	Axes of rotary inversion	Examples
1	4/mmm	C	3	4 ii, 1 iv	—	Zircon, cassiterite, rutile
2	42m	—	2	2 ii	1iv	Chalcopyrite, melilite
3	4mm	—	4	1 iv	—	Rare
4	422	—	—	4 ii, 1 iv	—	Phosgenite ($PbCl_2PbCO_3$)
5	4/m	C	1	1 iv	—	Scheelite ($CaWO_4$), wulfenite ($PbMoO_4$)
6	4	—	—	—	1iv	Rare
7	4	—	—	1 iv	—	No example

445

The tetragonal system includes all those crystals referred to three axes mutually at right angles, the parameters employed on the two horizontal ones being equal and either smaller or greater than that employed on the vertical axis. (For international symbols, ◊ *Symmetry*; ◊ *Axes, crystallographic*.)

Tetrahedrite. A copper ore mineral, $(Cu, Fe)_{12}Sb_4S_{13}$, found with other copper ores in ◊ *hydrothermal veins*. (◊ Appendix.)

Texture. Texture is the relationship between the grains of minerals forming a rock. Many petrologists use the words 'texture' and 'structure' almost interchangeably, and terms which are used for texture by one author are used for structures by another and vice-versa. Our selection is therefore somewhat arbitrary. (The following terms and remarks refer mainly to igneous rocks; for sedimentary textures, ◊ *Sedimentary rocks*; for metamorphic textures, ◊ *Metamorphic rocks*.) Texture depends on four factors:

(1) GRAIN SIZE. The implications of terms such as coarse, medium, and fine as applied to the grain size of igneous and metamorphic rocks (◊ *Igneous rocks*) are complicated, but the terms ◊ *aphanitic*, ◊ *phaneritic*, and ◊ *hyaline* may be used without defining absolute limits.

(2) GRAIN SHAPE. Three equivalent sets of terms are used to describe grain shape:

CIPW	Rosenbusch	Rohrbach	
Euhedral	Idiomorphic	Automorphic	Grains displaying fully developed crystal form
Subhedral	Hypidiomorphic	Hypautomorphic	Grains showing some trace of a crystal form
Anhedral *	Allotriomorphic	Xenomorphic	Grains showing no development of crystal form whatsoever

* May be applied to the shape of mineral grains of igneous, sedimentary or metamorphic rocks.

(3) DEGREE OF CRYSTALLINITY. Rocks which are entirely glass are said to be holohyaline. Those which are entirely crystalline are said to be holocrystalline. Those which contain both crystals and glass are variously termed hemicrystalline, hypocrystalline, hyalocrystalline, hypohyaline, merocrystalline. ◊ *Devitrified* glasses yield textures such as ◊ *variolitic*, ◊ *spherulitic*,

◊ *hyalopilitic*, and ◊ *microlitic*. ◊ *Intersertal* texture and hyalo-ophitic texture are varieties of ◊ *ophitic texture* in which plagio-clase laths are embedded in glass rather than augite.

(4) CONTACT RELATIONSHIPS OF THE GRAINS. (a) ◊ *Granular textures*. It is convenient to classify granular textures according to the grain shape terms in (2) above. Thus panidio-morphic-granular (automorphic-granular and idiomorphic-granu-lar are synonymous) implies that all the grains are euhedral. Similarly, hypautomorphic-granular (hypidiomorphic-granular) implies a mixture of euhedral, subhedral, and anhedral grains, and xenomorphic-granular (allotriomorphic-granular) signifies that a rock is made up of anhedral grains; aplitic texture (◊ *Aplite*), ◊ *saccharoidal* texture, and mosaic texture are synonyms. (b) ◊ *Porphyritic* textures. ◊) *Glomeroporphyritic*; *Vitrophyre*. The suffix -phyre implies a porphyritic texture.
(c) *Intergrowth* textures. ◊ *Granophyric*; *Graphic*; *Poikilitic*; *Ophitic*; *Myrmekitic*; *Symplectic*.
(For other textural terms, including metamorphic textures (in-dicated by 'M'), ◊ *Blasto-*; *-blastic* (M); *Crystalloblastic* (M); *Crystalline rock*; *Devitrification*; *Dyscrystalline*; *Equigranular*; *Eucrystalline*; *Eutaxitic*; *Felsitic*; *Granulitic texture*; *Helicitic* (M); *Lepidoblastic* (M); *Nematoblastic* (M); *Palimpsest structure* (M); *Perlitic texture*; *Pilotaxitic*; *Trachytic texture*. Cf. ◊ *Fabric*; *Structure* (1).)

Thallasso- (prefix). Pertaining to the sea.

Thallassocratic sea level. The sea level at the maximum of a marine transgression into a continental area.

Thallassostatic. A term applied to rivers which show evidence of adjustment of their base levels to changing sea levels.

Thamnasterioid. ◊ *Coelenterata* and Fig. 23.

Thanatocoenosis. An assemblage of fossils consisting of the re-mains of organisms which were not associated during life. The remains have usually been brought together after death by current action.

Thanetian. A stratigraphic stage name for the European Upper Palaeocene (◊ *Tertiary System*).

Theca, Thecal aperture. ◊ *Chordata* (Graptolithina) and Fig. 18.

Theralite. An ◊ *alkali gabbro* similar to ◊ *essexite*, without olivine, but containing nepheline to the exclusion of analcite and a higher ratio of nepheline to feldspar. It is equivalent to the volcanic rock nepheline tephrite. The amphibole barkevicite is commonly present. A theralite containing porphyritic barkevicite is termed lugarite. Olivine theralites are not uncommon; olivine-rich examples are distinguished under the name kylite. Luscladite lies between essexite and kylite. (◊ *Teschenite*.)

Thermo- (prefix). Heat.

Thin section. Most minerals are transparent when examined in sufficiently thin slices. Mineralogists and petrologists make use of such slices to study the optical properties of minerals, and to observe the interrelationships of minerals, using a microscope with an analyser and polariser. In order to standardise conditions for optical work, the standard thickness of a 'thin' section is 30μ (·03 mm.). The usual technique for making a thin section involves cutting a slice with a diamond saw, mounting the slice on a glass plate, and grinding to the correct thickness by means of succes-sively finer and finer grades of abrasive. The thickness is checked by optical means. Finally the section is covered with a cover glass. (◊ *Optical properties of minerals*.) The first thin section was made by ◊ *Sorby* in 1849.

Thixotropic. A term applied to certain types of solid/liquid systems which are effectively solid when stationary but which become mobile liquids when subjected to shearing stresses. Once started, they will flow down a slope as long as the velocity is sufficient to maintain the minimum shearing stress needed for liquidity. Certain clay–water systems form sediments which are thixotropic, and these give rise to certain ◊ *sedimentary structures*. It is pos-sible that some ◊ *magmas* consisting of crystals in a melt behave thixotropically when extruded as a lava. Volcanic ash–water systems are also frequently thixotropic.

Tholeiite. An important type of ◊ *basalt* consisting of basic plagio-clase and pigeonite (a pyroxene), with interstitial glass or quartz-alkali feldspar intergrowths.

Thorax. ◊ *Arthropoda* (Trilobita) and Fig. 3.

Three-point problem. A geometrical method of finding the ◊ *dip* and ◊ *strike* of a plane surface, given the altitudes above a datum plane at three known positions. Altitudes may be determined from an outcrop or from the depth to the plane in a borehole. The method depends upon the assumption that the dip and strike of the plane are constant over the area under consideration. From the point of view of natural geological problems this is only true to a first approximation over relatively small areas.

Throw. The measure of the vertical displacement between the up-thrown and downthrown sides of a ◊ *fault*.

Thrust. A low-angle reverse ◊ *fault*.

Thrust sheet. A ◊ *nappe*.

Thuringian. A stratigraphic stage name for the European Upper ◊ *Permian*.

Thurnian. A stratigraphic stage name for the British mid Lower Pleistocene (◊ *Tertiary System*).

Tiffanian. A stratigraphic stage name for the North American equivalent of the European Lower ◊ *Thanetian* (◊ *Tertiary System*).

Tiglian. A stratigraphic stage name for the British Lower Pleistocene (◊ *Tertiary System*).

Till (Tillite). A synonym for Boulder Clay. ◊ *Drift* (3).

Tinguaite. An obsolete name for ◊ *phonolites* occurring as intrusive bodies.

Tinstone. The popular name for ◊ *cassiterite* – SnO_2 – the main ore mineral of tin.

Tioughniogan. A stratigraphic stage name for the North American mid Middle ◊ *Devonian*.

Titanite. ◊ *Sphene*.

Tithonian. A stratigraphic stage name for the Middle and Upper ◊ *Kimmeridge* + ◊ *Portland* + ◊ *Purbeck*, or, by some authors, the Portland + Purbeck (cf. ◊ *Volgian*) (Jurassic System).

Toadstones. A local mining term from Derbyshire, England, subsequently used elsewhere, for various types of ◊ *intrusions*, ◊ *extrusive rocks*, and ◊ *pyroclastic* rocks. They include basic lavas and ◊ *sills*, and various kinds of bedded pyroclastic material. The word is derived from the German 'Tödstein' meaning 'dead stone', from the absence of ore deposits in the rocks. (Cf. ◊ *Trap*.)

Toarcian. A stratigraphic stage name for the top of the European Lower ◊ *Jurassic*.

Tombolo. A ◊ *spit* or ◊ *bar* which joins an island to the mainland or to another island.

Tonalite. ◊ *Diorite*.

Tongrian. A stratigraphic stage name for the European Lower Oligocene (◊ *Tertiary System*).

Tool marks. ◊ *Sedimentary structures*.

Topaz. ◊ *Aluminium silicates* and Appendix.

Topazfels. ◊ *Pneumatolysis* (Greisening).

-topic (suffix). A suffix suggested to signify sedimentary texture, to be comparable with the igneous suffix -morphic and the metamorphic suffix ◊ *-blastic* (◊ *Texture*); e.g. porphyrotopic – large dolomite crystals in a calcite matrix (analagous to porphyroblasts).

Topotype. Any specimen obtained from the same locality as the holotype. (◊ *Type*.)

Top-set beds. ◊ *Delta*.

Tor. A term originally used in south-west England for residual masses of rock, usually capping hills. A tor commonly appears as a pile of rock slabs or a series of slabs standing on end, according to whether the dominant joint system is horizontal or vertical. Weathering proceeds most actively along joint planes, thus reducing an originally solid mass first to piles of slabs and ultimately to a heap of loose boulders. Most tors in south-west England are found on the granite masses of Devon and Cornwall, although locally there are non-granite features named as tors. It is clear on the granite masses that the tors represent regions of un-kaolinised (◊ *Hydrothermal processes*) granite and are residuals left by differential ◊ *denudation* of the harder and softer materials. The term 'tor' has been applied to several other kinds of residual rock mass and, as mentioned above, to non-granitic features; some of these are only remotely related to the typical tors and caution should be exercised in widening the use of this term.

Torbanite. Boghead ◊ *coal*.

Torrejonian. A stratigraphic stage name for the North American equivalent of the West European ◊ *Montian* (◊ *Tertiary System*).

Torridonian. A stratigraphic division of the Scottish ◊ *Precambrian*, the unmetamorphosed equivalent of ◊ *Moinian*.

Tortonian. A stratigraphic stage name for the European Middle Miocene (◊ *Tertiary System*).

Toscanite (Rhyodacite). ◊ *Rhyolite*.

Touchstone. A hard, black, very fine-grained stone (◊ *basalt* or ◊ *chert* – notably the variety lyddite) used for determining the fineness (i.e. the purity) of gold and silver. The method involves comparing the ◊ *streak* of a gold or silver object of unknown purity on the touchstone with the streak produced by specimens of known fineness (touch needles).

Tourmaline. ◊ *Cyclosilicates* and Appendix.

Tourmalinisation. ◊ *Pneumatolysis*.

Tournaisian. A stratigraphic stage name for base of the European Lower ◊ *Carboniferous*.

Trace elements. ◊ *Geochemistry*.

Trace fossils. ◊ *Sedimentary structures* resulting from biological activity. Trace fossils include:

(1) Burrows and other excavations.

(2) Tracks, trails, footprints and resting marks.

(3) Evidence of feeding activity including patterns of shallow grooves left by mud surface feeders, e.g. gastropods.

(4) Trails of ◊ *faecal pellets*.

Any of these features can be preserved as ◊ *casts* or ◊ *moulds*. The marks made by shells rolling or bouncing on soft sediment are not included (◊ *Sedimentary structures*).

The study of trace fossils has become of great significance in the last few years, particularly in the field of ◊ *palaeoecology*. In a classic example in Northern Iraq the local ◊ *Ordovician* beds have been divided into three groups based on assemblages of trace fossils. (◊◊ *Fossils and Fossilisation*.) Trace fossils have considerable value as a means of determining the ◊ *way-up* of strata. It should be emphasised that trace fossils are almost entirely facies fossils, in that they are restricted to a limited environment. It is unfortunately rare to be able to find the actual organism responsible for making a particular trace.

Trachyandesite (Latite). A volcanic rock intermediate in character between ◊ *trachyte* and ◊ *andesite*; the volcanic equivalent of ◊ *monzonite*. The essential characteristic is the occurrence of both andesitic feldspar and alkali feldspars. Mugearites are properly included here, but are described under ◊ *andesite*. (For banakite, ◊ *Trachybasalt*.)

Trachybasalt. The volcanic equivalent of syenogabbro, a variety of ◊ *monzonite*. Undersaturated trachybasalts are tephrites and basanites (◊ *Alkali basalts*). Banakites (sometimes inaccurately classified as trachyandesites) are saturated or oversaturated varieties. Mugearites (◊ *Andesite*) are wrongly classified here. The essential characteristic is the occurrence of both basic feldspars and alkali feldspars (or feldspathoids).

Trachydolerite. A term used of rocks intermediate in character between ◊ *dolerites* and alkali dolerites. (Cf. ◊ *Trachybasalt*.)

Trachyte, Trachytic texture. Trachytes are fine-grained, ◊ *alkali*, ◊ *intermediate* igneous rocks. They range from slightly oversaturated types, containing less than 10% quartz, to undersaturated types, containing feldspathoids. Feldspars are often the dominant minerals – sanidine, albite, or a sodi-potassic type. The ferromagnesian minerals present may be ◊ *biotite*, ◊ *hornblende*, ◊ *augite*, ◊ *aegirine*, or ◊ *riebeckite*. ◊ *Glass* is rather uncommon. A characteristic feature of trachytes is the packing together of lath-like feldspars in parallel alignment, due to flow in the molten rock. This texture is called trachytic, but it is not confined to trachytes. Trachytes are not commonly ◊ *porphyritic*, although examples containing very large ◊ *phenocrysts* do occur.

Undersaturated trachytes containing dominant nepheline are called phonolites, and those containing dominant leucite, leucitophyres; other feldspathoids may be present, e.g. nosean, sodalite, haüyne (haüynophyre) etc. Aegirine-augite is the commonest ferromagnesian mineral in these rocks, but sodic ◊ *amphiboles*, and even melanite ◊ *garnet* or phlogopite ◊ *mica* may occur. Olivine phonolites are called kenytes.

The composition of the trachytes closely parallels the equivalent syenites:

	Potassic	Sodi-Potassic	Sodic
Oversaturated Saturated	←——QUARTZ TRACHYTE——→ ⎫ ←————TRACHYTE(s.s.)————→ ⎭		Keratophyre
Undersaturated	Leucitophyre	Leucite phonolite or Nepheline leucitophyre	Phonolite

Keratophyre is not now used for a sodic non-feldspathoidal trachyte (◊ *Rhyolite*). Biotite trachyte is called domite after its occurrence in the Puy-de-Dôme, France. Tinguaite is an obsolete name for phonolites occurring as intrusive bodies. Trachyphonolites, or phonolitic trachytes, are types containing exceedingly small amounts of feldspathoid.

Trachytes grade into ◊ *rhyolites* with an increase in quartz, and into ◊ *alkali basalts* and other feldspathoidal lavas by an increase in the feldspathoid to feldspar ratio and the development of more calcic plagioclase. Trachytes, and phonolites and leucitophyres, are the volcanic equivalents of syenites and feldspathoidal syenites respectively. Melanocratic trachytes may be regarded as the volcanic equivalent of some ◊ *lamprophyres*.

Trachytes occur mainly as lava flows of small extent, owing to their high viscosity, or as small ◊ *dykes*.

Traction. The process by which material is rolled by a transporting agent along a surface, e.g. the bed of a river, a bare land surface. (Cf. ◊ *Saltation*; ◊ *Suspension*.)

Transgression, marine. The invasion of a large area of land by the sea in a relatively short space of time (geologically speaking). It is generally considered that the sea advances over a more or less ◊ *peneplained* surface. Although the observable result of a marine transgression may suggest an almost 'instantaneous' process, it is probable that the time taken is in reality to be measured in millions of years. The plane of a marine transgression is a plane of ◊ *unconformity*. The reverse of a transgression is a ◊ *regression*.

Transition series. A group of beds linking two series of sediments formed in contrasting environments. For example, in Britain, the Downtonian Beds form a transition between the marine ◊ *Silurian* and the continental facies of the ◊ *Devonian* (the Old Red Sandstone).

Transport, tectonic. When considering the deformation of a rock it is often convenient to regard one part of the rock as having been displaced relative to another part. The direction of this relative movement is called the direction of tectonic transport and defines the a-tectonic axis. (◊ *Axis, tectonic.*) The scale of tectonic transport can vary from a few millimetres (as in the microfolds of a schist) to tens of kilometres (as in major recumbent folds).

Transportation of sediment. The main agencies by which sedimentary materials are moved are: gravity (◊ *Gravity transport*); running water (rivers and streams); ice (glaciers) (◊ *Glaciers and glaciation*); wind; the sea (currents and ◊ *longshore drift*).

Running water and wind are the most widespread transporting agents. In both cases, three mechanisms operate, although the particle size of the transported material involved is very different, owing to the differences in density and viscosity of air and water. The three processes are: rolling or ◊ *traction*, in which the particle moves along the bed but is too heavy to be lifted from it; ◊ *saltation*; and ◊ *suspension*, in which particles remain permanently above the bed, sustained there by the ◊ *turbulent flow* of the air or water. In the traction and saltation zones the particles are subjected to ◊ *laminar flow* of the fluid.

The maximum size of particle transported by water is considerably larger than for wind, and the total load is also greater. During transport there is considerable sorting (◊ *Grading*) of the particles and there is also a certain amount of mutual abrasion (attrition) which results in a general decrease in particle size. Transport of sediment is often intermittent, depending upon the medium attaining the necessary minimum velocity, e.g. floods and sandstorms. Water also transports material in solution.

Trap. (1) An obsolete name for a compact, fine-grained igneous rock (a ◊ *lava* flow, ◊ *sill*, or ◊ *dyke*) or occasionally an altered variety of these.

(2) A structure in which ◊ *oil* and/or gas may collect.

Travertine. A kind of ◊ *calc tufa* deposited by certain hot springs in volcanic regions.

Tremadocian. A stratigraphic stage name for (1) the top of the British Upper ◊ *Cambrian*, (2) the basal ◊ *Ordovician* in Europe and elsewhere. For the position of the Tremadocian, ◊ *Cambrian*.

Tremolite–actinolite. A series of minerals of the ◊ *amphibole* group, $Ca_2(Mg,Fe'')_5(Si_8O_{22})(OH,F)_2$, found in metamorphosed impure limestones. (◊ *Asbestos* and Appendix.)

Trempealeauan. A stratigraphic stage name for the top of the North American Upper ◊ *Cambrian*.

Trentonian. A stratigraphic stage name for the top of the North American Middle ◊ *Ordovician*.

Tri- (prefix). Threefold.

Triangular diagram. A graphical method of expressing the composition of a particular material in terms of three 'components' of which it consists. These three 'components' may comprise the whole of a system or part of a more complex system. Sometimes the 'components' are composite, e.g. one component may be labelled 'alkalies' and represents $K_2O + Na_2O + Li_2O$. The 'components' need not be chemically defined: it is possible, for example, to express the composition of a sediment in terms of the 'components' sand, clay, and pebbles.

The method depends upon the fact that for any point within an equilateral triangle, the sum of the perpendicular distances from the point to the sides is a constant. If this constant length is made 100 units, the three perpendiculars can then be selected so as to represent the percentages of each constituent. The vertices represent 100% of a constituent, the sides mixtures of pairs of constituents.

In Fig. 145, the three constituents are A,B,C; point X represents 50% A, 30% B, 20% C. By plotting series of analyses on such diagrams, trends, relationships, and groupings often become apparent. By adding a third dimension to the triangle, to produce a triangular prism, an extra variable (commonly temperature, indicated by an arrow in the figure) can be introduced. This is valuable when dealing with systems involving silicate melts. It is possible to represent these three-dimensional plots in two dimensions using 'temperature contours'.

Trias. An abbreviation of ◊ *Triassic*.

Triassic System. The system named, by von Alberti in 1834, from the three-fold division of the period which can be made at the type locality in Germany. The period extends from 225 to 195

454

m.y., a duration of 30 m.y. It marks the beginning of the Mesozoic Era. Widespread continental conditions which persist from the preceding ◊ *Permian* Period make the lower boundary difficult to interpret. The base of a 'zone' containing Lystrosaurus (a

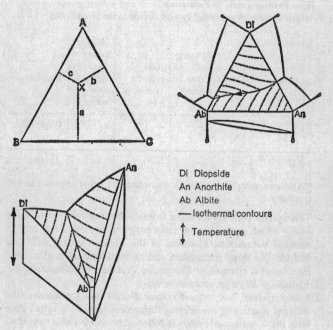

Di Diopside
An Anorthite
Ab Albite
—— Isothermal contours
↕ Temperature

145. Triangular diagrams.

reptile) has been suggested but its general utility is doubtful. The upper limit of the Trias is marked by the sharp change in conditions at the start of the ◊ *Rhaetic*. The base of the Rhaetic is a marine transgression with fossiliferous marine sediments above contrasting with continental sediments below.

The Trias is represented in Great Britain entirely by continental deposits but marine intercalations occur elsewhere. In Great Britain the Triassic sediments are largely unfossiliferous, the only fossil remains being those of reptiles (mainly as footprints). In Europe, where marine sediments occur, the fauna was largely one of lamellibranchs (◊ *Mollusca*), crinoids (◊ *Echinodermata*) and richly ornamented ammonoids (◊ *Mollusca*). In the Trias the earliest known dinosaur remains occur.

◊ *Evaporites* are found in the period and are of considerable economic importance.

(For Permio-Trias, ◊ *Permian*.)

Triclinic system. A ◊ *crystal system* divided into two classes:

	Symbol	Centre	Planes of symmetry	Axes of rotation symmetry	Axes of rotary inversion symmetry	Example
1	$\bar{1}$	(C)*	—	—	1	Plagioclase, kyanite
2	1	—	—	—	—	Axinite

* A unique centre of symmetry is equivalent to a 1-fold inversion axis of symmetry.

This system includes all those crystals that are referred to three axes – of which none are at right angles to any other and having unequal parameters. Elements of the triclinic system will thus include the three parameters and three interaxial angles. (For international symbols, ◊ *Symmetry*; ◊◊ *Axes, crystallographic*.)

Tridymite. ◊ *Silica group of minerals*.

Trigonal system. A ◊ *crystal system* divided into five classes (the modern practice of considering this system separately rather than with the ◊ *hexagonal system* is followed here). (◊ Table opposite.) Trigonal crystals are usually referred to the same set of axes as for the hexagonal system (the Bravais Axes). However, an alternative set of axes (due to Miller), consisting of three non-orthogonal axes, equally inclined to the three-fold axis of symmetry (or the three-fold inversion axis) and making equal angles with each other, is available. They are, in fact, parallel to three edge-directions of a rhombohedron, and when using them the characteristic feature is not the axial ratio but the axial angle. (For international symbols, ◊ *Symmetry*; ◊◊ *Axes, crystallographic*.)

Trilling. A crystallographic triplet. (◊ *Twinned crystal*.)

Trilobita. ◊ *Arthropoda*.

Trimorphism. ◊ *Polymorphism*.

Tripoli. A variety of ◊ *diatomite* containing radiolaria (◊ *Protozoa*).

	Inter-national symbol	Centre	Planes of symmetry	Axes of rotation symmetry	Axes of rotary inversion	Examples
1	$\bar{3}m$	C	3	3 ii	1 iii	Calcite and related carbonates; corundum; haematite
2	3m	—	3	— 1 iii	—	Tourmaline
3	32	—	—	3 ii, 1 iii	—	Quartz; cinnabar
4	$\bar{3}$	C	—	—	1 iii	Dolomite; willemite (Zn_2SiO_4) dioptase ($CuSiO_3.H_2O$) phenacite (Be_2SiO_4)
5	3	—	—	— 1 iii	—	Doubtful examples only

Trivium (Echinodermata). ◊ Fig. 45.

Trochoid. ◊ *Coelenterata* and Fig. 23. (Cornute is a synonym.)

Troctolite. ◊ *Gabbro.*

Trondhjemite (Trondjemite). A potash feldspar-free ◊ *granodiorite*, usually very low in ◊ *ferromagnesian minerals.*

Trough bedding (Trough cross-bedding). ◊ *Cross-bedding.*

Trough line, plane. ◊ *Fold.*

Tubercle (Mollusca). ◊ Fig. 105.

Tufa. ◊ *Calc tufa.*

Tuff. ◊ *Pyroclastic rocks.*

Tundra. A Lappish term for ◊ *soils* produced under conditions of extreme cold. They are often waterlogged because of a layer of ◊ *permafrost* at shallow depth.

Turbidite, Turbidity current. It is a matter of common observation, and an easily demonstrated fact, that slurries of sediment and water behaves as a discrete fluid phase when poured into fresh or sea water. It is now generally accepted that such slurries can be generated in large quantity in various types of basin and will flow down a slope at remarkably high speeds, covering distances of tens of kilometres. The movements of these masses of slurry are termed turbidity currents, or density currents, and the sediment

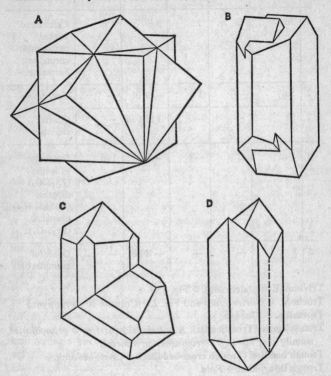

146. Twinned crystals. (A, B) Interpenetration twins; (A) Cubes; (B) Feldspar (Carlsbad habit); (C, D) Rotation twins; (C) Geniculate (rutile); (D) Butterfly (gypsum).

deposited as a result of such a current is termed a turbidite. Although many turbidites are greywackes (◊ *Arenaceous rocks*) the two terms are not synonymous. Turbidites display a wide range of ◊ *sedimentary structures*, including ◊ *graded bedding*, load, flute, and groove casts, and 'flame' structures. The major environment for turbidites is a ◊ *geosyncline*; but modern turbidity currents are known to move down continental slopes. Turbidity currents are capable of eroding the floor of the basin or slope and are responsible for transferring large quantities of shallow-water sediment into deeper zones. The recognition of the work of turbidity currents has revolutionised ideas on rates of sedimentation and the depths of water in which sediments were deposited.

Turbulent flow. Any flow which is not ◊ *laminar*, i.e. the stream lines of the fluid, instead of remaining parallel, become confused and intermingled.

Turgite. ◊ *Limonite*.

Turonian. A stratigraphic stage name for the European mid Upper ◊ *Cretaceous*.

Twinned crystal. A crystal made up of two parts which are orientated differently but in a regular fashion related to the crystallography of each portion. Two kinds of twin may be recognised:

(1) INTERPENETRATION OR CONTACT TWINS (◊ Fig. 146, A, B). These involve the idea of two simple crystals interpenetrating in such a way that one individual appears to have been rotated 180° about a twin axis relative to the other.

(2) ROTATION TWINS (◊ Fig. 146, C, D). These are conventionally derived by imagining the following operations: A simple untwinned crystal is divided along a plane – the twin plane – which may or may not be a plane of ◊ *symmetry*, but which is invariably parallel to a possible crystal face. At right angles to the twin plane is the twin axis (which cannot be an even-fold axis of symmetry) about which one half of the crystal is rotated through 180° relative to the other half. The two halves of the crystal are then united along a composition plane, which may or may not be the same as the twin plane.

This 'mechanical' explanation of twinning is purely a geometrical device; in nature, twinning arises as a result of a change in the direction of growth of a crystal lattice (◊ *Unit cell*), due to local conditions. Repeated twinning is possible – giving triplets (trillings), fourlings, sixlings, eightlings, etc. Small-scale repeated twinning on a series of parallel composition planes, e.g. in the plagioclase feldspars, is sometimes called polysynthetic twinning. A characteristic feature of twin crystals is the occurrence of re-entrant angles. The operation of twinning is often responsible for the appearance of additional pseudo-symmetry elements. Many twins mimic crystals belonging to higher symmetry classes, e.g. triplets of orthorhombic aragonite appear to be hexagonal. Twin crystals should be distinguished from ◊ *parallel growth*.

Certain types of twin have acquired popular names, e.g. geniculate, butterfly, cruciform.

Type. The convention of the 'type specimen', 'type species', 'type genus', etc. has arisen in palaeontology in order that the description of a form or group can be unequivocally tied to an actual specimen or specimens. Today, a 'type' (or 'types') is automatically designated in any description of a new species or new genera, etc. The concept has been extended to minerals and even to rocks. The following terms are commonly employed:

HOLOTYPE. A single specimen selected in order to show the main character of a species.

PARATYPE. Additional specimens selected with a holotype to show additional character of the species which are not shown by the holotype.

SYNTYPES. A series of type specimens of equal status, selected to show the range of variation within a species.

LECTOTYPE. Where a type has not been nominated in the past, it is now the practice to select a lectotype from the range of originally described material (if available). (The prefix 'lecto' may be used with the preceding three terms to give lectoholotype, lectosyntype, etc.)

NEOTYPE. A specimen or range of specimens chosen from available material when the original type material has been lost or destroyed. A neotype may be a cast of the original material. (Neoholotype, neosyntype, etc. are also used.)

TOPOTYPE. Any specimen collected from the locality from which the type specimens were obtained.

GENOTYPE. The type species of a genus. The holotype of this species is referred to as the genoholotype.

Type locality. (1) The locality from which a type specimen comes.

(2) A locality chosen to be a standard for comparison of a stratigraphic unit. This is the type section of the unit.

Ugrandite. ◊ *Garnets.*

Uintan. A stratigraphic stage name, the North American equivalent of the European ◊ *Bartonian* (Tertiary System).

Ukrainian. A stratigraphic division of the Russian ◊ *Precambrian* Pre-Saamian.

Ulsterian. A stratigraphic stage name for the North American Lower ◊ *Devonian.*

Ultrabasic rocks (Ultramafites). Igneous rocks consisting essentially of ◊ *ferromagnesian minerals* to the virtual exclusion of quartz, feldspar, and feldspathoids. Originally defined as rocks containing less than 45% silica, this artificial boundary has now been abandoned as a result of the recognition of many rock types which, while undoubtedly ultrabasic by other tests, contain more than 45% silica, e.g. a rock consisting entirely of hypersthene would contain between 43 and 53% silica, and one entirely bronzite, between 53 and 55%. (◊ *Pyroxenes.*) Minerals containing chromium, and platinum-group elements, are almost confined to ultrabasic rocks.

The commonest ultrabasic rocks are plutonic (i.e. coarse-grained), and there are relatively few volcanic (i.e. fine-grained or glassy) examples. Several types, however, occur in minor intrusives such as ◊ *dykes* and ◊ *sills*, but are not usually distinguished by separate names. ◊ *Porphyritic* types are rare. The three main classes of ultrabasic rocks are as follows:

(1) Peridotite – consisting predominantly of olivine with or without other ferromagnesian minerals.

(2) Perknite – consisting predominantly of ferromagnesian minerals other than olivine.

(3) Picrite – consisting of 90% or more ferromagnesian minerals and up to 10% feldspars. By an increase in feldspar this type grades into gabbroic types. This is the only ultrabasic group to include volcanic types (e.g. ◊ *oceanite* and ◊ *ankaramite*).

Most ultrabasic rocks occur in association with basic types, especially in ◊ *layered igneous structures*, and there are few examples of purely ultrabasic masses. (For volcanic ultrabasics, ◊ *Basalts*.) The following are the main plutonic and hypabyssal types (other types are described by adding a mineral name to a

461

group name, e.g. hornblende-peridotite, augite-picrite): ◊ *Cortlandite*; ◊ *Dunite*; ◊ *Glimmerite*; ◊ *Harzburgite*; ◊ *Hornblendite*; ◊ *Kimberlite*; ◊ *Lherzolite*; ◊ *Olivinite*; ◊ *Pyroxenite*; ◊ *Sapelite*; ◊ *Saxonite*; ◊ *Scyelite*; ◊ *Serpentinite*; ◊ *Websterite*; ◊ *Wehrlite*. Rocks consisting entirely of ◊ *chromite* and ◊ *ilmenite* are regarded as ultrabasic.

Ultramafic. (1) ◊ *Ultrabasic.*

(2) Hypermelanic. (◊ *Colour index.*)

Ultramarine. ◊ *Feldspathoids* (Complex).

Ultra-metamorphic rock. Rock which has undergone the highest grade of metamorphism, forming under the highest possible temperatures and pressures short of actual melting. Minerals are formed which have a close-packed structure of maximum density. ◊ *Charnockite* and ◊ *eclogite* are regarded as typical ultra-metamorphic rocks.

Ulvöspinel. ◊ *Spinel group of minerals.*

Umber. A naturally occurring pigment consisting mainly of hydrated iron oxides, manganese oxides, and sometimes clay. Raw umber is a brownish colour which becomes a characteristic red-brown when heated (burnt umber).

Umbilicus (Mollusca). ◊ Figs. 100, 104 and 105.

Umbo (Brachiopoda). ◊ Fig. 11.

Umptekite. An ◊ *oversaturated* ◊ *syenite*.

Unakite. A rock of granitic characteristics containing much ◊ *epidote*. The primary origin of the epidote is in doubt, and it may have developed as a result of ◊ *metamorphic*, ◊ *metasomatic*, or ◊ *hydrothermal* action.

Unconformity. It is very difficult to define unconformity in a single comprehensive sentence. The concept is a complex one and needs to be treated from various viewpoints. A number of terms have been used at various times, either as synonyms for unconformity or for different kinds of unconformity. There are three major aspects:

(1) Time. An unconformity develops during a period of time in which no sediment is deposited. This concept equates deposition and time, and an unconformity represents unrecorded time.

(2) Deposition. Any interruption of deposition, whether large or small in extent, is an unconformity. This aspect of unconformity pre-supposes a standard 'scale' of deposition which is complete. Major breaks in sedimentation can usually be demonstrated easily, but minor breaks may go unrecorded until highly detailed investigations are made.

(3) Structure. Structurally, unconformity may be regarded as planar structures separating older rocks below from younger rocks above, representing the 'break' as defined in (1) and (2) above. A plane of unconformity may be a surface of weathering, erosion or denudation, or a surface of non-deposition, or possibly some combination of these factors. It may be parallel to the upper strata, make an angle with the upper strata, or be irregular. Subsequent earth movements may have folded or faulted it.

Unconformity indicates a change, either temporary or permanent, in conditions. It may involve recognition of an ◊ *orogenic* period, a marine ◊ *transgression* or ◊ *regression*, a ◊ *facies* change, a climatic change, and even sometimes a faunal change. Unconformities have been used to delimit stratigraphic systems or sub-divisions of systems, but this is now regarded as unsatisfactory. It is clearly impossible to know the size of the break represented, and in any case, when strata are discovered which fill the gap, difficulties arise in assigning them to one or other of the divisions. In fact, a better dividing line may become apparent when a full sequence is studied.

Four types of unconformity may be recognised (◊ Fig. 147):

(1) Angular unconformity. The lower, older series of beds dip at a different angle to the younger, upper beds. This also includes the case where unfolded, younger, strata rest upon folded, older, strata.

(2) Parallel unconformity. The lower and upper series of beds dip at the same amount, and in the same direction.

(3) Non-depositional unconformity. A minor type of unconformity, representing a short, often local, period of non-deposition of sediment. Such breaks can only be detected either because a detailed sequence of fossil zones can be established, or because some physical feature, e.g. a bored surface, ◊ *washouts*, or an intraformational conglomerate, can be recognised. The terms non-sequence and diastem are used for this type of unconformity.

(4) Heterolithic unconformity. This type arises when sediment is deposited on top of intrusive igneous rocks or metamorphosed rocks which have been exposed at the surface by weathering and erosion. It is debatable whether this term should be extended to the case where sediments have been deposited on the top of lava flows. A buried landscape is a type of heterolithic unconformity.

The following terms have been used to describe the relationship of the beds above a plane of unconformity to those beneath:

Overstep. The unconformity which develops during a marine

transgression so that the younger series rests upon progressively
older members of the underlying series.

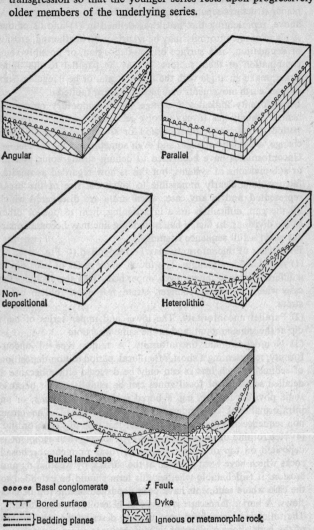

Angular	Parallel
Non-depositional	Heterolithic

Buried landscape

‚‚‚‚‚‚ Basal conglomerate		*f* Fault	
⊤⊤⊤⊤ Bored surface		◼ Dyke	
⇌⇌⇌ Bedding planes		Igneous or metamorphic rock	

147. Types of unconformity.

Overlap. This occurs when progressively younger members of the upper series rest upon the older series.

Offlap. This is the reverse of overlap, i.e. the lower beds of the

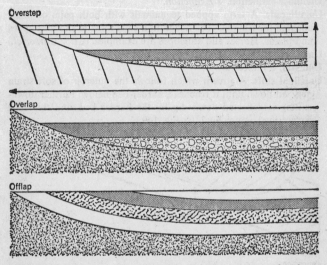

148. Terms describing the relationship of beds in an unconformity.

upper series extend further than the younger ones – a marine regression.

Onlap. This appears to be synonymous with overlap, or sometimes overlap plus overstep.

Non-conformity. As a synonym for unconformity or angular unconformity, this now seems to be obsolete.

Dis-conformity. This has been used in many senses; as a synonym for unconformity, in the sense of parallel unconformity as defined above, or as any kind of unconformity other than an angular one. Its use could well be abandoned.

S. I. Tomkieff, 'Unconformity – an Historical Study', *Proceedings of the Geologists' Association*, Vol. 73, Pt 4, 1962.

Undercutting. Erosion of material at the foot of a cliff or bank, e.g. a sea cliff, or river bank on the outside of a meander. Ultimately, the overhang collapses, and the process is repeated.

Underfit. A synonym of ◊ *Misfit*.

Undersaturated. A term used to describe igneous rocks deficient in

SiO$_2$, so that ◊ *unsaturated* minerals develop. The term is often applied to ◊ *feldspathoidal* rocks. (◊ *Oversaturated; Saturated.*)

Underthrust. ◊ *Faulting.*

Undulose extinction. ◊ *Extinction.*

Unguligrade. A term applied to animals with both pairs of limbs adapted for running. The adaptation suits them to life on steppes and plains. Horses and antelopes are examples. (Cf. ◊ *Cursorial.*)

Uni- (prefix). One.

Uniaxial. ◊ *Indicatrix.*

Uniclinal shift. A stream which follows the outcrop of soft strata may cut down until it reaches a harder bed below. Further downcutting is deflected down dip by the harder bed, and this shift is described as uniclinal. (◊ Fig. 149.)

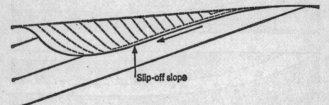

Slip-off slope

149. Uniclinal shift.

Uniclinal structure. Beds having a regular regional dip. Homocline is essentially a synonym. Neither term is synonymous with monocline (◊ *Folding*).

Uniformitarianism. The concept, fundamental to geology, that processes which operate at present also operated in the past, and produced the same results. These processes need not have operated at the same rate, nor at the same intensity. The saying 'The present is the key to the past', which is commonly offered as a 'definition' of uniformitarianism, is a considerable oversimplification. Acceptance of the principle of uniformitarianism does not exclude the possibility of a process which operated in the past not now being observable with current techniques; neither does it imply that all processes operating now operated in exactly similar ways in the past. It is fairly generally accepted that the degree of correlation between ancient and modern processes decreases as the time interval increases, e.g. in eras prior to the emergence of land vegetation, weathering and erosion must have been different in character and intensity compared with today.

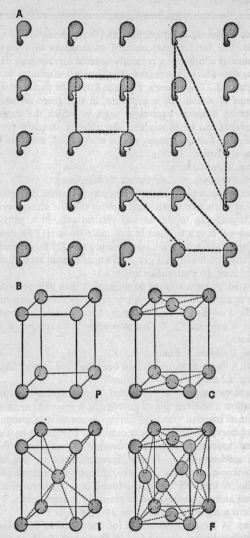

150. Unit cells. (A) Two-dimensional array showing alternative equivalent choices for the unit cell; (B) Simple orthorhombic lattices (Bravais lattices); P, primitive lattice; C, end-centred lattice; I, body-centred lattice; F, face-centred lattice.

Uniserial. ◊ *Chordata* (Graptolithina). (◊ *Echinodermata*.)

Unit cell. The fundamental element of structure in a crystalline substance. It consists of a regularly ordered arrangement of atoms (the lattice) which is repeated exactly in all directions and conforms with the ◊ *symmetry*. To some extent the choice of a group of atoms of a unit cell is arbitrary, in that there are usually a number of different equivalent ways in which the choice can be made. It can be shown that, by utilising standard symmetry operations, 14 simple lattices can be constructed; these are known as Bravais lattices. (◊ Fig. 150.)

Unit form. ◊ *Parameters*.

Univalve. Synonym of Gastropoda (◊ *Mollusca*).

Universal stage. A piece of apparatus which permits the rotation of a ◊ *thin section* or crystal grain about two (or more) horizontal axes, in addition to the normal vertical axis, of a petrological microscope. It can be used in two main ways: (1) To determine the optic characters of a mineral accurately; (2) To determine the orientation of the crystal grains in an orientated section of a deformed rock. (◊ *Petrofabric analysis*.)

Unsaturated. A term applied to minerals which cannot develop in stable equilibrium in the presence of excess silica. Such minerals include ◊ *olivine*, ◊ *feldspathoids*, ◊ *corundum*, ◊ *spinels*, and certain ◊ *garnets*. (Cf. ◊ *Oversaturated*; ◊ *Saturation*; ◊ *Undersaturated*.)

Upthrow, Upthrust. ◊ *Fault*.

Upwarp. A large area which has been uplifted, especially in the form of a broad, shallow anticline.

Uralian. A synonym of ◊ *Stephanian* (Carboniferous System).

Uralitisation. The alteration of an original ◊ *pyroxene* in an igneous rock to a mass of (usually) fibrous ◊ *amphibole*, commonly a type of hornblende, which was originally thought to be a distinct mineral species, 'uralite'. It commonly forms during the low-grade metamorphism of basic igneous rocks, or late stage ◊ hydrothermal alteration. (Cf. *Propylitisation*; *Saussuritisation*.)

Uraninite. A uranium ore, UO_2, found in ◊ *hydrothermal* veins, granites and pegmatites, and as grains in conglomerates. ◊ *Pitchblende* is a massive, impure type. (◊ Appendix.)

Uriconian. A stratigraphic name for the volcanic ◊ *Precambrian* underlying the ◊ *Longmyndian* in Shropshire, England.

Urtite. ◊ *Alkali syenite*.

Uvala. The coalescence of two or more ◊ *dolines*. (◊ *Karst scenery*.)

Vadose water. Water which occurs between the ground surface and the ◊ *water table*, i.e. the unsaturated zone. However it has been suggested that vadose water is equivalent to ◊ *meteoric water*, and the term may be ambiguous. (Cf. ◊ *Phreatic water*.)

Valanginian. A stratigraphic stage name for the base of the European Lower ◊ *Cretaceous*.

Valentian. A synonym of ◊ *Llandoverian*.

Valley bulging. ◊ *Fold*, Fig. 75.

Valley fill. Unconsolidated material partly or wholly filling a valley.

Valley profile. (1) ◊ *Long-profile*.

(2) The transverse profile of a valley, which can also be important, as in the comparison between the U-shaped cross-section of a glaciated valley and the V-shaped valley of a newly initiated stream.

Valve. One or more of the major skeletal units which make up a complete shell, e.g. a gastropod has a single-valved shell, lamellibranchs and brachiopods have two-valved shells (◊ *Brachiopoda*; *Mollusca*).

Varenius, Bernhard (1622–50). German scientist who produced in his *Geographia Generalis* in 1650 the first ever world-wide comprehensive description of the Earth's landforms.

Variation diagram. A graph or other similar diagram in which chemical features of a series of rocks are plotted against each other, or occasionally against some other property, e.g. colour index, specific gravity. The commonest type is the Harker diagram, in which percentages of elements (as oxides) are plotted against silica percentages. The diagrams are designed to reveal genetic and other relationships between rocks of a series. (◊ Fig. 151.) (◊ *Triangular diagram*.)

Variolite. A rock, originally glassy, which has developed ◊ *variolitic structure* throughout, leaving no interstitial glass. (◊ *Devitrification*.)

Variolitic structure. A type of ◊ *spherulitic texture* often occurring in glassy basic rocks (tachylites), at the chilled margin of ◊ *dykes* or ◊ *sills*. The radiating fibres making up the spherulite are generally plagioclase.

Varix (Mollusca). ◊ Fig. 100.

Varve. In its broadest sense the term is applied to the layer of sediment deposited in a single year. In practice, however, the use of the term is almost always confined to sediments deposited in glacial melt-water lakes. These consist of a coarser layer, representing summer deposition, and a finer layer, representing winter

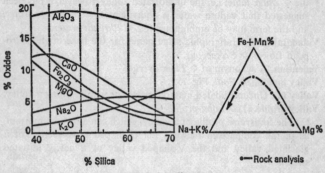

151. Variation diagrams.

deposition. Counting and correlation of these varves (notably by De Geer and his pupils) has led to the development of a detailed Pleistocene chronology for the Northern Hemisphere.

Vein. (The term lode is almost synonymous.) A vein is a tabular or sheet-like body of minerals which has been intruded into a joint or fissure, or system of joints and fissures, in rocks. Most veins are directly or indirectly of igneous origin, although in certain circumstances fissures may become filled as a result of sedimentary processes, e.g. calcite veins filling fissures in limestone. Veins are distinct from igneous ◊ *dykes*, although the term is often applied to the smaller intrusive tongues and fingers of igneous rocks. The most important use of the term vein, however, is in connection with ◊ *ore bodies*, and much of the terminology associated with veins derives from mining practice. (Because many veins occupy faults, much of the terminology of veins has been used, and defined, under ◊ *Faults*.) A veinlet is a small vein.

A saddle-reef or saddle-vein is a small ◊ *phacolith*-like body formed at the crest of an anticline. Ladder veins are those occupying a series of horizontal fractures commonly controlled in extent by lithology, e.g. horizontal mineralised joints in a dyke. A banded or crustified vein is a ◊ *multiple* and/or ◊ *composite* vein in which the various ore and ◊ *gangue* minerals are arranged in layers parallel to the walls of the vein, and represent successive periods of emplacement. Crustification may be symmetrical or asymmetrical. Gash veins are en échelon veins generally occupying tension gashes. (◊ *Joints*; ◊◊ *Replacement* (5), (6); *Stockwork*.)

Vent. The opening through which a ◊ *volcano* ejects igneous material.

Venter (Mollusca). ◊ Fig. 105.

Ventifact. A pebble faceted by the abrasive effects of wind-blown sand. The facets develop along the length of the pebble in a 'down-wind' direction. Initially three facets are present, two abraded and one residual from the original surface of the pebble, the pebble is then called a dreikanter. Ultimately, only two abraded faces are left, when it is called a zweikanter. (◊ *Wind erosion*.)

Venus' hair. Hair-like crystals of ◊ *rutile*, usually of a golden-brown colour, occurring embedded in quartz crystals.

Vermiculite. A group of ◊ *layer-lattice minerals*, especially certain chlorites, hydromicas, and clay minerals, which expand greatly when heated to give a light, cellular material much used for thermal insulation, packaging material, etc.

Vertebrates. ◊ *Chordata.*

Vesicular structure, Vesicles. Small spherical or ellipsoidal cavities found in volcanic ◊ *lavas*, which are produced by bubbles of gas trapped during the solidification of the rock. Where these are filled with a secondary mineral they are called ◊ *amygdales*. Lavas which contain abundant vesicles are pumice (acid lava) and scoriae (basic lava). Such froth-like material is often ejected from the volcano as ◊ *pyroclastic* material.

Vesuvianite. ◊ *Idocrase.*

Vibration directions. When a ray of light strikes an ◊ *anisotropic* medium, it is split into two rays:

(1) The ordinary ray (ω). The ray which passes through an anisotropic mineral as it would through an isotropic substance (e.g. a plate of glass). Thus such a ray perpendicular to the mineral passes through undeviated.

(2) The extraordinary ray (ε). This ray is produced and refracted, even with normal incidence, when light passes through an anisotropic mineral. When the mineral is rotated the ray describes a circle around the ordinary ray.

The two rays can clearly be seen by placing a clear cleavage rhomb of ◊ *iceland spar* over a small spot and observing the double image produced, one of the points moving in a circular path around the other as the cleavage rhomb is rotated.

The two rays are polarised in mutually perpendicular directions, corresponding to the two vibration directions of the mineral. The two rays arise because, in an anisotropic medium, the wave front is not perpendicular to the ray, except in certain directions. In an isotropic medium, the wave front is perpendicular to the ray in all directions.

Since the ordinary and extraordinary rays , $\omega \varepsilon$(respectively) correspond to different refractive indices, they also correspond to different velocities, and either ω or ε may be the fast ray. The polarisation of the ordinary and extraordinary rays has been used as a means of producing ◊ *polarised light* in the Nicol prism. Because of the differing velocity of the two rays there is a retardation between the light passing along the two paths. This retardation (R) is equal to the path length (d) multiplied by the difference in refractive index of the two rays $(n_1 - n_2)$:

$R = d(n_1 - n_2)$. If d $= 1$ then R $=$ the ◊ *birefringence*.

(◊ *Pleochroism*; *Polarisation colours*; *Axes, optic*.)

Villafranchian. A European stratigraphic stage name more or less precisely equivalent to ◊ *Calabrian* (Tertiary System).

Vindobonian. A stratigraphic stage name for the European Middle Miocene (◊ *Tertiary System*).

Virgation. A term used to describe a fan- or sheaf-like arrangement of mountain ranges.

Virgence. The converging or diverging of 'fans' of axial plane ◊ *cleavage*.

Virgilian. A stratigraphic stage name for the top of the North American ◊ *Pennsylvanian*.

Virglorian. A stratigraphic stage name, the Western European equivalent of ◊ *Anisian* (Triassic System).

Virgula (Chordata). ◊ Fig. 18.

Virgulian. A stratigraphic stage name for the Upper ◊ *Kimmeridgian* (Jurassic System).

Viséan. A stratigraphic stage name for the top of the European Lower ◊ *Carboniferous*.

Vishnevite. ◊ *Feldspathoids*.

Vitrain. ◊ *Coal*.

Vitreous. ◊ *Lustre*.

Vitric. Literally, glassy. (◊ *Pyroclastic rocks*.)

Vitrinite. ◊ *Coal*.

Vitrite. ◊ *Coal*.

Vitro- (prefix). ◊ *Glassy*.

Vitrophyre. A ◊ *porphyritic* texture in which the ◊ *groundmass* is ◊ *glassy*.

Vogesite. ◊ *Lamprophyre*.

Volatiles. Elements and compounds which are dissolved in a silicate melt and which would be gaseous at the temperatures involved were it not for the elevated pressures and the solvent effect of the magma. The commonest volatiles are water and carbon dioxide; others include chlorine and hydrochloric acid; fluorine, hydrofluoric acid, and many fluorides, including those of boron and iron; sulphur, sulphur oxides, and in special circumstances, hydrogen sulphide; exceptionally, nitrogen, ammonia, and borates may occur. The volatile constituents of magma are termed 'juvenile' if they are original constituents and 'resurgent' if they have been introduced as a result of contamination of the magma by material from country rocks.

Volatiles produce three major effects upon magma: (a) They lower the viscosity of the melt, thus making it more mobile. (b) They depress the freezing point of the melt, thus allowing it to remain liquid to a much lower temperature and prolonging the period of crystallisation. (c) They react with earlier formed minerals to produce various alteration products. (◊ *Hydrothermal processes*; *Pneumatolysis*; *Serpentinisation*; etc.)

The crystallisation of a melt commonly results in the formation of a volatile-rich residuum, from which ◊ *pegmatites* and ◊ *mineral deposits* arise.

It is estimated that the water content of a typical granite magma is about 5%; this amounts to 53×10^6 tons per cubic mile. Basic magmas are probably somewhat lower in volatiles.

Volatiles are important constituents of volcanic rocks, and modern volcanoes provide a direct source of volatiles for study. The commonest volcanic gas is undoubtedly steam (both juvenile and resurgent), followed closely by carbon dioxide. Most of the volatiles mentioned above also occur and to these may be added the chlorides of ammonium, iron, potassium, and other elements, and borates and boric acid. Sublimation of some materials, e.g. sulphur, occurs, while others crystallise out from solution in hot water. Emission of volatiles on a large scale is most noticeable during the waning stage of vulcanicity. The eruption of steam and hot water via the vents known as ◊ *geysers* is well known, while hot springs and pools, known as fumaroles and solfataras, are common. Many fumaroles develop a deposit of material crystallised from solution, e.g. the borates mentioned above, siliceous sinter and travertine ($CaCO_3$).

Gases escaping from lavas cause ◊ *vesicles* or *amygdales* to form and may also introduce secondary minerals such as ◊ *zeolites* into these cavities.

Volcanic. One of three groups into which H. H. Read has divided rocks. The volcanic assemblage includes all extrusive rocks and the associated intrusive ones. The group is dominantly basic, strictly magmatic, and usually but not necessarily associated with ◊ *orogeny.* (◊ *Plutonic*; *Neptunic.*) The term is used loosely as a synonym of 'fine-grained and/or glassy'.

Volcanic bombs. ◊ *Pyroclastic rocks.*

Volcanic breccia. (◊ *Pyroclastic rocks.*) Consolidated, early-formed volcanic material, brecciated by a later explosive volcanic eruption. (Cf. ◊ *Agglomerate.*)

Volcanic products. Volcanic products may be classified under three main headings: gaseous products including water (◊ *Volatiles*), liquid products (other than water) (◊ *Lava*), and solid products (◊ *Pyroclastic rocks*). There is an overlap to some extent between the last two, since some of the latter have solidified from a molten liquid.

Volcanoes and vulcanicity. A volcano is a vent or fissure in the Earth's crust through which molten magma, hot gases, and other fluids escape to the surface of the land or, in certain cases, the bottom of the sea. At the present time volcanoes are confined to certain limited areas of the Earth's surface; in the past, however, vulcanicity has occurred at a large number of sites, and volcanic rocks are a part of the geological column in most parts of the world. There is a generally clear association of vulcanicity with ◊ *orogeny*, and at the present time volcanoes are chiefly located along the great circum-Pacific orogenic belt, the Caribbean orogenic belt, the Alpine–Mediterranean–North African–Himalayan belt which stretches down through Indonesia, and the Rift Valley regions of Africa. There are also a large number of submarine volcanoes in certain regions of the oceans, such as the Mid-Atlantic Ridge, and certain zones in the Pacific, notably the Hawaiian area and the New Zealand–Tonga region.

Volcanoes may be classified broadly into central types – where the products escape via a single pipe (vent) – and fissure types – where the products escape from a linear vent or crack. A number of different types of central volcanic eruptions have been recognised, some undoubtedly resulting from variation in the gas content of the liquid magma and the composition of the magma, both of which have a pronounced effect on viscosity. It has also been suggested that several of the major types may in fact represent stages in an evolutionary sequence, with volcanoes commencing as one type and progressing through several stages until they eventually become quiescent. The following are the main types of volcanic eruptions and their characteristics:

HAWAIIAN. Very large open craters with quiet evolution of mobile lava and no explosive activity, although lava fountains may be flung up on to the surface of the lava lake by jets of gas.

STROMBOLIAN. Characterised by continuous small explosive eruptions; a rather more viscous lava which develops a crust under which gas accumulates until sufficient pressure exists to explode the solidified crust away.

VULCANIAN. Eruptions take place at longer intervals than those of the Strombolian type, owing to the fact that the lava is considerably more viscous and that the crust that develops is thicker and requires a higher gas pressure to destroy it. This is the type of eruption with which many volcanoes begin.

VESUVIAN. In this type long intervals, measured in tens of years, occur between eruptions, which are extremely violent, with large quantities of material being flung out by explosive activity. The lava itself is highly viscous, and the plug of solidified material which develops allows an enormous gas pressure to build up. The most violent type of Vesuvian eruption is sometimes described as Plinian.

PELÉAN. These develop as a result of the eruption of very viscous lavas, and are characterised by the eruption of ◊ *nuées ardentes* and 'spines' (see below).

Many volcanoes develop parasitic or daughter cones on the flanks of the main mountain (◊ Fig. 152), and these may become so extensive that the main crater becomes extinct; this has happened with Etna, which is probably the wreck of a shield or Hawaiian-type volcano.

The more viscous lavas tend to become disrupted by the explosive phase and give rise to fragmental or ◊ *pyroclastic* deposits (volcanic ash, tuffs, etc.). Less viscous lava types give rise to the characteristic ◊ *lava* flows. A classification of the main forms of volcano can be attempted on this basis. The characteristics of explosive activity are explosion vents (sometimes called maars) and ash or cinder cones. Many volcanoes have composite cones consisting of ash and pyroclastic deposits together with lavas. The effusive forms consist of lava domes, either growing by internal activity or external outpourings – the so-called shield volcanoes – and lava plateaus which are characteristic of fissure eruptions. In many cases explosive activity will give rise to a ◊ *caldera*, a

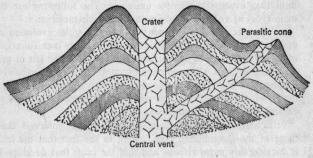

152. A volcano.

name usually applied to rather large-scale depressions formed either by explosion or collapse. In general there is a correlation between the relative acidity of lava and the viscosity: the more acid the lava the greater the viscosity. The most acid lavas often give rise to quite small conelets, sometimes called driblet cones, the viscous lava only flowing a few feet from its source. Very small lava flows are sometimes called hornitos. Occasionally in very viscous lava volcanoes, a spine of solidified material is pushed out of the crater by gas pressure, but in most cases this is rapidly weathered away. The most basic lavas, the ◊ *basalts*, have low viscosities and flow many miles, some of the Hawaiian flows reaching more than thirty miles from their point of eruption. Fissure eruptions are almost exclusively of basic lava. Volcanoes do not go on erupting indefinitely, and over a period of years degenerate and cease erupting lava and pyroclastic material; however in many cases this stage is followed by a period of ◊ geysers, hot springs and fumarole (solfatara) eruption (◊ *Volatiles*), and this slowly dies away as the underground source of heat decays. In New Zealand and Italy the underground heat of volcanic regions is being used as a source of energy.

Volgian. A stratigraphic stage name for the European Upper ◊ *Kimmeridge* + ◊ *Portland* + ◊ *Purbeck* (cf. ◊ *Tithonian*). Other usages also exist.

Vraconian. A stratigraphic stage name for the ◊ *Albian*/◊ *Cenomanian* transition beds in Europe (Cretaceous System).

Vugh. A cavity in a rock usually with a lining of crystalline minerals, a term used especially of cavities in mineral veins. (◊ *Geode*; *Druse*.)

Waalian. A stratigraphic stage name for the European Lower Pleistocene (◊ *Tertiary System*).

Wacke. A poorly sorted, ill-graded ◊ *arenaceous* sediment. The term is properly used only in the term greywacke and its use separately in any other sense is not recommended.

Wad. A soft black earthy mass of hydrated manganese oxides, usually containing iron oxides and occasionally barium or cobalt as well. Probably a kind of residual deposit.

Wadi. A valley with an intermittent stream, especially as developed in semi-desert areas.

Wall rock. The ◊ *country rock* of a vein or lode (◊ *Ore*).

Waltonian. A stratigraphic stage name for the British Lower Pleistocene (part ◊ *Ludhamian*) (◊ *Tertiary System*).

Wasatchian. A stratigraphic stage name for the North American equivalents of the ◊ *Cuisian* + ◊ *Ypresian* (◊ *Tertiary System*).

Wash. A term used to describe loose debris, commonly applied to alluvial material, especially in the dry beds of intermittent streams, or loose material on desert pavements. The term is normally used with an explanatory prefix, e.g. outwash, downwash.

Washout. A channel cut into sediment and filled by later material. The best-known examples come from coal fields where streams eroded channels through the rotting vegetation into the underlying sediments, which were subsequently filled with sand, etc. Washouts normally develop in shallow water or swampy conditions, e.g. deltas, coastal swamps, shallow lakes, and lagoons. The infilling material commonly contains fragments of the eroded bed. (◊ Fig. 153.) Washouts can be used as a ◊ *way-up criteria*.

Waterfall. A point in the course of a river where the water descends more or less vertically. Waterfalls may be produced by: (1) Differential erosion of a soft rock lying below a hard rock. (2) A mass of hard rock lying across the course of the river. (3) The knickpoint of a ◊ *rejuvenated* stream. (4) The damming of a stream by a landslip. Waterfalls are temporary features and represent a maladjustment of the river to the structure. A perfectly ◊ *graded* stream will have eliminated such irregularities.

Water gap. A pass through high ground occupied by a river or stream. (Cf. ◊ *Wind gap.*)

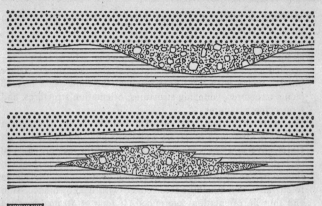

 Channel-fill

153. The formation of a washout.

Watershed. The region of higher ground which lies between two non-communicating drainage systems. It may be regarded as the boundary between two adjacent drainage systems and is synonymous with the term 'divide'.

Water table. The plane which forms the upper surface of the zone of ◊ *groundwater* saturation. Above the water table the rocks are

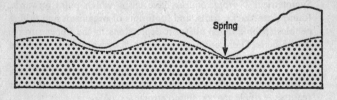

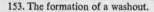 Water Table

154. A water table.

unsaturated. Should the water table rise so that it intersects the ground surface a ◊ *spring* results. The level of the water table measured from the ground surface (standing water level) is controlled partly by topography, partly by the nature of the near-surface rocks, and partly by the local climatic conditions. (◊ Fig. 154.)

Waucoban. A stratigraphic stage name for the North American Lower ◊ *Cambrian*.

Wave-cut notch/platform. ◊ *Marine erosion*.

Way-up criteria. Any geological phenomena which enable the original orientation of a rock mass to be determined. Other phrases which have been used in much the same sense include facing structures, younging structures, orientation structures, and sequencing structures.

The most important use of such criteria is in determining sequences in layered rocks, i.e. sediments and interbedded lavas. Unless inversion of the sequence is suspected for tectonic reasons, or layers are vertical, the normal principle of ◊ *superposition* should be assumed to apply. Also strata containing a reasonable fossil fauna can usually be correlated with a known succession and thus an orientation determined (◊ *Law of Strata Identified by Fossils*). Fossils in the position of growth may also be used, e.g. corals, sponges, plant roots, etc. Certain types of shells although not growing in a fixed orientation tend to lie on the sea floor in their most stable orientation, e.g. Productid brachiopods lie on bedding planes with their convex surfaces uppermost.

The most reliable criteria are ◊ *cross-bedding* and ◊ *graded bedding*. In an ◊ *unconformable* contact between folded and/or metamorphosed beds, and unfolded and/or unmetamorphosed beds, the latter are the younger.

Symmetrical ◊ *ripple marks* have cusps which point upwards. Rain prints, tracks, trails, and footprints of organisms and falling pebbles (pebble dints) all form depressions in the upper surface of the beds. Desiccation cracks taper downwards and are filled with sediment from the overlying beds (◊ *Sedimentary structures*). ◊ *Washouts* and channels cut into the underlying beds. Erosion on a small scale may also remove the tops of other features – ripple marks, slump structures, etc.

In volcanic rocks, tongues of lava may penetrate underlying sediments but care must be taken to distinguish this from deposition of sediment on an irregular surface of the lava, particularly when observing vertical sections. ◊ *Vesicles* in lava flows are generally concentrated at the top of a flow. A weathered or eroded upper surface may sometimes be recognised. Pipe ◊ *amygdales*, however, are more abundant in the lower parts of the flows, and if they are branched they bifurcate downwards. Pillow ◊ *lavas* have a rather rounded bulbous top and a more pointed, tapering, lower extremity. However, if the rocks have been extensively deformed these criteria may be difficult to establish.

Fossil 'spirit levels' occur when a cavity in a rock (a vesicle, fossil brachiopod or mollusc, or a solution cavity) becomes partly filled with fine sediment or material deposited from solution. The level surface of such partial filling defines an original horizontal plane. (◊ Fig. 155.) (◊ *Trace fossil*.)

Wealden. A stratigraphic name for the local fresh-water facies of the British ◊ *Neocomian* (Cretaceous System).

Weathering. The process by which rocks are broken down and decomposed by the action of external agencies such as wind, rain, temperature changes, plants, and bacteria. Weathering is the initial stage in the process of ◊ *denudation*. An essential feature of the process is that it affects rocks *in situ*; no ◊ *transportation* is involved. This is the factor which distinguishes it clearly from ◊ *erosion*. There are two main types of weathering:

(1) MECHANICAL. This is brought about chiefly by temperature changes, e.g. the expansion of water on freezing, in pores or cracks of the rocks; the differential expansion of the rock, or rock minerals, when strongly heated by the sun (insolation). This latter process tends to cause thin sheets of rock to split off (exfoliation – onion-skin weathering). The action of plant roots penetrating into cracks in rocks is frequently powerful enough to cause splitting.

(2) CHEMICAL. This is mainly brought about by the action of substances dissolved in rain water. They are usually acidic in character and leach rocks quite actively – a typical example is the solution of limestone by water containing dissolved CO_2.

$$2CaCO_3 + H_2O + CO_2 = 2Ca(HCO_3)_2$$
  Limestone Water   Calcium bicarbonate (soluble)

Under tropical conditions, intensive extractive leaching removes most elements from the rock except aluminium and iron, which remain as hydrated oxides in the form of ◊ *laterite and bauxite.*

Climatic factors play a large part in weathering. Chemical weathering is virtually absent in arid regions, while effective freeze-thaw action is confined to cold temperate and subarctic climates.

Websterite. An ◊ *ultrabasic rock* consisting of ◊ *hypersthene* and ◊ *diopside.*

Wehrlite. An ◊ *ultrabasic rock* consisting of ◊ *olivine* with accessory ◊ *augite* (diallage).

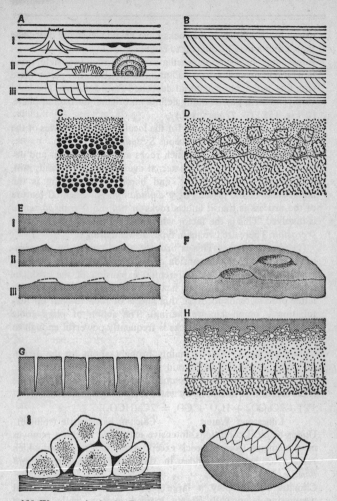

155. Way-up criteria. (A) Fossils in position of growth: i, tree roots, footprint; ii, brachiopod, coral colony; iii, simple corals; (B) Cross-bedding; (C) Graded bedding; (D) Included fragments; (E) Ripple marks: i, symmetrical; ii, asymmetrical; iii, eroded; (F) Rain prints; (G) Desiccation cracks (sun cracks); (H) Lava flow with scoriaceous tops and pipe amygdales; (I) Pillow lavas in section; (J) A fossil 'spirit level'.

Weischelian. A stratigraphic stage name for the top of the European Upper Pleistocene (◊ *Tertiary System*).

Weiss indices. ◊ *Indices, crystallographic.*

Welded. A term currently used almost exclusively to describe certain pyroclastic rocks in which the particles have been caused to cohere by the heat retained by the particles and the associated gas immediately after deposition. (◊ *Pyroclastic rocks* (Ignimbrite; Welded tuff).)

The use of the word as a synonym for indurated, coherent or compacted and applied to sediments as a term in ◊ *diagenesis* seems undesirable, as, although the process of induration of sediments may involve cohesion of particles, this appears to be a pressure phenomenon rather than a thermal one.

Welded tuff. ◊ *Pyroclastic rocks* (Ignimbrites).

Well-logging. The term applied to a variety of techniques of measurement of physical properties (by lowering suitable instruments) in a borehole to provide information about the geological strata traversed, or the directional attitude of the borehole itself. Logs may be run to give continuous and detailed information throughout the full depth of a borehole, or they may be restricted to selected zones of particular interest for potential production of oil, gas, or water. In zones of poor core recovery, well-logging has great advantages in providing information of use in solving problems of correlation between adjacent wells, because of the completeness of the well log record. Techniques employed include:

SPONTANEOUS OR SELF-POTENTIAL (S.P.) LOG. The measurement of variations in potential due to the natural currents which flow in the circuit formed by less permeable strata such as shale, the drilling fluid in the borehole, and more permeable strata such as sandstone.

RESISTIVITY LOG. The measurement of the resistivity of strata by means of electric currents applied via a multi-electrode sonde. The correlated interpretation of S.P. and resistivity logs can provide information about the ◊ *porosity* and ◊ *permeability* of strata, and about the nature of the pore fluid.

GAMMA RAY LOG. The measurement of the natural gamma ray activity of the strata by means of a Geiger counter or, preferably, a scintillometer.

NEUTRON LOG. The measurement of induced gamma ray activity due to the capture of neutrons emitted from a suitable source. Gamma ray and neutron logs provide information about the nature of the strata penetrated and give a measure of the hydrogen-bearing fluid content.

NUCLEAR MAGNETISM LOG. The measurement of the magnitude of the voltage due to proton precession, induced by the temporary application of a strong magnetic field, gives a measure of the free fluid index, and hence of the formation porosity. Information about the nature of the pore fluid can also be obtained by recording the relaxation time of the proton precession signal.

ACOUSTIC VELOCITY LOG. The measurement of the travel time of acoustic energy from a source via the strata to a receiver yields the velocity of travel, which varies with the relative path lengths through solid and liquid materials, and hence a measure of formation porosity can be obtained.

DIPMETER. The correlation of S.P. or resistivity values from three electrodes set at 120° to each other enables the dip and strike of a formation to be calculated by the solution of a ◊ *three-point problem*. The orientation of the dipmeter is determined by means of a photoclinometer attached to the instrument.

TEMPERATURE LOG. The variation of temperature in a borehole due to differential rates of heat exchange between drilling fluid and formations can yield information about the nature of the strata. A temperature log can also locate the source of gas flow, zones of lost circulation of drilling fluid, and the position of cement set between casing and the borehole wall.

PHOTOELECTRIC LOG. Changes in the opacity of drilling fluid, determined by photo-electric means, can locate the source of influx of formation water into a borehole.

BOREHOLE SURVEY. The deviation of a borehole from the vertical can be determined by recording the positions of a freely suspended pendulum bob, or the reference ball-bearing in a spherical level, azimuth control being provided by a magnetic compass, gyroscope, or orientated drill pipe.

CALIPER LOG (SECTION GAUGE). The variation in position of three spring-loaded flexible arms pressing against the borehole wall can be made to alter inductance or resistance components in an electrical circuit in such a way that the current variations caused can be calibrated to yield a continuous log of variation in borehole diameter. This information may be required in the interpretation of other logs which are sensitive to deviations from constant bore, or in calculations required in various well completion and production processes.

FLUID LEVEL MEASUREMENT. In pumping wells, the fluid level may be determined by the seismic principle of the reflection of sound waves.

GUN SAMPLERS. To obtain small core samples from the wall of a borehole in intervals where coring did not take place during drilling, hollow captive bullets are fired into the strata by small explosive charges, the bullet with its contained core being retrieved by wires attached to the gun, which can store up to thirty bullets and thus sample several zones on a single traverse of the borehole. A related technique involves the firing of radioactive marker bullets into the wall of the borehole to provide fixed reference points in a formation at great depth where difficulties in the precise positioning of logging instruments may arise, owing to stress and temperature effects on the cables connecting with the recorders at the surface.

Wemmelian. A stratigraphic stage name for the top of the European Upper Eocene (◊ *Tertiary System*).

Wenlockian. A stratigraphic stage name for the European Middle ◊ *Silurian*.

Wentworth-Udden scale. ◊ *Particle size*.

Werner, Abraham Gottlob (1749–1817). German geologist who was one of the great teachers of the subject, largely responsible for spreading the doctrine that the Earth will only be understood by observation in the field and the laboratory. He developed his own geological column and system of stratification and produced the neptunean theory of the origin of the earth which was widely accepted for about fifty years. This theory of the origin of rocks in aqueous solutions was based upon the methods of observation which he preached, but his fieldwork was too parochial for him to always reach the right conclusions – for example he believed that volcanoes had their origin in the combustion of coal seams. His principal British disciple was Robert Jameson.

Westphalian. A stratigraphic stage name for the European mid Upper ◊ *Carboniferous*.

Whaleback dunes. ◊ *Dunes*.

Whin stone. A colloquial term covering any dark fine-grained igneous rock, e.g. ◊ *dolerite*, ◊ *basalt*, ◊ *andesite*. The name is derived from the Great Whin Sill of Northumberland and Durham.

Whiston, William (1666–1753). A mathematics professor who won many admirers for his suggestion that the Earth originated as a comet which melted into shape by approaching the Sun and took on its landforms with the changes caused as it travelled away from the Sun again. Paradise was situated below the Tropic of Capricorn and the earth only began to rotate after the Fall of Man – which was punished by the great Flood!

Whitbian. A stratigraphic stage name for the base of the British Upper ◊ *Lias* (Jurassic System).

Whitcliffian. A stratigraphic stage name for the top of the British Upper ◊ *Ludlovian* (Silurian System).

Whitehurst, John (1713–88). English clock-maker and 'Stamper of the Money-Weights' who was both a keenly observant field geologist who realised the igneous origin of certain rocks, and a fanciful theorist on the origin of earth – explaining everything from Biblical chaos and rainbows to the possibility of a second Flood.

Whitneyan. A stratigraphic stage name, the North American equivalent of the European ◊ *Chattian* (Tertiary System).

Whorl (Mollusca). ◊ Figs. 100 and 105.

Wildcat. A borehole (or more rarely a mine) sunk in the hope of finding oil (or ore) in a region where deposits of oil or metallic ores have not been recorded. As usually employed, the term means any highly speculative exploratory operation, generally in search of oil.

Wind erosion. The process of the wearing away of exposed rocks by the abrasive action of particles carried by the wind. For such erosion should take place the particles must be of a reasonable size (approximately $\frac{1}{4}$ mm. minimum) and must be propelled against the rock surface with as high a velocity as possible. Wind erosion only occurs where there is an absence of vegetation and/or moisture to bind the loose soil together; it is thus most efficient in arid climates – deserts, and semi-deserts. Under ordinary climatic conditions, particles picked up by winds are too small and have too low a velocity to do much erosion, and it is only when winds of Beaufort force 6 are exceeded that sandstorms develop during which much erosion occurs.

Owing to the drag effect of the ground surface upon wind, the maximum velocity and hence maximum eroding power occurs a short distance from ground level, varying from a few inches to several feet according to local conditions. This effect produces some of the spectacular erosion features of many desert areas. Windblown sand grains are especially efficient at etching out bands of different hardness in rocks. (Cf. ◊ *Ventifact*; ◊ *Yardang*; ◊ *Zeugen*.)

Wind gap. A pass through high ground not at present occupied by a river or stream. Most wind gaps are produced by one of the following mechanisms: (1) ◊ *river capture*; (2) ◊ *glacial breaching*; (3) accidental diversion, e.g. landslides; (4) climatic changes.

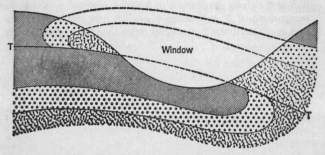

156. The formation of a window. T–T, thrust plane.

Window. An area where erosion has penetrated a thrust (◊ *Fault*) or recumbent ◊ *fold* to expose the rocks lying beneath (◊ Fig. 156). (Cf. ◊ *Inlier*.)

Witherite. A barium mineral, $BaCO_3$, found in hydrothermal veins associated with ◊ *barytes* and ◊ *galena*. (◊ *Carbonates* and Appendix.)

Wolfcampian. A stratigraphic stage name for the North American basal ◊ *Permian*.

Wolframite. An ore of tungsten, $(Fe'',Mn)WO_4$, found in ◊ *hydrothermal* and ◊ *pneumatolite* veins. (◊ Appendix.)

Wollaston, William Hyde (1766–1828). English chemist and mineralogist who is remembered among geologists for his invention of the goniometer and in the name of the Geological Society of London's highest award – the Wollaston Medal.

Wollastonite. A mineral of the ◊ *pyroxenoid* group, $CaSiO_3$, formed by the thermal metamorphism of impure limestone. (◊ Appendix.)

Woodward, John (1665–1722). English physics professor who was one of the first to recognise the value of systematic fieldwork and produced a catalogue of English fossils as a result of this. He suggested that the earth's stratification was the result of the Flood – that is that the Flood broke up the surface and redeposited it in layers according to the density of material – fossils being the organic remains deposited in this way.

Wulfenite. A mineral, $PbMoO_4$, found in the oxidised zones of lead deposits. ($\diamond$ Appendix.)

Würtherian. A synonym of $\diamond$ *Skythian* (Triassic System).

Wyomingite. A lava consisting of $\diamond$ *leucite*, $\diamond$ *phlogopite*, and a $\diamond$ *pyroxene* in a glassy groundmass. On the basis of the composition of the glassy groundmass (it has been reported as having the composition of $\diamond$ *orthoclase* + $\diamond$ *nosean*) it is probable that the rock is related to the $\diamond$ *phonolites* rather than the $\diamond$ *alkali basalts*.

Xeno- (prefix). Strange, foreign.

Xenoblastic. A textural term applied to metamorphic rocks. It is applied to mineral grains which have developed without showing crystal faces and corresponds to ◊ *allotriomorphic* in the igneous textural terminology.

Xenocryst. The term applied to crystals, often apparently ◊ *porphyritic*, in an igneous rock which have not formed in the place in which they are now found, i.e. they have been introduced into the melt from some extraneous source. These crystals may be derived from part of a body of magma which has undergone ◊ *differentiation* and which is crystallising material of different characters in different places.

Xenolith. An inclusion of pre-existing rock in an igneous rock. The fragment may be derived from the ◊ *country rock* or may be a fragment of an earlier solidified portion of the igneous rock having a slightly different composition. Xenoliths may show extensive reaction with the igneous rock and may have suffered partial ◊ *assimilation*. (Enclave is effectively a synonym.)

Xenomorphic. Having no crystal form. (◊ *Texture*.)

Yardang. A grooved or furrowed topographic form produced by wind abrasion acting upon weakly consolidated sediments. They are elongated in the direction of the prevailing winds and are usually strongly undercut. (◊ Fig. 157.) (◊ *Wind erosion.*)

157. A yardang.

Yeovilian. A stratigraphic stage name for the top of the British Upper ◊ *Lias* (Jurassic System).

Young, to (intransitive verb). A usage invented to avoid the awkward construction 'the bed of rock is orientated so that its youngest (i.e. uppermost) surface faces north' – which may be written 'the bed of rock youngs towards the north'. The term has been objected to as a 'barbarism', and the alternative 'to face' has been suggested, e.g. 'the bed of rock faces to the north'. It will be appreciated that this is ambiguous, especially to a person whose mother tongue is not English.

Ypresian. A stratigraphic stage name for the European Lower Eocene (◊ *Tertiary System*).

Zechstein. A European stratigraphic stage name equivalent to ◊ *Thuringian* (Permian System).

Zeolites. A group of tekto◊*silicates* containing true water of crystallisation; one of the few mineral groups displaying reversible dehydration. The cations present typically include Ca, K, Na, Ba. Zeolites show powerful properties of base exchange (◊ *Clay minerals*), and sodium zeolites were originally used as water softeners. Although the zeolites are classified as tektosilicates, the degree of bonding between the SiO₄ tetrahedra varies considerably, and results in the following three classes of zeolites:

(1) The fibrous group: natrolite, mesolite, scolecite.

(2) The platy group: heulandite, stilbite.

(3) The equant group: harmatome, chabazite (analcite, which is sometimes included in this group, is here regarded as a ◊ *feldspathoid*).

The structure of the zeolites is such that open channels occur through the lattice and this property has been utilised as a molecular sieve. The chemistry of the group is complex. Crystals are often twinned, which sometimes obscures the true symmetry.

Most zeolites are found in ◊ *amygdales* and other cavities in basic volcanic rocks, but they may also occur in other volcanic rocks, and not uncommonly as late-stage low-temperature minerals in ◊ *hydrothermal* veins. The evidence suggests strongly that they are true hydrothermal products, and are not a result of leaching of feldspars in the primary rock. Zonal arrangements of zeolites in thick piles of lava flows have been demonstrated in several localities, e.g. Iceland; Mull, W. Scotland; N. Ireland.

Zeugen. 'Rock mushrooms'; tabular masses of more resistant rock resting on undercut pillars of softer material. They are very often elongated in the direction of the prevailing wind. (◊ Fig. 158; ◊◊ *Wind erosion*.)

Zinc blende. ◊ *Sphalerite*.

Zinnwaldite. A lithium-bearing biotite mineral (◊ *Mica*; *Pneumatolysis, Greisening*).

Zircon. A zirconium mineral, $ZrSiO_4$, found as an accessory in the more acid igneous rocks and in detrital deposits. (◊ Appendix.)

Zirkel, Ferdinand (1838–1912). German professor who realised the value and importance to petrology of ◊ *Sorby*'s use of thin sections, publicising it in his teaching and in his books – particularly that on the classification of basalts and on the microscopic structure of minerals, which emphasised and explained the methods

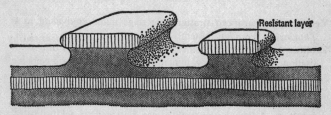

158. A zeugen.

used. Later he produced a complete text-book survey of the petrological nature of rocks, with a classification based upon mineralogical composition (cf. ◊ *Rosenbusch*).

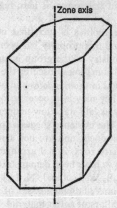

159. A zone axis.

Zoantharia. *Coelenterata* (Anthozoa).

Zoisite. ◊ *Epidotes* and Appendix.

Zone, crystallographic. A set of ◊ *faces* which meet in a series of parallel edges. The faces involved need not all belong to the same ◊ *form*. An axis parallel to these edges is known as a zone axis. (◊ Fig. 159.)

Zone, metamorphic. The concept of metamorphic zones was orginally evolved by Grubenmann, and developed extensively by Barrow in the regionally metamorphosed areas of the Scottish Highlands. (◇ Fig. 160.) Zones are effectively 'mapping units' defined by the appearance of an index mineral in rocks of appropriate composition. Thus Barrow and later workers, using metamorphosed argillaceous rocks, established six zones characterised by the following minerals:

Low grade	(1) Chlorite
	(2) Biotite
	(3) Almandine garnet
	(4) Staurolite
	(5) Kyanite
High grade	(6) Sillimanite

It has been shown that these Barrovian zones are not the only possible set, e.g. in the Buchan district of north-east Scotland a series of zones characterised by the occurrence of andalusite, cordierite, staurolite, and sillimanite are found. The Abukuma type has been described from Australia and Japan and appears to have features in common with the Buchan type; however almandine garnet replaces staurolite. Broadly speaking, the assemblage of rock types formed in a zone will constitute a metamorphic ◇ *facies* (or possibly a sub-facies).

The use of zones as 'mapping units' depends upon being able to locate the first appearance of the index mineral. Since this may be affected by the bulk composition of the parent rock, it is essential that the rocks used to define zones should be as near isochemical as possible. It has been found that the staurolite zone is apparently missing over fairly wide areas of Scotland because of the absence of parent rocks of appropriate composition.

Zone, stratigraphic. ◇ *Stratigraphic nomenclature.*

Zone fossil. A fossil species which characterises a particular horizon, and is restricted to it in time.

Zoned crystal. A crystal which does not have a uniform composition throughout its volume. It is usually a result of non-uniform distribution of cations in the atomic structure. In most cases the variation occurs in shells arranged concentrically around a point or axis. Sometimes the variation gives rise to visual features (e.g. colour banding in tourmaline – ◇ *Cyclosilicates*), while in other cases the change is only revealed by careful examination (e.g. zoning in plagioclase ◇ *feldspars* is often visible only in ◇ *thin*

493

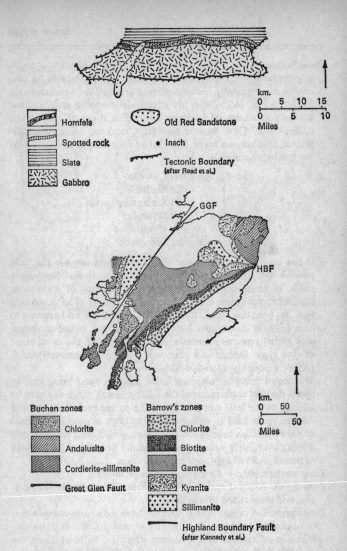

160. Metamorphic zones. (A) The thermal aureole around the Insch gabbro in Aberdeenshire, Scotland, showing the development of thermal metamorphic zones on the north side; (B) A simplified map of the distribution of the zones of regional metamorphism in Scotland between the Highland Boundary Fault and the Great Glen Fault.

section, either as a change in refractive index or as a change in ◊ *extinction* position). Zoning arises in mineral groups where continuous series of solid solutions exist between a set of end-

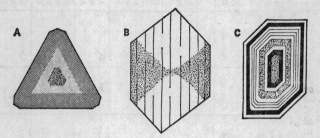

161. Zoned crystals. (A) The prismatic zoning of tourmaline; (B) The hour-glass zoning of pyroxene; (C) The concentric zoning of feldspar.

members, and usually reflects progressively changing conditions of crystallisation. (◊ Fig. 161.)

Zweikanter (German, two-edged). A wind-faceted pebble. (◊ *Ventifact*.)

APPENDIX: TABLE OF MINERALS

Crystal System: C Cubic, H Hexagonal, M Monoclinic, O Orthorhombic, Tet. Tetragonal, Tric. Triclinic, Trig. Trigonal. *Lustre:* Ad. Adamantine Gry Greasy, Met. Metallic, P'ly Pearly, Res. Resinous, S'ky Silky, Sub-met. Sub-metallic, Sub-vit. Sub-vitreous, Vit. Vitreous.

Name	Composition	Crystal System and Habit	Hardness Specific Gravity	Lustre	Colour	Streak	Cleavage	Fracture	Occurrence and Remarks
Actinolite	(see Tremolite)	(below)							
Aegirine (◇ *Pyroxenes*)	NaFe'' (Si₂O₆)	M.; prismatic, elongated	6 3·55	Vit.	Brown, green	White	Perfect (110)		Soda-rich igneous rocks
Analcite (◇ *Feldspath-oids*)	NaAl Si₂O₆ (H₂0)	C.; trapezohedral rhombohedra-2-3; also massive	5 2·3	Vit.	White pinkish	White	None		In amygdales, ◇ *zeolites*, as primary mineral in some undersaturated alkaline rocks
Anatase	TiO₂	Tet.; often acute pyramidal or tabular	5½ 3·9	Ad.	Brown, indigo-blue, black	Colour-less	Perfect (001) (011)		Joint planes and veins, in schists and gneisses
Andalusite (◇ *Aluminium silicates*)	Al₂SiO₅	O.; nearly square prisms, also granu-lar or massive	7½ 3·2	Vit.	Grey, pink, purplish-red	White	Good (110) Poor (100)	Uneven; tough	Metamorphic rocks. Variety ◇ *chiastolite* has patterned inclusions
Anhydrite (◇ *Evaporites*)	CaSO₄	O.; commonly fibrous, granular	3½ 2·9	Vit. /P'ly	White with grey, blue	White	Perfect (010) Good (100) (001)	Uneven	Sedimentary rocks, associated with gypsum

Name	Formula	Crystal system & habit	H / S.G.	Lustre	Colour	Streak	Cleavage	Fracture	Occurrence
	$(PO_4)_3$	combination, also massive	3·2		green, olive, brown, white, yellow	White	Very Poor (0001)	Conchoidal; uneven; brittle	igneous rocks especially pegmatites; metamorphic limestones
Aragonite (§ Carbonates)	$CaCO_3$	O.; pseudo-hexagonal twinned crystals common. Stalactitic	3½ / 2·94	Vit.	White, grey	White	Poor (010)	Sub-conchoidal; brittle	Sedimentary rocks. Passes into calcite with heat or pressure
Arseno-pyrite	$FeAsS$	O.; prismatic, massive	6 / 6	Met.	Silver-white, tin-white	Grey-black	Imperfect (101)	Uneven; brittle	Hydrothermal veins
Atacamite	$CuCl_2.3Cu(OH)_2$	O.; massive, lamellar	3 / 3·76	Ad. to Vit.	Green	Apple-green	Perfect (010)	Brittle	Oxidised zone of copper lodes
Augite (§ Pyroxenes)	$(Ca,Mg,Fe,Al)_2 (Al,Si)_2 O_6$	M.; short prismatic, often twinned crystals; also massive and lamellar	6 / 3-4	Vit.	Black, greenish-black	White	Good (110)		Ultrabasic and basic volcanic and plutonic rocks
Axinite (§ Cyclo-silicates)	$Ca_2(Mn, Fe')Al_2 (BO_3)(Si_4O_{12})(OH)$	Tric.; thin sharp-edged crystals	7 / 3·3	Vit.	Brown	Colour-less	Good (100)	Conchoidal; brittle	Contact metamorphism of limestone; and by pneumatolysis
Azurite (§ Carbonates)	$2CuCO_3.Cu(OH)_2$	M.; modified prisms; usually massive	4 / 3·8	Vit. to Ad.	Azure-blue	Light blue	Good (011)	Conchoidal; brittle	Associated with other oxidised copper minerals; copper ore
Barytes (Barite)	$BaSO_4$	O.; tabular; massive, fibrous columnar	3 / 4·5	Vit.	Colourless; white	White	Perfect (001, 210)	Uneven; brittle	Hydrothermal veins and replacements; nodular masses

Name	Composition	Crystal System and Habit	Hardness Specific Gravity	Lustre	Colour	Streak	Cleavage	Fracture	Occurrence and Remarks
Beryl (◊ Cyclosilicates)	$Be_3Al_2Si_6O_{18}$	H.; prismatic, often large crystals, also massive	8 / 2·7	Vit.	White, green, yellow, pale blue	White	Imperfect (0001)	Conchoidal or uneven; brittle	Acid igneous rocks, pegmatites. Gem varieties – emerald, aquamarine
Biotite (◊ Micas)	$K(Mg, Fe')_3 (AlSi_3)O_{10} (OH,F)_2$	M.; prismatic six-sided crystals	2½ / 3	Vit.	Black, dark-green	White	Perfect (001)		Igneous rocks of all kinds, and many metamorphic rocks
Bornite	Cu_5FeS_4	C.; cubes; usually massive	3 / 5	Met.	Purplish brown (tarnishes)	Pale grey-black	None	Conchoidal; uneven; brittle	Hydrothermal copper veins
Brookite	TiO_2	O.; thin tabular (010)	5½/6 / 4	Met. or Ad.	Brown to black	Colourless	None	Brittle	Hydrothermal veins in schists and gneisses
Brucite	$Mg(OH)_2$	Trig.; broad tabular; usually massive, foliated; fibrous	2½ / 2·39	P'ly or Vit.	White to pale-green, blue, grey	White	Perfect (0001)	Sectile	Contact-metamorphosed dolomitic limestones; serpentines
Calcite (◊ Carbonates)	$CaCO_3$	Trig.; nail-head and dog-tooth prismatic; twinning common; massive, stalactitic, granular	3 / 2·71	Vit.	Colourless or white; sometimes tinted by impurities	White	Perfect (10$\bar{1}$1)	Conchoidal	Common mineral: in sediments; altered basic igneous rocks; in veins, etc.; as stalactites, travertine, etc.
Cassiterite	SnO_2	Tet.; prismatic twinning frequent; massive, fibrous	6½/7	Ad.	Brown to black	White, greyish, brownish	Perfect (100) Imperfect	Sub-conchoidal; brittle	Hydrothermal veins; alluvial; in acid igneous rocks, especially ...

Mineral	Formula	Crystal habit	H / SG	Lustre	Colour	Streak	Cleavage	Fracture	Occurrence
Celestite (Celestine)	$SrSO_4$	O.; tabular; massive, fibrous	3/3½ 3·97	Vit.	White, with pale blue tint	White	Perfect (001) Good (210)	Imperfect conchoidal; very brittle	Chiefly in sedimentary rocks; but also in hydrothermal veins. Source of strontium
Cerussite (⟨⟩ *Carbonates*)	$PbCO_3$	O.; prismatic twinning common; granular, massive, compact	3 6·55	Ad.	White to grey – tinted by impurities	Colourless	Distinct (110) (021)	Conchoidal; very brittle	In oxidised zone of lead veins; valuable ore when found in quantity
Chalcocite	Cu_2S	O.; prismatic or pinacoidal; twinning frequent; usually massive	3 5·7	Met.	Blackish; lead-grey	Grey to black	Imperfect (110)	Conchoidal; brittle	Copper ore mainly in the enriched zone of sulphide deposits;
Chalcopyrite	$CuFeS_2$	Tet.; tetrahedral; twinning frequent; usually massive	4 4·2	Met.	Brass-yellow (tarnishes)	Greenish black	Imperfect (011)	Conchoidal; uneven; brittle	Wide occurrence; mainly hydrothermal and metasomatic veins. Main ore of copper
Chlorite (⟨⟩ *Chlorites*)	$(Mg,Fe')_{10} Al_2(Si,Al)_8 O_{20}(OH,F)_{16}$	M.; tabular; usually granular, scaly	2·5 3ap.	Vit. to Dull	Green with other tints	White to pale green	Perfect (001)		In igneous rocks from alteration of biotite, etc.; in schists
Chromite (⟨⟩ *Spinels*)	$FeCr_2O_4$	C.; octahedral; commonly massive, granular	5½/6 4·6	Met.	Black	Brown	None	Uneven; brittle	Ultrabasic igneous rocks and serpentines – often as small grains; ore of chromium
Chrysocolla	$CuSiO_3, 2H_2O$	O.; fine seams; usually massive, compact	3ap. 2·4 ap.	Vit.	Green to blue	White (when pure)	None	Conchoidal	Oxidised zone of copper deposits

Name	Composition	Crystal System and Habit	Hardness Specific Gravity	Lustre	Colour	Streak	Cleavage	Fracture	Occurrence and Remarks
Cinnabar	HgS	H.; tabular, rhombohedral or prismatic, usually massive, granular	$2/2\frac{1}{2}$ 8·1	Ad. (Dull when massive)	Cochineal red – sometimes brownish	Scarlet	Perfect $(10\bar{1}0)$	Sub-conchoidal; sectile	In veins and impregnations where volcanic activity has taken place; most important ore of mercury
Cobaltite	$CoAsS$	C.; pyritohedra or cubes, usually massive, granular, compact	$5\frac{1}{2}$ 6·2	Met	Silver-white (tarnishes)	Grey-black	Perfect (100)	Uneven; brittle	Hydrothermal veins with other cobalt and nickel minerals
Copper	Cu	C.; usually twinned; massive, thin sheets or dendritic	3 8·8	Met	Copper-red	Metallic	None	Ductile and malleable	Hydrothermal or metasomatic deposits; in cavities of basic igneous rocks; also zone of oxidation of copper veins
Cordierite (◇ Cyclo-silicates)	$Al_3(Mg, Fe)_2(Si_5 Al)O_{18}$	O.; pseudohexagonal twinned but usually massive	7 2·6	Vit	Various shades of blue, also greyish	Colour-less	Im-perfect (010) Poor (001) (100)	Sub-conchoidal; brittle	Hornfels; rarely in other metamorphic rocks
Corundum	Al_2O_3	Trig.; barrel-shaped, pyramidal; also massive, granular	9 4	Ad. to Vit.	Grey, blue, red to pink, green, yellow to brown	White	None Partings 0001 $01\bar{1}2$	Conchoidal; uneven	Value as abrasive and gemstone (ruby, sapphire), etc., wide occurrence – metamorphism of shales and limestones; veins—some igneous rocks

Covellite	CuS	H.; hexagonal plates; commonly massive	1½/2 4·7		Indigo-blue	Black	Perfect (0001)		Zone of secondary enrichment in copper lodes
Cryolite	Na_3AlF_6	M.; twinning common; usually massive	2½ 2·97	Vit.	Colourless to white; reddish-brownish tinge	White	None	Uneven; brittle	Uncommon – few localities of importance; pegmatite vein in Greenland
Cuprite	Cu_2O	C.; octahedral, sometimes dodecahedral; also massive earthy	3½/4 6	Ad. Sub-met.	Shades of red	Brownish red	Imperfect (111)	Conchoidal; brittle	Zone of weathering of copper lodes; sometimes an important ore of copper
Diamond	C	C.; octahedron most common; twinning frequent	10 3·5	Ad.	Colourless to white; also yellow, black, green, blue, etc.	White	Perfect (111)	Conchoidal; brittle	Gems and abrasive; ultrabasic igneous rocks, alluvial deposits; the latter is now the most important source
Diopside (⟨⟩ *Pyroxenes*)	$Ca(Mg,Fe")$ Si_2O_6	M.; prismatic; usually massive, granular	6 3-3	Vit.	White, green	White	Good (110)		Igneous rocks, metamorphosed impure dolomites
Dolomite (⟨⟩ *Carbonates*)	$CaMg$ $(CO_3)_2$	Trig.; rhombohedral crystals common; twinning frequent, massive; granular	3-5/4 2·85	Vit. Dull	White, often tinged with yellow, brown, red	White	Perfect (1011)	Conchoidal brittle	Sedimentary rocks, alteration of limestone; also in metalliferous veins as gangue
Enstatite	(see Hypersthene (below))								
Epidote (⟨⟩ *Epidotes*)	$Ca_2Fe"Al_2O.$ $Si_2O_7.$ $SiO_4(OH)$	M.; elongated (b-axis) also massive, granular	6 3-4	Vit.	Various greens	White	Perfect (001)	Uneven	Metamorphosed impure calcareous rocks and lime-feldspar – rich igneous rocks; hydrothermal veins

Name	Composition	Crystal System and Habit	Hardness Specific Gravity	Lustre	Colour	Streak	Cleavage	Fracture	Occurrence and Remarks
Fluorite (Fluorspar)	CaF_2	C.; octahedrons, dodecahedrons; also massive, coarsely or finely granular, compact	4 3·18	Vit.	Colourless but tinted with impurities giving many colours	White	Perfect (111)	Conchoidal; brittle	Blue John – blue ornamental variety; vein mineral; also in pneumatolytic deposits, cassiterite veins, pegmatites
Galena	PbS	C.; massive, also granular	2½ 7·5	Met.	Lead-grey	Grey	Perfect (100)	Sub-conchoidal; brittle	Most important ore of lead; hydro-thermal veins, replacements
Glauconite (◇ Micas)	$K(Fe^{\cdot},Mg,Al)_2$ Si_4 $O_{10}(OH)_2$	M.; small granules	2 2·7	Dull	Green-blackish/yellowish	Pale green	Perfect (001)		Marine sedimentary rocks; modern sedi-ments; also as grain in limestone, chalk, marl, sandstone
Glaucophane (◇ Amphiboles)	$Na_2(Mg,Fe^{\cdot})_3(Al,Fe^{\cdot\cdot})_2$ Si_8O_{22} $(OH,F)_2$	M.; prismatic; sometimes fibrous, massive, granular	6 3·2	Vit.	Blue, bluish-black	White to blue-grey	Perfect (110)		Metamorphic rocks – e.g. glaucophane schists – igneous rocks, particularly soda-rich granites
Gold	Au	C.; commonly dendritic or in alluvial grains or scales	3 17 (max.)	Met.	Gold-yellow when pure – white/copper tints when alloyed	Yellow	None	Ductile, sectile, malleable	Hydrothermal veins; placer deposits and conglomerates

				Met.	Black–iron-grey	Black and shining	Perfect (0001)	None	Greasy feel; metamorphic rocks, crystalline limestones; also in igneous rocks, veins, pegmatites
Graphite	C	H.; commonly foliated, massive, columnar, earthy	1 2·1						
Gypsum (◊ *Evaporites*)	$CaSO_4.2H_2O$	M.; tabular; twinning common; also massive, granular, fibrous	1½/2 2·3	Vit. p'ly	Colourless-white; tinted pink, yellow grey	White	Perfect (010)		Evaporite; in clays and limestones; associated with sulphur
Halite (◊ *Evaporites*)	NaCl	C.; cubes, massive granular	2½ 2·16	Vit.	Colourless, white—various tints when impure	White	Perfect (100)	Conchoidal; brittle	Evaporite. Soluble in water; salt taste
Haematite	Fe_2O_3	Trig.; tabular; micaceous compact, fibrous, granular, reniform	5½ 5·2	Met.	Red to black; steel-grey	Red-brown	None	Sub-conchoidal	Important ore of iron. Occurs as an accessory in igneous rocks; in hydrothermal veins and replacements; in sediments
Hemimorphite	$Zn_4Si_2O_7(OH)_2.H_2O$	O.; thin tabular; frequently twinned; massive, granular, mammillated, stalactitic	5 3·45	Vit.	White with faint yellow, green, blue	White	Perfect (110)	Uneven; brittle	In oxidised zone of zinc deposits; associated with smithsonite
Hornblende (◊ *Amphiboles*)	$NaCa_2(Mg,Fe'')_4(Al,Fe''')(Si,Al)_8O_{22}(OH,F)_2$	M.; prismatic; twinning common; also in blade-like forms, massive	5½ 3·2 ap	Vit.	Black, greenish-black	White/grey	Perfect (110)	Uneven	Widespread in metamorphic and igneous rocks

Name	Composition	Crystal System and Habit	Hardness Specific Gravity	Lustre	Colour	Streak	Cleavage	Fracture	Occurrence and Remarks
Hypersthene-Enstatite (⟨⟩ *Pyroxenes*)	$(Mg,Fe'')SiO_3$	O.; massive, irregular grains	$5\frac{1}{2}$ 3·6 ap	Vit.	Brown to black; greenish; yellowish; colourless; grey	White/grey	Good (210)	Uneven; brittle	Continuous series; occurs in basic and ultrabasic rocks low in calcium; also in some metamorphosed rocks
Idocrase	$Ca_{10}Al_4$ $(Mg,Fe'')_2$ $(Si_2O_7)_2$ $(SiO_4)_5$ $(OH)_4$	Tet.; prismatic or pyramidal; massive, granular	$6\frac{1}{2}$ 3–4	Vit.	Brown; green; yellowish	White	Poor (110)	Uneven	Contact metamorphism of impure limestones. Name 'vesuvianite' also used
Ilmenite	$FeTiO_3$	Trig.; thick tabular or prismatic; usually massive, or sand grains	$5\frac{1}{2}$ 4·7	Submet.	Iron-black	Black	None	Conchoidal	Accessory mineral in basic igneous rocks; as detrital deposits; in veins
Kaolinite (⟨⟩ *Layer-lattice minerals*)	Al_4Si_4 $O_{10}(OH)_8$	Tric.; fine powdery clay, pseudomorphs after feldspar	2 2–6	Dull	White, greyish	White	Perfect (001) when detectable		Hydrothermal alteration and weathering of feldspars, etc.
Kyanite (⟨⟩ *Aluminium silicates*)	Al_2SiO_5	Tric.; blade-like; often flexible and bent	4 & 7 3·63	Vit.	Patchy; blue-grey, green, white	White	Perfect (110) Good (010)		Metamorphism of aluminium-rich rocks; in quartz veins
Lepidolite (⟨⟩ *Micas*)	$K_2(Li,Al)_5$ (Si_6Al_2) $O_{20}(OH,F)_4$	M.; massive, granular	3 ap 2·85	P'ly	Colourless, pink, purple	White	Perfect (001)		In pegmatites; an ore of lithium; also in tin veins

Mineral	Composition	Crystal system; habit	H / SG	Lustre	Colour	Streak	Cleavage	Fracture	Occurrence
Leucite (◊ Feldspathoids)	$KAlSi_2O_6$	C.; trapezohedra	6 / 2·47	Vit.	White, ashy grey	Colourless	Imperfect (110)	Conchoidal; brittle	Primary constituent of undersaturated volcanic rocks rich in K
Limonite	Hydr.Fe_2O_3 & $FeO(OH)$	Massive, compact, granular, botryoidal, stalactitic, etc.	3–5½ / 4 ap.	Earthy Sub-met.	Black, yellow, brown, reddish, etc.	Yellow			Mixture of goethite, lepidocrosite, etc. Weathering of iron minerals
Magnesite (◊ Carbonates)	$MgCO_3$	Trig.; rhombohedra; massive, fibrous, granular, compact	4 / 3	Vit.	White or colourless yellow/brown tinge	White	Perfect (10$\bar{1}$1)	Conchoidal	Irregular veins in serpentine; replacement of dolomite, limestone
Magnetite (◊ Spinels)	Fe_3O_4	C.; octahedral, sometimes do-decahedral; also massive, granular	6 / 5·2	Met.	Iron-black	Black	None	Sub-conchoidal	Igneous rocks; in contact-meta-morphic deposits; replacement deposits; placer deposits
Malachite (◊ Carbonates)	$Cu_2CO_3(OH)_2$	M.; commonly massive, encrusting, with botryoidal or mammillary surface	3½/4 / 4	S'ky/Dull	Bright green	Pale green	Perfect (201) Fair (010)	Uneven	Oxidised zones of copper deposits; rarely as a cementing material in sand-stones; copper ore
Marcasite	FeS_2	O.; tabular; often repeatedly twinned; also radiating, spear-shaped, and cockscomb aggregates	6/6½ / 4·9	Met.	Pale bronze–yellow	Greyish	Imperfect (101)	Uneven; brittle	Low-temperature near-surface deposits in sedimentary rocks; in limestone often in concretions or as replacements

Name	Composition	Crystal System and Habit	Hardness / Specific Gravity	Lustre	Colour	Streak	Cleavage	Fracture	Occurrence and Remarks
Microcline (◊ *Feldspars*)	$KAlSi_3O_8$	Tric.; prismatic; twinning common; massive, granular	6 / 2·56	Vit.	White, grey, pink (bright green – amazonite)	White	Perfect (001) Good (010)	Conchoidal, uneven	Acid igneous rocks, pegmatites; also metamorphic rocks
Molybdenite	MoS_2	H: Plates, scales	1 / 4·7	Met.	Silver-grey	Blue-grey	Perfect (0001)		Hydrothermal veins. Marks paper
Monazite	$(Ce,La,Y,Th)PO_4$	M.; small flattened or elongated crystals; twinning common; also massive or as detrital grains	$5/5\frac{1}{4}$ / 5·2 ap	Res.	Yellow to reddish brown	White	Imperfect (100)	Uneven	Accessory mineral in acid igneous rocks, in pegmatite dykes; commercially found in detrital sands
Muscovite (◊ *Micas*)	$KAl_2(AlSi_3O_{10})(OH,F)_2$	M.; tabular; usually platy, massive or flakes	$2\frac{1}{2}$ / 2·85 Vars	Vit./P'ly	Colourless – or pale brown, green	White	Perfect (001)		Granite pegmatite; schist, greisen; detrital mineral. Plates flexible and elastic
Nepheline (◊ *Feldspathoids*)	$NaAlSiO_4$	H.; prisms, also irregular massive	6 / 2·6	Vit.	White-grey, brown, yellowish	White	Imperfect ($10\bar{1}0$)	Sub-conchoidal	In undersaturated soda-rich volcanic and plutonic igneous rocks, and in pegmatites
Olivine (◊ *Olivines*)	$(Mg,Fe'')_2SiO_4$	O.; usually granular massive	$6\frac{1}{2}$ / 3–4 ap	Vit.	Olive-green; also grey-white, yellow, brown, black	Colourless	Imperfect (010)	Conchoidal	Essential mineral in peridotites; in basic igneous rocks; variety 'forsterite' in metamorphosed dolomitic limestone

Mineral	Composition	Crystal form / habit	H / SG	Lustre	Colour	Streak	Cleavage	Fracture	Occurrence
Opal (⟨⟩ *Silica*)	$SiO_2 . nH_2O$	None; amorphous; massive, compact	5½/6½ 2·2	Vit.	White but with blend of colours brown, red, etc.	White	None	Conchoidal	Fissures and cavities of many rocks deposited by silica-bearing waters. Apart from gem-stone, variety di-atomite is commer-cially important
Orpiment	As_2S_3	M.; commonly foliated, massive	1½/2 3·48	P'ly	Lemon-yellow	Yellow	Perfect (010)	Sectile	Oxidised zones of arsenic minerals in veins and hot spring deposits
Orthoclase (⟨⟩ *Feldspars*)	$KAlSi_3O_8$	M.; short prismatic crystals; twinning common; also massive, granular	6 2·56	Vit. to P'ly	White to pink, also greenish-grey	White	Perfect (001) Good (010)	Conchoidal to uneven	Essential constituent of acid igneous rocks; common in metamorphic rocks
Phlogopite (⟨⟩ *Micas*)	$KMg_3(AlSi_3)O_{10}(OH,F)_2$	M.; pseudohexa-gonal prisms but lamellar plates more common	2½ 2·8	P'ly/Sub-met.	Pale yellow to brown	White	Perfect (001)		Metamorphosed dolomites, ultra-basic igneous rocks
Plagioclase (⟨⟩ *Feldspars*)	$Na(AlSi_3O_8) - Ca(Al_2Si_2O_8)$	Tric.; prismatic, flattened, poly-synthetic twinning; also massive, granular	6 2·7 ap	Vit.	White, grey	White	Perfect (001) Good (010)	Uneven	Metamorphic and igneous rocks according to variety; see entry for variety names
Psilomelane	$(Ba,H_2O)_2Mn_5O_{10}$	M.; always massive often botryoidal, stalactitic	5½ 4·2 ap	Sub-met.	Iron-black	Brownish black	None visible		Secondary mineral in association with manganese deposits. Residual deposits due to weathering

Name	Composition	Crystal System and Habit	Hardness Specific Gravity	Lustre	Colour	Streak	Cleavage	Fracture	Occurrence and Remarks
Pyrite	FeS_2	C.; cube and pyritohedron with striated faces, also massive, granular	6/6½ 5	Met.	Brass-yellow	Brown-black	Poor (001)	Conchoidal; brittle	Most widespread sulphide mineral; accessory mineral in igneous rocks; in ore veins; by contact metamorphism; anaerobic sediments; magmatic segregation
Pyrolusite	MnO_2	T.; usually massive, fibrous, reniform; pseudomorphs common	2/6 4 ap	Met.	Iron-grey	Black Bluish-black	None	Uneven; brittle	Associated with other manganese and iron oxides; found in bog and lake deposits
Pyromorphite	$(PbCl) Pb_4 (PO_4)_3$	H.; prismatic, often aggregated; also massive, globular, botryoidal	3½/4 7	Res.	Green, yellow, brown	White, Yellow-white	None	Uneven; brittle	Oxidised zones of lead deposits
Pyrrhotite	$Fe_{1-x}S$	M.; commonly massive, granular	4 var. 4·6	Met.	Coppery bronze-brownish; reddish	Dark grey-black	None	Uneven; brittle	Basic igneous rocks, pegmatites, contact metamorphic deposits
Quartz (⟨·⟩ *Silica*)	SiO_2	Trig.; hexagonal pyramid/prisms; faces irregularly developed; also massive, granular	7 2·65	Vit.	Colourless – white; coloured by impurities	White	None	Conchoidal	Acid igneous rocks; many metamorphic rocks; as a veinstone; in sedimentary rocks

Realgar	AsS	M.; usually massive, granular	1½/2 3·6	Res.	Red	Reddish-orange	Good (010) Poor (101)(100)	Sectile	Associated with orpiment in hydrothermal sulphide veins; also as deposit from hot springs and vulcanicity
Rhodochrosite (♦ Carbonates)	$MnCO_3$	Trig.; usually massive, encrusting, botryoidal	4 3·7 ap	Vit.	Pink to red; brownish-yellow	White	Perfect (10$\bar{1}$1)	Uneven; brittle	Veinstone in lead and silver-lead ore veins; also in metasomatic deposits
Rhodonite (♦ Pyroxenoids)	$MnSiO_3$	Tric.; usually massive	6 3·5	Vit.	Flesh-red-black, green, yellow where impurities	White	Perfect (110)	Conchoidal	Hydrothermal or metasomatic veins – as veinstone in lead and silver lead veins. Sometimes used ornamentally
Riebeckite (♦ Amphiboles)	$Na_2Fe'''_3Fe''_2Si_8O_{22}(OH,F)_2$	M.; prismatic crystals; also as aggregates and radiating tufts	6 3–4	Vit.	Blue to black	White/blue grey	Perfect (110)		Occurs in acid igneous rocks – soda-rich granite, etc.; also in metamorphic rocks
Rutile	TiO_2	T.; prismatic acicular – striated – twinning common; also massive	6·5 4·2	Ad. to Vit.	Reddish-brown, yellowish, black	Pale brown grey-black	Imperfect (110)	Uneven	Accessory mineral in igneous rocks, in pegmatites, metamorphosed limestones and veins with quartz
Scheelite	$CaWO_4$	Tet.; pyramidal; usually massive, granular, reniform	5 6	Vit.	Yellowish-white, brownish	White		Uneven; brittle	Pneumatolytic and hydrothermal veins, and in contact

Name	Composition	Crystal System and Habit	Hardness / Specific Gravity	Lustre	Colour	Streak	Cleavage	Fracture	Occurrence and Remarks
Serpentine (⟡ *Layer-lattice minerals*)	$Mg_6Si_4O_{10}(OH)_8$	M.; crystals unknown; occurs massive, fibrous or granular	3 / var. 2·55	Gr'y	Green, yellow, brown	White	None visible	Conchoidal; tough	Metamorphosed basic or ultrabasic igneous rock. Used ornamentally. Fibrous variety – asbestos
Siderite (⟡ *Carbonates*)	$FeCO_3$	Trig. Rhomb. crystals, massive, granular, nodular, oolitic, etc.	4 / 3·8-4	Sub-vit.	White, yellow, brown	White	Perfect (1011)	Conchoidal	Sedimentary iron ore; hydrothermal veins; nodules in clays
Sillimanite (⟡ *Aluminium silicates*)	Al_2SiO_5	O.; needle-shaped crystals or fibrous aggregates	7 / 3·24	Vit.	White, brownish greenish	White	Good (010)	Uneven	High-grade metamorphosed aluminium-rich rocks
Silver	Ag	C.; usually arborescent or massive, dendritic	3 / 10·5	Met.	Silver-white (often tarnished)	Silver-white	None	Malleable; ductile	Oxidised zone of ore deposits or in hydrothermal veins
Smithsonite (⟡ *Carbonates*)	$ZnCO_3$	Trig.; rhombohedra; massive, encrusting, earthy, botryoidal, stalactitic	4 / 4·4	Vit.	White-greenish, brownish, grey	White	Good (1011)	Uneven; brittle	Oxidised zone of ore deposits. Ore of zinc
Sodalite (⟡ *Feldspathoids*)	$Na_8(AlSiO_4)_6Cl_2$	C.; dodecahedra; massive	6 / 2·3	Vit.	Blue-white, pink, yellowish	White	Poor (110)	Conchoidal; uneven	In soda-rich igneous rocks particularly nepheline syenites

Name	Formula	Crystal form	H / SG	Lustre	Colour	Streak	Cleavage	Fracture	Occurrence
Sphalerite (Zinc blende)	ZnS	C.; tetrahedra and dodecahedra; twinning and distortion frequent; also massive, botryoidal, fibrous	4 4·1	Res. to Ad.	Black, brown; also yellow, white	White/reddish brown	Perfect (011)	Conchoidal; brittle	Main ore of zinc; in metasomatic deposits with galena, in hydrothermal lode and vein deposits, in replacement deposits
Sphene	$CaTiSiO_5$	M.; wedge and lozenge shaped; massive	5½ 3·5	Ad.	Brown, also grey, green, yellow	White	Imperfect (110)	Conchoidal; brittle	Accessory mineral in acid igneous rocks; in metamorphosed limestones
Spinel (◇ Spinels)	$MgAl_2O_4$	C.; octahedra, dodecahedra; often twinned	8 3·6	Vit.	Black brown, green, red, blue	White	None	Conchoidal; brittle	In ultrabasic igneous rocks; some metamorphic rocks low in SiO_2; alluvial sands
Staurolite (◇ Aluminium silicates)	$FeAl_4Si_2O_{10}(OH)_2$	O: prismatic crystals, cruciform twins	7 3·8	Sub-vit. dull	Brown	White	Fair (010)	Conchoidal	Medium-grade metamorphic rocks
Stibnite	Sb_2S_3	O: prismatic, bladed	2 4·6	Met.	Silver-grey	Grey	Perfect (010)	Hackly, brittle	Hydrothermal veins (low temp.)
Strontianite (◇ Carbonates)	$SrCO_3$	O.; prismatic; frequent twinning; massive fibrous, granular	3½/4 3·7	Vit.	Grey, yellowish, greenish	White	Good (110)	Uneven; brittle	In veins; nodules in limestones; as concretionary masses. Source of strontium
Sulphur	S	O.; pyramidal; massive, encrusting	2 2·07	Res.	Yellow-reddish or greenish tinge	Yellow-white	Imperfect (110) (111)	Conchoidal to uneven; brittle	In areas of recent volcanic activity and around hot springs; also in sedimentary rocks with gypsum and limestone

Name	Composition	Crystal System and Habit	Hardness Specific Gravity	Lustre	Colour	Streak	Cleavage	Fracture	Occurrence and Remarks
Talc (⟨⟩ Layer-lattice minerals)	$Mg_3Si_4O_{10}(OH)_2$	M.; massive; foliated; compact, fine-grained aggregates	1 2·7	P'ly	Pale green, white, grey, brown	White	Perfect (001)		In metamorphic rocks rich in magnesium; with serpentine
Tetrahedrite	$(Cu,Fe)_{12}Sb_4S_{13}$	C.; tetrahedra; often twinned; massive, granular or compact	4 4·8	Met.	Iron-black, steel grey	As colour	None	Subconchoidal; brittle	With other copper ores in hydrothermal veins
Topaz (⟨⟩ Aluminium silicates)	$Al_2SiO_4(OH,F)_2$	O.; prismatic; massive, granular	8 3·5	Vit.	Colourless, white with yellow, blue, pink shades	White	Perfect (001)	Subconchoidal	In cavities in acid igneous rocks; in greisens; in quartz veins; gemstone
Tourmaline (⟨⟩ Cyclo-silicates)	$Na(Mg,Fe')_3 Al_6(BO_3)_3 (Si_6O_{18}) (OH,F)_4$	Trig.; prismatic triangular cross-section; massive, columnar	7 3·1	Vit.	Usually black; also blue, green, red, colourless, yellow and other colours	Colourless	Poor (1120)(1011)	Subconchoidal	Pneumatolytic mineral in acid rocks and in schists, gneisses
Tremolite-Actinolite (⟨⟩ Amphiboles)	$Ca_2(Mg,Fe')_5 (Si_8O_{22}) (OH,F)_2$	M.; prismatic; radiating, fibrous; columnar	5½ 3·2	Vit.	White to green with increasing Fe'	White	Perfect (110)		In metamorphosed impure limestones
Uraninite	UO_2	C.; octahedra; usually massive, botryoidal, or dendritic, as var. pitchblende	5½ 8	Sub-met.	Black; brownish, greenish tinge	Brown-black	None	Conchoidal	In hydrothermal veins; granites and pegmatites; and as grains in conglomerates

Mineral	Composition	Crystal system; habit	H SG	Lustre	Colour	Streak	Cleavage	Fracture	Occurrence
Witherite (◇ *Carbonates*)	$BaCO_3$	O.; pseudo-hexagonal twinned crystals; massive, columnar, granular, tuberose, botryoidal	3½ 4·3	Vit.	White, yellowish, greyish	White	Distinct (010)	Uneven; brittle	Hydrothermal veins, associated with barytes and galena. Source of barium
Wolframite	$(Fe\cdot,Mn)WO_4$	M.; tabular; also massive, bladed	4½ 7·3	Sub-met.	Brown – blackish, reddish, greyish	Brownish-black	Perfect (010)	Uneven; brittle	Ore of tungsten; hydrothermal and pneumatolytic veins
Wollastonite (◇ *Pyroxenoids*)	$CaSiO_3$	Tri.; tabular; also massive, cleavable, fibrous	5 2·9	Vit.	White – greyish, red/brown tints	White	Perfect (100) Good (001) (102)		Contact metamorphism of impure limestones
Wulfenite	$PbMoO_4$	Tet.; square, platy; also massive, granular	3 6·8	Res.	Orange yellow – wax yellow, yellowish-grey, brownish	White	Distinct (101)	Sub-conchoidal	Oxidised zone of lead deposits
Zircon	$ZrSiO_4$	Tet.; prismatic; also in detrital grains	7½ 4·3	Ad.	Colourless; reddish brown, yellow, grey	Colourless	Imperfect (110)	Conchoidal	Accessory mineral of more acid igneous rocks; detrital deposits
Zoisite (◇ *Epidotes*)	$Ca_2Al_3Si_3O_{12}(OH)$	O: massive; bladed	6·5 3·3	Vit.	White, grey, pink, greenish	White	Perfect (001)	Conchoidal	Metamorphosed impure limestones

BIBLIOGRAPHY

(This list is not an exhaustive list of books consulted but rather a selection of those recommended for further reading.)

Reference and General

Holmes, A., *Principles of Physical Geology*, Nelson, 1966.

Read, H. H., and Watson, J., *Introducing Geology*, vol. 1, *Principles*, Macmillan, 1966.

Weller, J. M., ed., *Glossary of Geology and Related Sciences* and *Supplement*, American Geological Institute, 1962.

Geomorphology

Dury, G. H., *The Face of the Earth*, Penguin Books, 1959.

Thornbury, W. D., *Geomorphology*, Wiley, New York, 1969.

Mineralogy and Crystallography

Berry, L. G., and Mason, B., *Mineralogy: Concepts, Descriptions, Determinations*, Freeman, San Francisco, 1959.

Bishop, A. C., *Crystal Morphology*, Hutchinson, 1967.

Bragg, L., and Claringbull, G. E., *Crystalline State*, vol. IV, *The Crystal Structure of Minerals*, Bell, 1965.

Dana, J. D., *System of Mineralogy*, Chapman & Hall, London, Wiley, New York, 7th edn 1962.

Deer, W. A., Howie, B.A., and Zussman, J., *Rock-Forming Minerals*, 5 vols., Longman, 1966.

Rankama, K., and Sahama, T. G., *Geochemistry*, Chicago University Press, 1950.

Read, H. H., *Rutley's Elements of Mineralogy*, Murby, 26th edn, 1970.

Palaeontology

Moore, R. C., Lalicher, G. G., and Fischer, A. G., *Invertebrate Fossils*, McGraw-Hill, New York, 1952.

Moore, R. C., ed., *Treatise on Invertebrate Palaeontology*, 24 parts, Geological Society of America and University of Kansas Press, 1953ff.

British Museum of Natural History, *Palaeozoic Fossils, Mesozoic Fossils* and *Tertiary Fossils*.

Rocks

Hatch, F. K., Rastall, R. H., and Greensmith, J. T., *Petrology of the Sedimentary Rocks*, Murby, 1965.

Hatch, F. K., Wells, A. K., and Wells, M. K., *Petrology of the Igneous Rocks*, Murby, 1961.

Holmes, A., *Nomenclature of Petrology*, Murby, 1920.

Pettijohn, F. J., *Sedimentary Rocks*, Harper, New York, 1957.

Stratigraphy

Gignoux, M., *Stratigraphic Geology*, Freeman, San Francisco, 1956.

Kummel, B., *History of the Earth*, Freeman, San Francisco, 1970.

Phanerozoic Time-Scale: A Symposium, Quarterly Journal of the Geological Society, 120S, London, 1964.

Structure

Billings, M. P., *Structural Geology*, Prentice Hall, New York, 1942.

Hills, E. S., *Elements of Structural Geology*, Methuen, 1963.

Sitter, L. V. de, *Structural Geology*, McGraw-Hill, New York, 1956.

Spencer, E. W., *Introduction to the Structure of the Earth*, McGraw-Hill, New York, 1969.

MORE ABOUT PENGUINS
AND PELICANS

For further information about books available from Penguins please write to Dept EP, Penguin Books Ltd, Harmondsworth, Middlesex UB7 0DA.

In the U.S.A.: For a complete list of books available from Penguins in the United States write to Dept CS, Penguin Books, 625 Madison Avenue, New York, New York 10022.

In Canada: For a complete list of books available from Penguins in Canada write to Penguin Books Canada Ltd, 2801 John Street, Markham, Ontario L3R 1B4.

In Australia: For a complete list of books available from Penguins in Australia write to the Marketing Department, Penguin Books Australia Ltd, P.O. Box 257, Ringwood, Victoria 3134.

METALS IN THE SERVICE OF MAN

William Alexander and Arthur Street

Metals in the Service of Man is a fascinating record, illustrated with photographs and diagrams, of the part played in life and civilization by the many metals in common use. This new edition was partly rewritten to take account of the important developments of the last ten years in research and industrial technique. This particularly affected the sections on iron and steel, aluminium, the minor metals, and corrosion. A chapter on nuclear energy explains how new metals have been developed to meet the needs of a new industry.

The book also contains an analysis of the earth's crust, an explanation of the methods by which we get our metals, and discussions of the alloying of metals and of their inner structure. The authors finally look to possible future developments.